IRELAND'S VANISHING NATURE

PÁDRAIC FOGARTY, a professional ecologist from Castleknock in Dublin, holds degrees in Environmental Protection, Environment and Geography. Having served as chairman of the Irish Wildlife Trust, he now works as its campaigns officer and editor of the magazine *Irish Wildlife*.

Stay up to date with the author at:
 @whittledaway
 www.whittledaway.com

Follow the Irish Wildlife Trust at:
 @irishwildlife
 Irish Wildlife Trust

50% of royalties from this book will be donated to the Irish Wildlife Trust

This book is dedicated to all the people who work in environmental groups in Ireland – whether it's a national conservation organisation or a local Tidy Towns group. There are more of them than we appreciate – from counting birds to collecting litter, the voluntary work done by low-key but inspiring individuals leaves us all in their debt.

The book is also dedicated to my wife, Annika, whose words of wisdom helped me find my voice, and my children, Max and Maia, who I hope will be able to show their grandchildren a healthier environment than the one we have today.

IRELAND'S VANISHING NATURE

PÁDRAIC FOGARTY

The Collins Press

First published in 2017 by
The Collins Press
West Link Park
Doughcloyne
Wilton
Cork
T12 N5EF
Ireland

Reprinted 2017, 2021

© Pádraic Fogarty 2017

All photographs © Pádraic Fogarty unless otherwise credited.

Pádraic Fogarty has asserted his moral right to be identified as the author of this work in accordance with the Irish Copyright and Related Rights Act 2000.

All rights reserved.
The material in this publication is protected by copyright law. Except as may be permitted by law, no part of the material may be reproduced (including by storage in a retrieval system) or transmitted in any form or by any means, adapted, rented or lent without the written permission of the copyright owners. Applications for permissions should be addressed to the publisher.

A CIP record for this book is available from the British Library.

Hardback ISBN: 978-1-84889-310-8
PDF eBook ISBN: 978-1-84889-617-8
EPUB eBook ISBN: 978-1-84889-618-5
Kindle ISBN: 978-1-84889-619-2

Typesetting by Carrigboy Typesetting Services
Typeset in Garamond, Frutiger and WolfSpirit
Printed in Poland by Białostockie Zakłady Graficzne SA

Front: all images from Shutterstock except (*clockwise from top*): Sturgeon (W. J. Gordon, *Our Country's Fishes*), badger (The Graphics Fairy), yellowhammer (National Library of Ireland), bee (Vintage Illustrations). *Spine*: Beetle (Vintage Illustrations). *Back*: Bee (Vintage Illustrations). Illustrations on pp. 36 and 208 (www.vecteezy.com); p. 133: Huish, Robert, *The Wonders of the Animal Kingdom*, Thomas Kelly, London, 1830.

Contents

Abbreviations used		vi
1	Not as green as we'd like to think	1
2	The quest to catch the last fish	36
3	The whittling-away of our iconic landscapes	78
	Killarney National Park	86
	Wicklow Mountains National Park	95
	Connemara	104
	Glenveagh National Park	113
	The Burren	124
4	Extinct: Ireland's lost species	133
5	Culling: the urge to kill animals	173
6	The battle to save the bogs	208
7	The myth of Ireland's 'green' farming	247
8	A future for wildlife and people	286
Endnotes		326
Further Reading		350
Acknowledgements		351
Index		352

Abbreviations used

ASI	Area of Scientific Interest
BAG	Burren Action Group
BIM	An Bord Iascaigh Mhara
BNPSA	Burren National Park Support Association
ESRI	Economic and Social Research Institute
EPA	Environmental Protection Agency
EU	European Union
ICABS	Irish Council Against Blood Sports
IFI	Inland Fisheries Ireland
IFA	Irish Farmers' Association
IPCC	Irish Peatland Conservation Council
ISS	Irish Seal Sanctuary
IUCN	International Union for Conservation of Nature
IWT	Irish Wildlife Trust
MPA	Marine Protected Area
NHA	Natural Heritage Area
NPWS	National Parks and Wildlife Service
OPW	Office of Public Works
RBCT	Randomised Badger Culling Trial
RSPB	Royal Society for the Protection of Birds
SAC	Special Area of Conservation
SPA	Special Protection Area
UNESCO	United Nations Educational, Scientific and Cultural Organization
WMNP	Wicklow Mountains National Park
WWF	World Wide Fund for Nature (formerly the World Wildlife Fund)

'Ireland has a remarkable heritage of historic values and scientific sites, varied in type, widely spread over the face of this country. This heritage is of growing importance for purposes of education and recreation, not only in the daily lives of Irish people, but also in the development of tourism and the pursuit of historical and scientific research.

This heritage is, however, being steadily whittled away by neglect and natural forces on the one hand, and on the other hand by human exploitation, pollution, and other aspects of modern development. This could represent a serious loss to the nation, since the heritage, properly cared for, and suitably developed, can bring many and continuing benefits to present and future generations.'

An Foras Forbartha, June 1969[1]

1

Not as green as we'd like to think

> 'Sustainability has always been at the source of who we are. Think about it. Our climate has always been this mild; our landscape this lush. Our fields have always been this green, and windswept and rainwashed. Our seas have always teemed with fish.'
>
> From a promotional video for Irish food, 2013[1]

THERE IS A CORNER OF CONNEMARA, to the west of Galway city, that is a paradise for nature lovers. Within a walkable radius there is an ancient oak woodland, a rocky shore, an expanse of blanket bog and that rough 'West of Ireland' farmland that is pockmarked with small wetlands and rocky outcrops. No matter the time of year, nature's ebb and flow is laid bare for those who care to look. In January small flotillas of Brent geese bob on the water near the saltmarsh, and are replaced in summer by Arctic terns dive-bombing in the surf for sprat and sand eels. The winter gales disgorge the flotsam of the deep sea while in April and May the oak forest is bejewelled with bluebells, anemones, ramsons and celandines. Throughout the year the setting sun silhouettes the drying wings of cormorants against the breaching humpbacks of the Burren hills that enclose the south side of the bay. Kestrels patrol rough grass near the sea in search of rodents. Stoats dart out of stone walls and zip across roads. There's no end to the exploring and discovering. While walking with my children one February we nearly stood on the silky nest of a brood of the rare marsh fritillary butterfly. The tiny black caterpillars were just emerging on the edge of the grassy bog track and, so far as we knew, this was the first time they were recorded from this part of the country. During the low tides of summer I pull on a mask and snorkel and peer down the trunks of kelp forests, their strap-like 'leaves' swaying in unison at the whim of the current. Beneath lie giant pink sea urchins the size of oranges, well-armoured spider crabs tip-toeing, and occasionally a spotted-dogfish quartering the depths, weaving its way through the shifting stems. All this within a radius of a kilometre or so. A little further in the car even greater adventures await. The mountains of the Twelve Bens and Maumturks provide some of the best hillwalking to be had in Ireland. The Aran Islands, fragments of limestone pavement that have been marooned from the Burren by rising seas millennia ago, are bestowed with vertiginous cliffs, rare flora and a colony of lolling seals. To the north the limestone basin of Lough Corrib is fringed with luxuriant native woodland and famed for its clear waters and run of trout. Every visit is different from the last. No matter how often I make the dash from Dublin there are always new things to see, new perspectives, new adventures. Seeing the mountains of Connemara with a rare dusting of snow was as if a new landscape had been created from the old. The warming sun melting the crust served only to focus my appreciation on the present. The islands beyond Connemara are jewels in a sea that sometimes glitters and sparkles in bands across their golden beaches. A marriage of people, time and the forces of nature have made Connemara the way it is. At first glance it seems perfect. But all is not as it seems. Its undoubted beauty is on the wane, subtly, imperceptibly, but surely, like a once grand castle falls into ruin one brick at a time.

On a grey, cold, dry April morning in 2013, I went walking along one of the bog tracks that wind their way across the peatlands which cloak the ridge between Galway Bay and Lough Corrib to the north. Beyond I could see the dark curtains of pine plantations and a stiff wind blew across the otherwise treeless landscape. For most of the year, at least until the heather blooms in late summer, the bog is rather colourless and drab, but on this day a pall seemed to bring the grey skies even closer to the ground. As I went on I realised from the acrid whiff that this was not morning fog but smoke; the bog was on fire. The flames were working their way slowly across the surface, leaving blackened fingers of heather stems and scorched peat in their wake. I continued on my way and passed two men leaning on a car, chatting. I don't know why I didn't ask them why the bog was on fire but I regretted it later. I also regretted not calling the emergency services, or complaining to the local wildlife ranger. This was, after all, a protected area during the height of the bird-nesting season. Any nests of pipits or skylarks on the open bog, or those of wrens or stonechats that may have been in the low shrubs of willow and furze, would have been destroyed. Frogs could not have survived while any caterpillars of the marsh fritillary that we had so diligently logged the year before had now surely perished. Perhaps I didn't complain to anyone because it was a Sunday and I wasn't really in the mood to be making a fuss. Perhaps it was because I am not a native in this area and so might have caused hassle for myself with the locals. Perhaps I had seen so much of our countryside on fire that year, and the year before, that I thought there was nothing unusual about fires in the landscape. Perhaps I also thought that complaining to either the Gardaí or the National Parks and Wildlife Service (NPWS, the state agency charged with the conservation of our natural heritage) would just further my frustration as action is rarely taken except where private property or lives are threatened. This is not because of disinterest or lack of will but because these agencies have never been given the resources to deal with wildfires. The Gardaí stand out in having no unit devoted to wildlife crime, unlike most forces elsewhere in Europe. The NPWS have dedicated staff but the bar for a successful criminal prosecution is extremely high.

By the time I returned home, about 2km from where I had been, a dim glow was visible against the monochrome sky, and shortly afterwards I could hear sirens pulsating in the still air. Back home in Dublin the 'gorse fires', as the media referred to them, were reported on national news as they had threatened homes. People had been evacuated to the sidelines until the flames were brought under control. The report didn't mention that the fire had probably been started deliberately and that it had effectively destroyed an area that was of national importance for nature conservation. But then the state broadcaster, RTÉ, had no environmental correspondent at the time while the national print media had few reporters dedicated to such issues. Having been spoilt during the 1970s and 1980s with high-profile figures like Éamon de Buitléar and Gerrit van Gelderen who spoke out on environmental issues, today Ireland lacks such voices. Commentary, such as it is, tends to be restrained in tone, something that stands in stark contrast with the UK where vigorous debates concerning nature conservation are played out in print and across the web. The fire, in the end, did not damage anyone's home or property and luckily no one was injured.

A year later I returned to the bog. Scorched earth was still in evidence but there were also signs of life. The green shoots of purple moorgrass were probably growing more vigorously, unhampered by a blanket of dead vegetation that would otherwise have formed a crust over the peat surface. Bracken, well known for its resilience to fire, was also doing well. The heather, meanwhile, was showing little sign of recovery. Fire is not a part of natural processes in Ireland, like it is in Australia or parts of the United States, and while some plants will survive, even thrive, from fire, overall the community of plants that is left is much poorer than before. Insects and amphibians meanwhile are ill-equipped to survive the heat and flames, and an impoverished habitat is left behind. Records show that this bog had been burnt before. Certainly turf had been cut on it for a long time, but determining the health of this Natural Heritage Area (NHA) is difficult because there is no regular system of monitoring and reporting for these protected sites. There is no management plan, just like there are few management plans for the vast majority of the 600 or so NHAs across Ireland. In fact, the majority of NHAs are not even really protected under law at all, being only 'proposed'. This NHA – the Moycullen Bogs – was described in 2005 as:

> an extensive area of lowland blanket bog in an area of high landscape beauty. The site supports a diversity of habitats

[and] supports *Irish Red Data Book* [i.e. threatened] species red grouse and several additional notable species of fauna including Irish hare, common frog, snipe, curlew, fox, kestrel and lapwing.

Peat cutting (both mechanical and hand) is the dominant land use at present, while grazing pressure by donkey, cattle and ponies is low but locally damaging ... There are a number of quarries within the site ... A golf course has been constructed on the north side of Lough Inch and a small pitch and putt course has been established on the southern shores of the lake. There has been some burning of the bog surface in the recent past and conifer plantations have been planted in the centre and eastern area of the site. Due to the proximity of the site to Galway City there is increasing pressure from housing development (typically single dwellings). Development of wind energy installations is also a potential threat.[2]

But many of the birds that were once common on the bog or hills, like the red grouse, curlew and lapwing, are barely hanging on as the vegetation changes from mostly heather to mostly moorgrass (this is most striking in summer where hills that once were flush with purple flowers are now green). As I write this, reports of fires across the hills of Ireland are once again in the media, including from parts of Connemara. Back in 2013 a battle was raging over the conservation of a small number of bogs in midland counties, which were home to the last remnants of this habitat type. But even here full legal protection was not sufficient safeguard in preventing further damage. In other words, although we have a large number of protected areas in Ireland, many of them remain little more than lines on maps.

Connemara is a major tourist attraction in a region with few other economic opportunities; one website dedicated to local tourism declares: 'Connemara, situated at the very edge of Europe, on the west coast of Ireland, is one of the most beautiful, unspoilt places it's possible to find.'[3]

The sentiment that large parts of Ireland are 'unspoiled' is commonplace – but how true is it? In fact, while Connemara has a lot, it has also lost a lot.

Even beyond this relatively small area of bog bigger changes have occurred. During the nineteenth century the fishing village at Claddagh, close to the centre of Galway city, was described in Peadar O'Dowd's book on the enclave as 'one of the leading fishing stations in the country'. In 1829 there were 2,420 open sailboats and 1,199 row boats employing 7,305 fishermen. On shore there were men and women involved in curing the fish, gutting and packing the catch as well as barrel and net making. In total 8,700 people were employed in the industry in the Galway district.[4] The quantity and variety of fish in Galway Bay at the time is mindboggling from today's perspective:

> a great quantity and variety of fish are taken in the bay; amongst which the following are to be had daily in the public market, during their proper seasons, viz. herrings in great quantities, turbot of the largest and finest kind, sole, plaice, cod, haddock, hake, whiting, ling, mullet, black and white pollack, mackerel, bream, eel, gurnet, and several other varieties, with abundance of lobsters, crabs, cray-fish, oysters, shrimps, cockles, mussels, &c and at very reasonable prices.[5]

O'Dowd describes in his book how 'a single rowing boat could often return with a catch of 20,000 herrings in one night's work [...] It even mentions that a Claddagh man by the name of Seán Mac an Ríogh once landed 40,000 herrings in one haul.'

The Claddagh fishermen operated from open boats with oars and sails and fought tenaciously against the introduction of bottom trawling, then a technological innovation, rightly predicting how it would destroy the fishing grounds. Fishing in Galway Bay today involves a handful of boats potting for crabs and lobsters. Looking out on the bay it is hard to imagine the level of activity it once sustained both above and below the waves.

Conifer plantations smother landscapes, communities and wildlife.

On land the once great forests that cloaked Connemara in oak and pine are reduced to tufts of trees on islands in lakes. Even in the most scenic parts of the mountains, plantations of conifers have been planted where native forests would have brought sustainable incomes and habitat for wildlife. During the 1980s disastrous European farming policies saw sheep crowded onto the hills, causing erosion and loss of vegetation. Farming as a way of life across the region is dwindling, dependent upon state subsidies and the sometimes contradictory instructions from government officials that go with it. The result is wetlands filled in, patches of scrub removed, and flower-rich hay meadows 'improved'. Once-common farmland birds like the corncrake and the grey partridge died out in the early 1990s. Victorian lodges hark to the days when Connemara's salmon and trout fishing were the envy of Europe. The Galway fishery was described in one guidebook as 'remarkable compared with that on any similar piece of water in the United Kingdom'.⁶ Lough Corrib still attracts anglers and is internationally renowned for its brown trout fishing. But it has suffered in recent years from a number of invasive species, including zebra mussels, curly waterweed (a plant) and roach (a fish). It is believed that pollution resulted in the extinction of the Arctic char, a relative of the salmon, during the 1980s. The ecological status of the lake was assessed in 2015 as varying from 'poor' in its upper regions to 'moderate' nearer the city.⁷ During the 1980s numbers of sea trout returning to rivers in Galway and Mayo collapsed due to lice infestations from salmon farms at sea.

Irish people tend to see our country as 'green' and 'unspoiled'. The Wild Atlantic Way, a scenic drive and arguably our most successful tourism initiative in decades, is rooted in the quality of our countryside, where 'its landscape, flora, fauna, and sheer size have inspired everyone from W. B. Yeats to John Lennon'.⁸ State agencies charged with marketing our food and drinks industry heavily promote our 'green' image abroad to generate income and jobs at home. And Irish people love nature. A 2015 poll by the European Commission found fully 97 per cent of Irish people agreed with the statement that 'we have a responsibility to look after nature' while 90 per cent agreed that 'our health and well-being are based on nature and biodiversity'.⁹ We are a less-urbanised society than many European countries, with strong connections to the land, so perhaps we are in tune with our landscape more than most. Even city dwellers usually only need to go back a generation or two to find their links with rural life. We are told that our environment is in good shape by various state agencies, our air is clean and that our countryside is unspoiled. We are told by chefs that our seas are teeming with fish. But these claims are rarely challenged or subject to scrutiny.

I remember my childhood in the 1980s and the fascination I had with nature. It was instilled in me by my parents and a particular teacher who was especially enthusiastic about wildlife. She encouraged us never just to walk to school but to be alert to what was happening along the way. Despite growing up in the suburbs of Dublin I became aware of when buds appeared on trees, what bird was singing or from which direction the wind was blowing. We were encouraged to get down on our hands and knees to look at things up close. Nature has astonishing detail, whether in the bark of trees, the crenulations and texture of leaves, or the hairs on a bumblebee's legs that allow it to carry barrels of pollen between flowers. There was no talk of evolution, no talk of ecology but there was an understanding that everything is connected. At the time the Amazon was being felled at a frightening rate. An area the size of Belgium (it was always Belgium) was being cleared every year to make way for ranches. We were told that the meat from these ranches was ending up in burgers sold in McDonalds so even at the age of ten I could make a connection between our individual actions and awful consequences that were happening a world away. I vaguely remember the English naturalist David Bellamy, with his mumbled speech and crusty appearance, warning us that what we were doing to our bogs was a travesty. I had never been to a bog. I didn't know what one looked like. I don't remember ever learning about bogs in school except perhaps that we were (and still are) the only country that generates electricity from burning turf. There was a sense of achievement that as a nation we had managed to find a profitable use for unproductive wasteland. I knew that my grandparents cut turf on a bog in Meath and some friends took deliveries of hand-cut turf in the autumn. We were warned sternly (to no avail) not to climb the heaps that filled their gardens. Today, less than 1 per cent of the midlands' raised bogs remain intact – a figure that continues to shrink.[10] I remember also stories of dead lakes in Cavan. Our water quality had become so bad in some areas that mats of green algae blanketed lakes during the summer, sucking the oxygen out of the water and suffocating fish. This was a result of run-off from agriculture, poorly constructed septic tanks and discharges from industry. Thirty years on, despite some improvement, nearly half of our water is still polluted.[11]

The Environmental Protection Agency (EPA) was established in 1992 by the state to monitor our environment and has a reputation for independence and professionalism. Every four years it publishes its 'state of the environment' report, an excellent document, very readable and accessible to non-specialists who are interested in the health of the nation. Its introduction states that 'Ireland's environment is generally good', despite acknowledging 'this has to be qualified'.[12] If we look internationally, the Environmental Performance Index (EPI), produced annually by the Yale Centre for Environmental Law and Policy in the United States, placed Ireland at a respectable nineteenth out the 150 or so countries that they looked at in 2016. But are we getting the full story? Until recently, the EPA, for example, did not assess the state of our seas – until 2016 it was simply a missing chapter from their report. This is not necessarily their fault; the seas around Ireland are a kind of no-man's land, and the EPA's remit does not extend past the estuaries of our major rivers. Their 2012 report stated boldly that 'humanity is dependent on nature for survival' but yet gave little prominence to the declining state of much of our wildlife. Yale's EPI, meanwhile, awards Ireland's high ranking based on a strong performance in five areas: human health, air quality (which is among the best in Europe), water and sanitation (i.e. access to same), water resources (which looks at the standard of wastewater treatment and not the quality of water in our rivers and lakes), and forests. We have weaker scores for agriculture and energy/climate while we barely register a score for either fisheries or biodiversity. Our high score in the 'forests' category is based on an increase in tree cover, consisting of conifer plantations that most ecologists would not equate with the rich tapestry of life to be found in a forest. Nature is the very foundation of life, yet it is something our politicians rarely speak about. What information we have shows that nature is in a bad way. A 2013 report from the NPWS showed that of fifty-eight important habitats only five were in 'good' condition, while of sixty-one rare species just half were 'favourable'.[13]

Since 2003 I have been a volunteer with the Irish Wildlife Trust (IWT), one of a handful of environmental organisations that have been active in Ireland since the late 1970s. From 2008 to 2012 I was its chairman and I continue to be its campaign officer and editor of its magazine *Irish Wildlife*. I have also spent time with the Irish Peatland Conservation Council, an organisation based in County Kildare, where I worked for a short time on a conservation project restoring a bog that had previously been cut for turf. For a few months in 2007 I volunteered with BirdWatch Ireland as part of a national census of red grouse. For hours at a time a friend and I hauled ourselves through the knee-high heather of the Wicklow Mountains with a 3kg loudspeaker blasting out the grouse's mating call (the idea being that the disgruntled male would reveal itself by calling back).

Red grouse was once a common bird across Irish hills but is now considered critically endangered due to habitat loss. Nevertheless it is still legal to hunt it during certain months. During my time with the IWT I have been a part of campaigns to improve the health of our seas, end the culling of badgers, stop hare coursing, ban the pesticides that are killing off bees, reintroduce species that have gone extinct due to human pressures, make farming more environmentally friendly and for wildlife crimes to be investigated. We have had some success, usually working with other like-minded groups, but mostly it is a frustrating business. Very few Irish people are members of environmental organisations. The biggest organisation of any kind in the UK is the Royal Society for the Protection of Birds (RSPB), which has over a million members. A similar-sized organisation in Ireland, correcting for the difference in population, would have at least 71,000 members. I estimate that combining the members of all environmental groups in Ireland might get to a third of that figure. And the UK has many prominent organisations other than the RSPB that are concerned with wildlife issues, such as the Wildlife Trusts (membership 800,000), Greenpeace, or the World Wide Fund for Nature (WWF). In fact, Ireland is the only European country with no branch of either Greenpeace or WWF. Are Irish environmental organisations failing to connect with people? On visiting colleagues in Ulster Wildlife, a similar outfit to the IWT in Northern Ireland, I listen in awe and envy when told of their burgeoning membership of 12,000 people and the staff who are dedicated to their campaigns. During my term as chairman of the IWT we had two paid staff which was supplemented at times by interns or people on work experience programmes.

In June 2013 the UK's Department for Food and Rural Affairs announced that there would be a pilot cull of badgers in restricted areas of England (West Somerset and West Gloucestershire). Since the 1980s they have been implicated in the spread of tuberculosis in cattle, a disease with serious repercussion for farmers. Despite spending a colossal £40 million on one of the biggest and most thorough field-based scientific studies ever carried out on wild animals, a study that concluded that culling could make 'no meaningful impact' on bovine tuberculosis (bTB) levels in cattle,[14] the British government decided that landowners in the affected areas could deal with the problem through 'free shooting'. In other words, farmers were free to shoot badgers on their land at will. The uproar and level of protest that ensued has never been witnessed on this side of the Irish Sea for a wildlife issue. Mass protests were organised including a march on the Houses of Parliament attended by thousands of people. The cull was debated on local and national media outlets. The level of personal invective directed at the UK's then-Environment Secretary, Owen Patterson, during the debate was on a level not seen in Ireland even during our most heated conflagrations. The anti-culling lobby had well-known faces behind it, particularly ex-Queen guitarist Dr Brian May but also leading voices of authority such as Sir David Attenborough, *Guardian* columnist George Monbiot and wildlife TV presenter Simon King (search for 'stop the cull' on YouTube). Even non-wildlife personalities weighed in, such as Sir Roger Moore, Dame Judi Dench, Brian Blessed and Joanna Lumley. The matter was brought to the courts by The Badger Trust, a campaign group, but the effort was unsuccessful and the cull went ahead. When the shooting started, however, things really heated up. Squadrons of volunteer activists in the affected areas mobilised their forces. Some travelled great distances to disrupt the cull while it was happening, staying up all night, outdoors in wintery conditions, and noisily, but peacefully, scaring off the badgers before they fell into the crosshairs. It was a tense and dangerous situation and one which thankfully passed without injury, at least to people. The cull was unexpectedly extended after the target number of animals to be shot was not reached, with Secretary Patterson bizarrely claiming that the badgers had 'shifted the goalposts'. It finally came to an end in the week before Christmas 2013. In total it is estimated that 1,862 badgers were shot, and that most of these were not carrying bTB.[15] In late 2016 culling continued in England, in an extended zone, with both sides digging their heels in.

Its effectiveness remains to be seen but so far has proved to be useless in fighting bTB. That politicians should overlook scientific advice in order to be seen to be doing something is neither new nor unique to the UK. But what astounded me most was the reaction of the British people. While ultimately they failed to stop the cull they left nobody in any doubt that it was proceeding only to satisfy the vested interests of the National Farmers Union whose members tend to be Tory supporters. Prime Minister David Cameron's 2010 election pledge to be the 'greenest government ever' was called into question, and the claim was largely dropped in the run-up to the 2015 election campaign (which Cameron went on to win). Badgers in England are not an endangered species, their population is well counted and assessed[16] and they occur in the highest densities from any of the countries within their range.[17] English farmers face a huge challenge in confronting bTB and the pressure to cull continues. However, the badger itself is not at risk and non-cull disease control approaches are being vigorously adopted elsewhere in the UK. The protest may well ultimately be seen as a success.

But contrast the 1,862 badgers that were killed in the English cull against the 7,000 that are snared and shot in Ireland *every year* by Department of Agriculture contractors under licence from the

NPWS. In 2013 the total number of badgers that have been killed in this way exceeded 100,000 since the cull started back in the 1980s. Badgers are culled throughout the year, even during the breeding season when cubs are left in their underground setts to starve. There is no census of badgers in Ireland so it is likely that badgers are now rare or have disappeared entirely from many areas of the lowlands where culling has been going on for years. The NPWS assessed badgers as being of 'least concern' in their conservation evaluation of Irish mammals[18] but this was based on limited hard data. Meanwhile, information received by environmental groups in 2016 showed that the NPWS have no role in overseeing the culling and were rubber-stamping licences despite having serious concerns related to animal welfare and long-term conservation effects. Combine the annual removal of 7,000 badgers a year over three decades with roadkill and illegal persecution and badgers in Ireland may well be under serious threat. In 2011 the IWT started a petition to end the cull and we were sure that when people became more aware of what was happening we would be inundated with signatures. We weren't. In fact it took us two years to reach our target of 10,000 signatures despite a pretty sustained campaign by us through social media. A similar petition in the UK overflowed with 100,000 signatures after its first week. Conn Flynn and Fintan Kelly, who did the bulk of the work researching the issue for the IWT, uncovered the fact that €70 million was spent on the cull programme in 2011. Perhaps when hospitals were being closed and special needs teachers being laid off to make cuts under Ireland's despised bailout we would start to question whether this money could have been better spent. There was great support from people we met from across the country, including many farmers who suspected that the badger was not responsible for the bTB problem. Our campaign attracted some media attention and we could see the attitude of the Department of Agriculture personnel, with whom we had been in reasonably regular contact, change from mild alarm that their arguments might be under threat at the start of our campaign, to one of relaxation as they realised that there would be no political pressure to re-evaluate the rationale for the cull.

Badgers are unlikely to go extinct in Ireland. There are lots of places away from intensive farmland in which they still thrive, not least our cities and some of the remaining fragments of natural woodland. They are hardy and adaptable and if they have survived this long they are probably here to stay. Yet a lot of taxpayers' money is spent on a programme that is likely to be wiping out badgers from much of the country while failing to reach its stated aim of eradicating bTB after nearly thirty years. Why does this not stir more debate?

In global terms, Ireland was never a particularly rich place for wildlife, at least in terms of the numbers of different plants and animals that live here. There are probably suburban gardens in Costa Rica that have a greater variety of species than in all the island of Ireland. For this I cannot blame the government. For a start, islands tend to have fewer species than continents. This is one of the basic laws of biogeography – the study of where plants and animals live. Britain, being a bigger island, has a far greater number of species across nearly all groups than Ireland. It is true of mammals, birds, flowering plants, trees, amphibians and reptiles. In turn, mainland Europe has more species again. So because of its small size and isolation Ireland was never going to be a hotspot for biodiversity.

In their favour, islands have a disproportionate number of unique species. In other words, they tend to have species that are not found anywhere else. This is true, for instance, of Madagascar with its lemurs, the Galapagos Islands with their flightless cormorants and giant tortoises, or New Zealand with its kiwis. These islands have been isolated for so long that evolution has had time to shape and form new species. In many of the small islands there have historically been no land predators, which has promoted flightlessness in birds, hence the dodo in Mauritius, the kiwi and extinct moas of New

Zealand and even the penguins in Antarctica. In Ireland's case successive ice ages in the last 100,000 years have caused ice sheets to descend from the north like the bar on a giant 'etch-a-sketch'. Most animals and plants could not have survived such a dramatic and severe environmental disruption. So unfortunately Ireland doesn't have unique species simply because our wildlife has not had the millions of years of isolation required for the hand of evolution to fashion its creations. There are a small number of animals that show signs that they are evolving into distinct species and the hare, stoat and the dipper (a type of bird) are three that are officially considered subspecies, a kind of acknowledgement of their uniqueness but not different enough from their British or European cousins that they couldn't start a family if they were pushed.

While it is possible, and the subject of continuing debate, that some plant species survived the ice sheets, perhaps on mountain tops that jutted above the frozen mass, ice-age Ireland was more or less a lifeless and hostile place. Like the centre of Greenland today, there would have been no mammals, no birds and, in the main, no vegetation. During the height of the most recent ice age, a 1km-thick sterile sheet of ice covered all but the most southerly part of the island. South of this lay a frozen tundra-type landscape, perhaps not lifeless but certainly with a restricted number of plants and animals. When the ice began its retreat, about 13,000 years ago, only a very few large animals, adapted to the tundra habitat, remained. The wolf, the mountain hare and the stoat are among these pioneers. Bears, lynx, and wood mouse are also believed to have been present at this time. Birds, bats, invertebrates and plants colonised by air or by piggybacking on larger animals (i.e. sticking to fur or feathers). There is no evidence that an ice-free land bridge linked Ireland to mainland Europe in this time, unlike Britain which was a part of the continental land mass until perhaps 7,000 years ago, and for this reason it seems many animals such as the red deer, beaver, moose, dormouse and auroch (the antecedent to the domestic cow) were never to be found in Ireland.[19] Ireland's archaeological record is quite poor compared to Britain and it is hard to be sure about which animals did or did not make the journey. While woolly mammoths, reindeer and hyenas were in Ireland at one time or another there is no evidence that they survived the last ice age and were probably not seen by the earliest human arrivals. The magnificent Irish elk, with its 3.5m antlers, did survive until about 9,000 years ago but became extinct before the arrival of humans (assuming humans arrived a little later, something which itself is not certain). It may have been that a change in climate and the associated changes in vegetation, particularly the growth of high forest, played a significant role in the demise of the giant deer. All the same it is quite possible that new evidence will come to light, perhaps underneath a worked-over bog or in an as yet undiscovered cave, that will illuminate our knowledge of this poorly understood time.

Nevertheless, the arrival of humans, possibly around 9,000 years ago, would ultimately have profound implications for the developing ecosystem. There are many environmentalists who see human intrusion as inherently linked to the decline of nature and paradise lost. However, it is equally possible to look at humans as just another species of large mammal that was expanding its range with the retreat of the ice. Just like the wolves, the bears and the plants that were colonising new lands freed from the ice, humans were availing of new opportunities and were driven, not by a thirst for conquest, but a struggle for survival – finding food, shelter and rearing their young to reproductive age. Humans had the tools, in the form of boats, weapons and clothes, to cross the Irish Sea and establish settlements. Unlike today, people at that time were subject to attack by large predators (there were lions throughout Europe at the time). They probably subsisted on foraging rather than hunting large game (reading Jared Diamond's *The Rise and Fall of the Third Chimpanzee*, it is likely that hunter-gatherers relied on small game such as birds and fish, as well as foraging for shellfish and berries, rather than the traditional image of men arriving home with deer draped across their shoulders). The large number of coastal middens (prehistoric rubbish heaps) in Ireland seems to bear this out. Humans had evolved in Africa and spread slowly into southern Asia and Europe. Current theory has it that large animals in particular learned fear of humans and kept their distance, which is why elephants, rhinos and lions, etc. still inhabit much of Africa and Asia today. There is now a fair consensus that the arrival of humans in Australia and the Americas precipitated a mass extinction of large animals, probably because the fauna had no reason to fear people and so were easy targets. The reasons why large animals went extinct in Europe are less clear and while there is no doubt that hunting took place, climate change and habitat loss must also have played their part. It is therefore reasonable to assume that in the rapidly changing landscape of the European continent humans were no more of a disruptive presence than any other species in the complex web of life. In Ireland they would have been no different. Before long, though, people started changing the environment in profound ways.

As technology improved more trees were felled. The first species to disappear was the lynx although its record is scant in Ireland, just one bone (a femur) from a cave in County Waterford dated to 8,875 years ago. It is a shy woodland animal that today is confined to northern and mountain climates so it may have declined as a result of climate change or habitat loss. It may also have been hunted for its fur. Next to go was the brown bear, which survived in Ireland up to the Bronze Age, about 5,000 years ago. Bears are a greater threat to people than lynx and may have attacked the livestock of the early farmers. It can easily be imagined that fearful communities sent forth bands of hunters expressly to exterminate local bears.

Today bear densities in Europe vary significantly and have been influenced by habitat changes and hunting. Paddy Sleeman, from University College Cork, has studied the likely wildlife populations of Ireland in the Mesolithic period. He estimated that there was a bear density of only 1 per 100km^2. This would give an estimated population of no more than 850 bears in the whole of Ireland.[20] The population would have been vulnerable to hunting and a determined, even if uncoordinated effort by people to eradicate them entirely would have been relatively easy. In any case, there is no trace of bears in Ireland after the Neolithic period, about 5,000 years ago; a record from this time placed them in Lough Gur, in County Limerick. In Britain they may have hung on until the twelfth or even the fourteenth century. The brown bear may have been the first example of an animal that was exterminated from Ireland as a direct, perhaps the deliberate, result of human activity. But it wouldn't be the last.

From the time that people abandoned their hunter-gatherer lifestyles and became farmers there have been enormous changes in our landscape with widespread habitat change and a subsequent extinction of species. Such changes are always easier on an island than on a larger land mass because populations are invariably smaller and there are fewer refuges where animals can hold out. By my calculation just over 115 species have become extinct in Ireland as a result of human pressure, either deliberately – in the case of the wolf and the golden eagle – or inadvertently through over-exploitation or loss of habitat – in the case of cranes, bitterns or red squirrels. This figure is likely to be an underestimate as data is lacking on many groups of species. Meanwhile, other species were brought to the brink, and both the pine marten and red deer came close to being wiped out until timely laws intervened to protect them in the 1970s. The Killarney fern was nearly collected out of existence by wealthy Victorians during the 1800s. The last species to be officially extirpated was a small farmland bird, the corn bunting, which died out in the late 1990s. Today a number of species hang on by a thread, including the curlew, corncrake and the freshwater pearl mussel. The nightjar (an upland bird) and, at sea, the angel shark, may already have vanished.

Admittedly, this is a more complex picture than it first appears. Some, such as the red squirrel and golden eagle, have been reintroduced although they remain under threat. Others, such as the corn bunting and the corncrake, were abundant because of agricultural practices of the time but then went into decline when these changed abruptly in the 1970s and 1980s. They originally benefited from the changes to the landscape brought about by human presence as they are not forest creatures. Nevertheless, most of the species that have gone extinct have remained extinct and this loss has been driven by one equal in scale to the habitats upon which they depend.

The primeval oak woodlands, which can be glimpsed today in Killarney National Park and a handful of other scattered remnants, were felled with nothing more technologically advanced than an axe. By the middle of the eighteenth century Ireland was mostly a treeless landscape and this can be seen in many of the paintings from this period. Really old trees, which are a vital part of forest ecosystems, are today missing and only 1 per cent of our land area is covered by native woodland. Overall, we are a more wooded island today than a century ago, with trees covering about 12 per cent of the land, still low by European standards, but enough to be a more visible component of our landscape. However, this has not benefited our wildlife because much of this is serried ranks of non-native conifers. Dark and impenetrable, they are practically devoid of natural life and pollute our water through acidification and nutrient run-off. The first forests were replaced by farmland and enclosed with the familiar hedgerow field boundaries that define much of the landscape today. Until recently farming was a low-intensity practice, essentially organic, that accommodated a wide range of plants, birds and insects, alongside the crop or the animals that were being reared. This has been replaced with a system of intensification under a production-based regime of subsidies that pays little reward for environmental protection. Today's farms are not the wildlife-rich places they once were. Nature has been squeezed into the thin ribbons of treelines and hedgerows that are themselves under threat from a combination of one-off houses, wire fencing and lack of management. Meanwhile, to maintain the levels of production required to fuel the demands of our agri-food sector, tonnes of fertiliser, both artificial and animal manures, as well as chemical sprays are spread on the land. Too much of this finds its way into streams and rivers, causing pollution and depleting aquatic life. It combines with the discharges from thousands of malfunctioning septic tanks and wastewater treatment plant outfalls and accumulates in our estuaries, nearly half of which are polluted

(Galway Bay and Clew Bay are exceptional among those near large urban centres in their high standard).

By the Middle Ages those areas not used as farmland were bog or wetland. While turf has always been cut by people for fuel it was not until the establishment of Bord na Móna in 1946 that the large-scale destruction of our bogs occurred with industrial efficiency. Today the raised bogs that have been at the centre of so much acrimony are the final remnants, less than 1 per cent and shrinking, of this rich and important habitat. Meanwhile, our rivers were deepened and straightened and our wetlands drained in an effort to make agriculture more productive. Our uplands have fared little better. Overloaded with sheep, under a subsidy regime that paid out depending on the number of animals under the farmer's possession, the hillsides came close to total collapse in the 1990s. The sheep ate everything, leaving the underlying peat exposed and running into the nearest river during the frequent rains. Even though the upland forests had been long since stripped away, any remaining wildlife interest was lost under the trample of hooves. The red grouse is one of the few birds to be found in the artificially treeless uplands. Its historic range collapsed by 70 per cent in the forty years to 2000. Where once it was found in all upland and mountainous areas in Ireland it has now vanished from many of its haunts, including the Cooley Peninsula in Louth and the Dingle Peninsula in Kerry.[21]

For most of human history, people went about their business unaware that their actions could cause serious or irreversible harm to their environment. It seemed impossible that another species could be driven to extinction. It is understandable that earlier generations prioritised their economic development over what only seems from hindsight to be environmental concerns. It is totally rational that society chose to find a 'use' for the bogs, or that 'unproductive' land should be planted with tight-fitting and fast-growing trees. The word 'ecology' has only been with us since the 1960s while widespread concern for environmental matters among the general population only emerged in Ireland in the 1970s through the formation of environmental groups (such as the Irish Wildlife Federation in

1979 or the Irish Wildbird Conservancy in 1968). A Department of Environment within the government dates from only 1977. We have had nearly forty years, therefore, of an 'environmental awareness' – i.e. that our actions were having negative impacts on our land, air and water that needed to be addressed. Today, we make choices with a much fuller understating of the consequences.

A lot has changed for the better in the last four decades. For one thing, our planning system is much better at accommodating environmental concerns. We have a lot of information available to us over the Internet or, if needs be, the use of the Access to Information on the Environment legislation. People are better informed than ever before. And after losing so many of our animal species some started to return. In 2001 golden eagles once again became a feature of the Irish landscape. They had disappeared over 100 years before as a result of persecution. Their return not only saw the reinstatement of a magnificent bird that was an important element of our natural heritage, it also marked the re-establishment of top predators in Ireland. With the loss of our eagles, and the wolf before, our ecosystems had been decapitated. If the loss of one species has the potential to disrupt the balance of life, the loss of a whole layer in a food web can turn things on its head. Since 2001 there has been much to celebrate with the reintroduction of white-tailed eagles in Kerry and red kites (another bird of prey) in Wicklow, Down and Dublin. Thanks to the efforts of the Golden Eagle Trust, all three are now breeding again and this must surely give us hope that the path of nature is not fixed on a downward trajectory. In 2012 a pair of white-tailed eagles nested on an island in Lough Derg, near Mountshannon in County Clare and within sight of a pier in the centre of the town. Since then, volunteers have manned telescopes to allow visitors to get a better look and answer their questions. It was a public relations gift. The eagles got lots of publicity and when, in 2013, two chicks were reared for the first time, people flocked to see them. It was a boon for the local economy and proof positive that the eagles could provide an economic as well as a cultural and environmental dividend. But the reintroduction projects have been beset from the start by illegal poisoning and shooting. The Golden Eagle Trust knew this would be a problem. They have worked with local landowners to promote the rewards of successfully returning the eagles, and assuaging fears that they would be a threat to livestock. In February 2014 one of the two Irish-reared chicks was found shot in a field in Tipperary. Cruelly, the bird survived the injuries, being hit only in the wing and leg, but was left to starve and endure great suffering. In April 2015 a white-tailed eagle about to lay eggs was found dead on its nest in Connemara. It had been poisoned and was a major setback to the birds becoming established in this area. It wasn't only a blow for wildlife, the region as a whole has lost out from the tourism potential that the eagles would have generated.

Lacing the countryside with poison has been going on for decades and only became a prominent issue in light of the eagle deaths. The law was changed in 2008 to make it illegal to lay poisoned meat bait. Poison on the land affects not only eagles and red kites but also any other animal that comes into contact with the bait. To date, nobody has been charged with these crimes.

In my early days of studying ecology I would look out on the sea and think of it as a sort of wilderness expanse, little touched by human hand. But the seas around Ireland have been fished on an industrial scale for at least 150 years and to such an extent that overfishing and habitat loss have taken a heavy toll. Because this damage is more or less invisible, however, people look out from our coastline onto a scene that has not changed in thousands of years. Fishermen in particular are deeply unhappy with a system that has seen coastal fishing communities disappear. In 2014 RTÉ aired a TV programme by Hector Ó hEochagáin that attempted to summarise the issues facing the fishing sector. In it, the host described how, despite having productive waters teeming with fish, Ireland has traditionally turned its back on the sea. Successive governments failed to invest in a fishing fleet or the infrastructure required to bring fish to market. When we joined the European Economic Community (now the European Union) in 1973 there was a back-room deal that saw the seas' riches bartered away in exchange for more favourable terms under the Common Agricultural Policy. When the first Common Fisheries Policy was agreed between member states in 1982 Ireland became locked within a system, known as 'relative stability', whereby the size of the catches for each country would remain the same relative to each other. This size was based on a historic reference point so that countries, such as France and Spain, that had large fleets, would win a higher proportion of the total catch than those with small or underdeveloped fleets such as Ireland. Thus, since 1982, despite having 20 per cent of EU waters, Irish fishermen can only ever catch around 4 per cent of the fish. It was the worst of deals that haunts the industry to this day. How that 4 per cent is divided up is decided every year at the December Council of Ministers meeting in Brussels where a quota for each fish species is given out. The concerns of the Irish fishing industry generally centre on this annual availability of quota. In the programme Hector interviewed the then Minister for Agriculture, Food and the Marine, Simon Coveney, who was forthright in saying that, despite its inherent unfairness, 'relative stability' will never change because none of the other countries wants it to change. Government agencies will have their plans, politicians will wring their hands, and like the punter who lost the winning lottery ticket, the fishing industry will forever stare at the sea with a tear in its eye.

It is widely acknowledged that Ireland got an awful deal under the Common Fisheries Policy. Unlike France or Spain, Ireland never had a well-developed fishing industry. In his fascinating book on the history of our sea fisheries, *Troubled Waters*, historian Jim Mac Laughlin reveals how central authorities never seemed to take fishermen seriously, how tonnes of freshly caught fish would rot on piers for a lack of ice to preserve them en route to market, and how the uncertainty and seasonality of the catch, combined with a grinding poverty, meant that fishermen would frequently sell all their gear and turn to farming to make ends meet.[22] But there is another story that is less well known. There is only the slightest hint of it in Hector's programme when Minister Coveney defends a rules-based system because in its absence 'we would fish out our waters ... like what happened in the Irish Sea'. It was a telling comment and viewers were left guessing as to 'what happened in the Irish Sea'. Today the Irish Sea is an ecological ruin. The only significant fishery left is not for fish at all, but prawns, a crustacean which was once thrown overboard when fishing for whiting. Later the reverse was the case, as whiting were dumped (because of lack of quota) in the fishery for prawns (this practice of dumping unwanted fish, referred to as 'discarding', is thankfully coming to an end). The reason why prawns switched places with whiting has been attributed to the overfishing of large predators from the Irish Sea, such as cod and whiting, but also the other larger fish such as sharks and rays, leaving the prawns to increase greatly in number. In the 1980s whiting, as well as Irish Sea cod and sole, were regular mealtime features for Irish families. No more. According to the Marine Institute, the state body which gathers the science behind commercial fishing in Irish waters, all three species have 'collapsed' in the Irish Sea.[23] Indeed, these days, far from needing more quota, Irish fishermen struggle to catch the quota they have been given. As fisheries scientist Ed Fahy says on his blog www.eatenfishsoonforgotten.com: 'the industry in Ireland constantly bemoans "lack of quota", but the real problem is shortage of fish'.

The story of the Irish Sea, of an abundance of many different types of fish to only one or two, is one that has been repeated around the coast of Ireland. Until the arrival of steam trawling in the 1800s nearly all fishing was done in small boats close to shore, something that provided enormous employment in harbours dotted around the coast. But fishing as a way of life in these places has all but disappeared. There are now fewer ports, with much larger boats, heading much further out to sea, and with smaller crews. The story of our fisheries is one of seas that once really did teem with fish – colossal shoals of herring that burst nets, single cod that would feed a large family, carpets of oysters that spread out from every high tide, runs of migratory salmon that made the waters of our estuaries churn and boil. At the time it must have seemed that this incredible abundance was inexhaustible. But then came bottom trawling, first with sails, and then steam power, scraping the sea floor and gathering all in its wake. Ever bigger boats, with ever more sophisticated gear that allowed fewer fish to escape began to draw down what is, in theory, an infinitely renewable resource. Once one species became overexploited, the boats, usually backed by state grants, simply moved on to the next one. One by one, productive fisheries that supported livelihoods and local economies vanished. One testimonial recalls the memories of Gerry O'Shea, from Howth in County Dublin. He recounts how, when he first started landing at Howth around 1956/57, there was a good living to be made. 'I once landed over one hundred boxes of fish that included eight boxes of plaice, three boxes of prawn tails and the remainder of good cod.'[24]

That's eighty-nine boxes of cod on a good day. Very little now comes into Howth, except for prawns, and the fish shops that line the quay source their produce from around the world. Government agencies established to spend public money on developing the seafood sector never prioritised the sustainable management of this renewable resource, and have yet to produce a plan to restore this abundance.

Scuba-diving clubs still grope their way through the inky murk but the once gin-clear waters off Dublin and Wicklow have been churned up so that visibility is rarely more than a few metres. In Iceland it is still possible to snorkel in shallow water with cod that are over 2m long but the chances of an unexpected encounter with a giant off Howth or Skerries are slim. Under the Common Fisheries Policy Irish boats have to share waters with those of any other EU country that wants to fish here but that is only the case beyond 9.6km from the shore, the so-called six-mile limit. Inside this limit Irish boats have the water to themselves and our authorities can manage it as they see fit. It is here that traditionally the vast majority of fishing occurred before the advent of the bottom trawlers.

In 2015 the Irish catch was worth over €205 million but over half of this (€123 million) was made up of only three species: Dublin Bay prawns, mackerel and horse mackerel.[25] Prawns are now the principal source of income for the Irish Sea trawling fleet. As exploitation of prawns has increased dramatically, not only in the Irish Sea but in the Celtic Sea and the Atlantic, there are worrying signs that it too is now being overfished in places. Mackerel is an open-water fish that migrates through the cold waters of the North Atlantic. It is shared by the countries of the EU as well as Norway, Iceland and the Faroe Islands. In July 2010 the mackerel stock was certified by the Marine Stewardship Council (MSC) for its sustainability, a rare accolade and the only such certification in Ireland at that time. However, the shoals of mackerel, once confined to waters north of Ireland and Scotland, have since moved further north, perhaps as a result of climate change. The Icelandic fleet massively increased its take of the fish but the EU and Norwegian boats continued to take what they had always taken. The result is the unfolding tale of yet another decline and in March 2012, after only two years, MSC suspended its certification, citing the inability of all states targeting the fishery to agree on a quota.[26] The Marine Institute's scientific assessment of fish populations, published each year as *The Stock Book* has indicated that it is not being fished sustainably. Horse mackerel, meanwhile, will not be found on any supermarket counter or fine-dining menu. It is a bony fish and the 21,600 tonnes that were allocated in 2015 went to make fish meal, i.e. they were caught to be turned into food for other fish, such as farmed salmon, or were exported to West Africa. It requires on average around 2.5–3kg of these wild fish to produce 1kg of farmed salmon.

In April 2013, the business development manager of An Bord Iascaigh Mhara (BIM), the state body which promotes the seafood industry in Ireland, wrote an article for the *Irish Examiner*, entitled 'Ireland can become a global player in the seafood sector'. In it he said:

> It is not an exaggeration to say that Irish seafood is about to take off. The market opportunities are vast. Given this changing landscape, and as an island nation adjacent to the abundant fishing grounds, how can Ireland take advantage of

this tremendous global opportunity? Firstly, we need more raw material – we can achieve this in a number of ways, including the development of new salmon farms, harvesting new species such as boarfish, and by attracting foreign landings into our processing plants.[27]

More fish (raw material) will be needed in order to realise this golden windfall but the wild fish catch peaked in the early 1990s and has been on the decline since. Indeed, BIM reported a fall in sales in 2013 to €810 million, from €822 million the year before, attributed to what the chief executive described as 'one of our key issues – lack of supply'.[28] The idea of targeting new species seems optimistic given the scale of the exploitation that has already occurred. The announcement of an export market to China for the tiny boarfish, a species that had heretofore been cast overboard with the other unwanted catch (and what Minister Simon Coveney referred to on radio as 'trash fish'), seems only to underline the relative poverty of the unexploited ocean. Predictably perhaps, the boarfish stock has collapsed due to mismanagement. The other suggestion was the need to develop new salmon farms. To this end, BIM proposed a massive 15,000-tonne salmon farm to occupy a large area of Galway Bay near the tourist hotspot of the Aran Islands. While sounding like a sensible way to produce more edible protein from the sea, the reality is that raising salmon in cages at sea is beset with problems. They are implicated in passing parasitic sea lice to an already stressed and diminished wild fish population, the food required to feed all these salmon will come from wild fish and so further deplete stocks elsewhere around the world, and the pollution caused by antibiotics, pesticides and salmon faeces will have grave knock-on effects for the general health of the marine environment. In the face of mounting local opposition the plan was ignominiously dropped late in 2015.

Today, not only has the quantity of fish in the seas around Ireland greatly diminished but the potential for tourism is also much reduced. In a 1968 article in the British angling magazine *Fishing: Coarse and Sea Angling*, entitled 'That Fabulous Irish

Fishing', the author bemoaned the decline of sea bass at his native haunts in England. 'Most readers,' he enthused, were 'well aware of my pre-occupied love for Ireland, especially the south-west, where I have taken so many large bass from the beaches of Co. Kerry.'[29] A guidebook to the sea angling in Ireland which appeared around the same time declared 'Ireland's sea coast is one of the last left unspoiled in Europe ... For the coastal waters of Ireland abound with fish.'[30]

The author went on to describe some of the best bass angling to be had from Wexford to Cork and Kerry. But in the 1990s restrictions on bass fishing were imposed due to overfishing and even today the stock of the fish has yet to rebound. Ashley Hayden, a regular contributor to the angling literature and native of County Wicklow, summed up conditions in 2010: 'Due to a combination of factors which include commercial overfishing, mussel dredging, and unregulated whelk fishing, the once rich fishing grounds between Bray Head and Wicklow Head now lie denuded and degraded.'

All the same, it is estimated that in Wexford sea bass fishing was worth €1.4 million to the local economy in 2008.[31] This is a glimpse, surely a tiny fraction, of the potential that exists were fish stocks to recover fully.

Why does all this matter? Ireland, after all, is a decent country by international standards. Some may gripe but the United Nations ranks us seventh in the world in its Human Development Index and we are among the happiest nations in the European Union.[32] The same UN index puts us a much more lowly fifty-first in environmental quality. But if our environment is really so bad, why is it not more obvious to us? Why does it not seem to be impacting on economic indictors or our quality of life? There are certain sectors that feel it every day, particularly those who earn their living from the sea or the poorer land in upland and coastal districts. These communities have seen their livelihoods decline in lockstep with the wildlife around them. But in population terms this is a small number of people. In the cities and towns, or the factories, farms and offices across much of the country the fortunes of nature are one step removed from our own standard of living.

A growing mountain of scientific research is demonstrating that we are in the midst of an ecological catastrophe, principally from the twin evils of climate change and biodiversity loss. The latter is exacerbating and limiting our ability to adapt to the former. Nature provides us with our food, recycles most of our waste, pollinates many of our crops, provides us with building materials and textiles for clothing, protects our homes from flooding (when they haven't been built on a floodplain), and it purifies our air and water. It plays a crucial role in regulating our climate and so makes Earth habitable. No mean feat. And it does this at no cost. In 2008 the Department of Environment, Heritage and Local Government put the value of the services provided to us by biodiversity at €2.6 billion per annum.[33] This figure is surely too low as one cannot put a price on the benefits that a healthy outdoors brings to our mental health or general happiness. And yet we are watching nature diminish slowly, steadily, inexorably. Perhaps we are on the edge of an unforeseen calamity that will dramatically result in a decline in our living standards because of insecure food supplies, undrinkable water, unbreathable air and a grim landscape that leaves our souls malnourished. It's the kind of reality that is unfolding in parts of China and some other developing countries. However, most of the problems that China and others are facing are solvable, and so I don't believe that this dystopian nightmare is due on our shores any time soon. Rather, I think that the decline in nature has left us with a diminished country, one that could be so much better.

Ireland is a beautiful country that, in places, still has the power to take your breath away, but that is despite our best efforts, not because of them, and is partly a side effect of our cultural past. It certainly has little to do with wise management of our resources. In a recent poll by the travel gurus at Rough Guide, Ireland came in as the nineteenth most beautiful country in the world. Tourists say that one of the main reasons they come to Ireland is the value of our landscape, so if we don't look after it then we are eroding the basis of one of our most valuable industries. Scotland is estimated to earn £1.4 billion (€1.7 billion) from wildlife and adventure tourism every year and is marketing itself as Europe's No. 1 wildlife-watching destination. In fact, it is considered that 40 per cent of all spending there is wildlife or outdoor-adventure based.[34] No comparable figures are available from Fáilte Ireland, the state body that markets Ireland as a holiday destination, although the estimated €7.2 million earned from the Great Western Greenway cycle and walking route in Mayo gives us a glimpse of the potential. In the top twenty free visitor attractions there is only one National Park, Connemara, which comes in tenth.[35] In the USA there are three in the top ten (Denali, Yellowstone and the Grand Canyon). Surely we are not making the most of what are our greatest natural treasures. In Ireland only the Burren, within which there is a little known National Park, really succeeds in being a world-class outdoor attraction, something that has come about first through protest and rancour, followed by local support and genuinely sustainable initiatives. The success of the Burren is due to the careful work of grass-roots organisations that cooperate with farmers and other landowners in a sustainable manner. The Burren is a farming landscape but one that is nevertheless rich in wildlife, layered upon an ancient human presence and a savvy small business culture. Here, strict designations for nature conservation have been an opportunity to create a unique identity, one which has received international attention. Ireland has six National Parks, each of which are small kernels of government-owned land in wider areas of natural and cultural interest. Those at Killarney, Connemara, Wicklow Mountains and Glenveagh in County Donegal have their unique problems and are each failing to reach anything near their potential in terms of wildlife and economic opportunity. The newest national park, at Ballycroy in County Mayo, is at the centre of a unique 'wilderness' experiment which is exciting. It is intended to allow natural forces to reassert themselves and give visitors an opportunity to enjoy a unique outdoor experience. Will we see the reintroduction of long-lost species such as bears, wolves or lynx? The question remains to be answered. The paradox is that the Burren has succeeded across the whole landscape, not just within the boundaries of the National Park. What if our National Parks were managed, not as mere tourist attractions as they are now, but for their wildlife and landscape value? If the lessons learnt in the Burren were to be replicated across the other five regions mentioned and their wider cultural hinterlands, there would be immense dividends, both economically and ecologically.

Tourism, however, is only one of the economic sectors that would benefit from a prioritisation of nature protection. The communities that have suffered most directly from the loss of nature are surely fishermen. Fishing, after all, is not farming, but hunting wild animals. Fishermen depend upon healthy seas to make a living. Prognostications of the demise of fishing as a way of life are being realised all along our coast because the natural wealth of our seas has been depleted. A report by the New Economic Foundation found that restoring seas to full health, something that is still possible, would add £2.7 billion (€3.2 billion) in additional revenues and 100,000 extra jobs across Europe.[36] This does not take into account the additional revenue from sea angling or other water-based activities. And it is not just the fishing industry. Bringing wildlife back to our farmed landscapes would reduce flooding and improve water quality – two areas that are currently costing us many millions of euro. 'Natural' infrastructure is more effective, cheaper to install, easier to maintain, and lasts longer than concrete-based engineering solutions. A landscape with clean, freely flowing rivers, functioning wetlands and native woodlands would not only be more attractive but would provide us with a buffer against the unpredictable changes that are already occurring in our climate. Even intensive farming would benefit from a more wildlife-friendly approach. There are huge opportunities in becoming the 'green food island' referred to in marketing literature and providing high-end produce from farms that work to protect nature. Healthy hedgerows provide shelter for livestock from wind and rain, something which has a direct effect on the welfare of the animals. Dense, lush hedgerows keep neighbouring herds of cattle apart and so can help reduce the spread of disease. They are reservoirs of insects that pollinate flowering crops such as potatoes, rape and fruit trees. Protecting river catchments means clean water for animals is freely available, while the integration of woodlands and wetlands would reduce the extremes of flood and drought by storing water on the land for longer.

Then there are the benefits that cannot be measured. After a long day at work nobody chooses to go for a walk in an industrial estate. People seek out natural areas with more than just well-mown grass and uniform lollipop-type trees. We need nature to soothe our souls and stimulate our minds. Natural places with healthy ecosystems are places of adventure and learning, not just for children but adults as well. They exercise mind and body and make us happier and healthier. They are an antidote to our hyperconnected lives where we are bombarded with the demands and perceived wants of modern living. Increasing the space for nature around our homes reduces stress and even our energy bills (natural vegetation can shield our homes from cold winter winds). Children's imaginations are ignited by the messy tangle of branches and briars, and water spilling over rocks. This type of unregulated, unsupervised, uninstructed play is undoubtedly healthier for children than the regimented and controlled upbringing that is more and more commonplace. They are active for longer in a day, learn about real things in a real world and can independently assess risks. Such an upbringing is becoming rarer in Ireland, as it is in much of the developed world but there is a longing among parents and children to reconnect with this world. We as humans have an umbilical connection to nature. It is a connection that has existed for all of human history. The separated and electronic world that today we take for granted has only come about in a mere two generations.

Indeed, there are signs that changes are afoot. They are being led by small groups of foresighted individuals or merely nature asserting itself where human influence has faded. Our towns and cities are now more wildlife rich than much of our countryside. Sales of bird food are rocketing and our substantial parks are being subtly managed to encourage wildlife without restricting access to people. Out of the thirty-three species of land mammal to be found in Ireland, twenty-seven of them are seen regularly in and around Dublin, including our two types of seal. Beyond the city boundaries the reintroduction of eagles, the meticulous tending of the living landscape in the Burren, and the rewilding of parts of Mayo may not be perfect but they give us hope that change is possible. Most counties now have areas of forest that are freely open for all kinds of amenity use and are managed with nature in mind. We still need lots more forests – real forests, not just stands of conifers – but the state forest company, Coillte, deserves praise for recognising that forests are about a lot more than just timber.

This book is not a call for the 'rewilding' of our country, whatever that might mean. Humans are a part of nature and we must be made to feel welcome within it – not an observer or a custodian or a benefactor that 'allows' other organisms to exist. We must be able to feed ourselves and provide jobs for a growing population. The use of the land will be, as it always has been, an essential component of that. I dream of a country that can do that in a way that looks beyond short-term profits and perpetual predictions of untrammelled growth. I want to see nature restored to its full glory. This means reintroducing species that have been driven to extinction by our actions. It means a total reorganisation of the way we fish so that marine habitats and species can be restored to their former abundance. It means changing the way we farm so that water is protected, wildlife is encouraged rather than scorned, and that uplands are allowed to regain an element of their natural vegetation. It means protecting what is really important and valuing those things that do not appear on balance sheets. At the turn of the millennium the then minister in charge of natural heritage, Síle de Valera, granddaughter of Éamon, told the Dáil: 'We in Ireland are perhaps better off than many other countries in that we still have a great deal that is worth protecting, from the environmental perspective.'[37]

As the years pass the veracity of this statement wanes. The time is right for a new relationship with the natural world in Ireland because our relationship with the environment is showing the same cracks that were obvious in the financial world in 2008.

I yearn for politicians to talk about wildlife as though it was as important as health, education or roads. Some countries, such as France for instance, have a Minister for Ecology, and their politicians talk comfortably about the importance of nature to the country's well-being. Conflicts arise from time to time between people affected by laws that restrict activity, but the resources are provided to resolve these in a sensible way. Had this been the case in Ireland perhaps we would not have seen the anger over ending turf-cutting on protected areas, the continuing dismay of fishermen over the decline of their way of life, or the draining away of rural upland communities. In Ireland the model of nature protection follows the lines of 'designation – procrastination – fulmination – compensation'. This formula has left nature being seen as a problem, not a solution. Politicians, rather than leading, look for solutions from powerful lobbyists, and too often lack the will to defend the public interest. Underfunded conservation agencies are blamed for everything from flooding to destroying local traditions. Everyone pays a price, including future generations.

It is important to look to the past in order to appreciate what has been lost, but there is no point in trying to turn the clock backwards. This book aims to set out the scale to which we have lost nature and the effect that this has had on people and our landscape. But in doing so my hope is that we are spurred to action. Wildlife has enormous powers to rebound; the question is, do we have the desire to allow it happen?

2

The quest to catch the last fish

> 'I want to raise awareness about the fantastic, sustainable fish stocks we have right here, off our own shores, with plentiful supplies of mackerel, pollack, monkfish, skate, flounder, bass and even sustainable sources of tuna and Arctic charr.'
>
> TV chef Clodagh McKenna, *Fresh From the Sea*[1]

NAUTICAL CHARTS RETAIN a powerful aura of mystery and adventure. Even for landlubbers like myself, they evoke the era of exploration, with their unfamiliar contours and lines of latitude veering off into the unknown. In emphasising the sea rather than the land they present an odd inversion of our known world of roads and towns and little dots that mark the hilltops. The dots on these maps of the marine mark the unseen depths; their deepening blue contours seep into canyons only to be viewed in the mind's eye. Ever so slightly the curtain of blue is pulled back to reveal the valleys and plains of the ocean floor, the mountaintops that barely scrape the surface and reefs strewn with shipwrecks. Little dotted lines point to safe passage through deeper water while icons of anchors and lighthouses beckon vessels on stormy nights. Ireland's waters have been charted since the Middle Ages. Armadas sailed through them, early Christian monks rowed across their vast expanses to reach America, and Vikings mastered them in their quest for gold. Ireland's crenulated western coastline dissolves into the Atlantic swell and with its scattering of islands and skerries is on a literal and figurative edge.

The first people to arrive in Ireland came by boat and their prehistoric rubbish heaps (middens), full of fish scales and shells, speak of the importance of the sea to their survival. Life in these

Irishwomen photographed in the early twentieth century bringing fish home from market, near Galway. The large fish is a halibut, now probably extinct in Irish waters. COURTESY NATIONAL LIBRARY OF IRELAND

far-flung inlets and offshore islands today conjures a mix of freedom tempered by hardship and independence, yet subject to the vagaries of the wind and the whim of ocean swells. Then, as now, storms can bring life to a halt.

Roaringwater Bay lies east of Mizen Head, that finger of land that marks Ireland's most southerly point. Its wonderful name evokes the ceaseless currents that wash through the many islets and larger islands that enclose a body of water stretching from Brow Head to the Fastnet Rock in the south, to the cosy harbours of Baltimore and Ballydehob to the east. It has a long and colourful history that includes the Sack of Baltimore in 1631 that saw a raid by Algerian pirates from the distant Barbary coast. Villagers were locked in irons and carted off to a life of slavery on galleys or in the Sultan's harem. An English colony had been established here in 1605, pioneers lured by a rich fishery in pilchards. Pilchards, also known as sardines, are small fish, around 20cm in length, that shoal in open water. They feed on floating plankton and small crustaceans and were caught using seine nets. These nets are like curtains that float on the surface while a boat was rowed to encircle the shoal, something that no doubt required great skill and exertion. Baltimore was equipped, however, with fish 'palaces' and a strict system of licensing that excluded foreign vessels. The inhabitants were nearly all involved in the industry, which not only caught the fish but had adopted salting and packaging techniques that allowed their produce to be exported to France, Italy and even North America.[2] The pilchard fishery was so important to the south-west of Ireland that it accounted for 40 per cent of the customs receipts for the coastal region and is estimated to have employed 2,000 people on a seasonal basis. Descriptions of the fishery from the early seventeenth century are given in Jim Mac Laughlin's meticulously researched *Troubled Waters* from 2010. Here, an account emphasises the physical labour involved in finding and hauling in the fish, and the rich abundance of the shoals as they entered the bay. The fish are:

> Taken either by day or night, but mostly in the day, by means of hewers [a person who would shout out] placed on the adjacent high grounds above the bays. The nets are from 100 to 140 fathoms long, and from 6 to 9 fathoms deep; the net being shot or dropt into the sea, they surround the fish, having two boats to attend them, one of which is called the seine boate and the other the follower. The pilchards being thus enclosed between the two boats, by drawing both ends of the net or poles together, they begin to haul the net up, and bring the bottom and top of it together, this is called tucking the net; then by means of oval baskets which they call maons they empty the net of fish into their boats. The fish are brought out of the large baskets and are laid in the fish-house, which they call a palace.[3]

In 1623 it was reported that Ireland was exporting 20,000 tons of pilchard annually.[4] To put this figure in perspective, it is over four times the total landings for all species from the Irish Sea in 2014 and is exceeded today only by the individual open-water catches of mackerel, horse mackerel, and blue whiting.[5] Even if this figure is an exaggeration it gives some idea of the scale of shoals. However, it was not to last. The pilchard began its decline in the middle of the eighteenth century and by 1795 the Baltimore fishery was spoken of in the past tense. After this, shoals of fish continued to appear, albeit intermittently, well into the nineteenth century. An appearance off Kilmore Quay in County Wexford in 1835 was the first in living memory and 'people did not know their value and there was no means of saving them and great quantities were used as manure'. In 1879 the Baltimore Company was wound up after a number of unsuccessful years and then fell into a terminal decline. The pilchards didn't disappear entirely but subsequent appearances resulted in catches being dumped or used to fertilise fields because demand was not there to actually eat them. Today, there is no targeted fishery for pilchards and their occurrence in Irish waters is sporadic and of no commercial significance.[6]

Life in Baltimore and around Roaringwater Bay went on but far from being the prosperous centre of a thriving export industry it was described in 1833 as having 'no manufacture, and the town is very poor and does not seem to be improving'.[7] The Great Famine of 1845–52 hit particularly hard but a change of fortune was at hand when a wealthy English baroness, Angela Burdett-Coutts, joined forces with a local parish priest, Fr Charles Davis. They recognised that to raise the people out of the poverty in which they had become enmeshed they had to emulate the foreign boats that were profiting from fishing in the area. They had to abandon their open boats and subsistence ways for deep-sea vessels and market demand. In 1880 the Baroness established what would today be called a system of micro-financing, small loans at zero interest, and in 1887 the Baltimore Industrial School was established to develop the fishing industry for the long term. With the pilchard now a folk memory the attention of the boats was turned to mackerel. Mackerel is a much bigger fish than pilchard and can grow to half a metre, but like its diminutive predecessor it is oily, and so high in calorific value. It also shoals in open water, and seasonally migrates between deep waters of the open ocean in winter to warmer surface waters along the coast in summer.[8] The fishing started with the arrival of spawning shoals in mid-March and would last for about twelve weeks. The renaissance of Baltimore and the communities around Roaringwater Bay would be driven by mackerel – a story recounted by Séamus Fitzgerald in his study of the period from 1879 to 1913.[9] Mackerel was in great demand in the United Kingdom but also in North America where their own mackerel stocks had collapsed in 1888 due to overfishing. At its height in 1884, 7,200 tonnes of fish were landed in Baltimore. It was not the only port to benefit from this fishery, with nearby Kinsale and other harbours up the Cork and Kerry coast also taking their share. Nationally the mackerel fishery peaked in 1890 and was worth £212,672, nearly €27 million in today's money. However, after a few boom years landings began a steady decline. Changes in the distribution of the fish, fluctuating prices, a dependency on foreign entities to market and transport the fish meant that by 1910 the landings into Baltimore were no more than they were prior to the boom in 1880.

Today in Baltimore the ruins of the old fishing school crumble quietly away from the bustling town centre. Nearby the port is still home to fishing boats and according to the Sea Fisheries Protection Authority 1,500 tonnes of various fish and other sea life were landed in 2015 – no great change on a decade earlier but a fraction of the mackerel at its height.

The first hint that Roaringwater Bay and its islands were of scientific interest came in a 1972 report which anticipated the establishment of designated 'areas of scientific interest' throughout the country.[10] It coincided with a broader environmental awakening, both in Ireland and abroad, in the wake of that seminal photograph from 1972, known as the 'Blue Marble', which showed us Earth from space for the first time. Basking sharks, the second largest fish in the world, were seen regularly off Cape Clear island, mainly during the summer months, while schools of up to 250 harbour porpoise, a diminutive dolphin, were also regularly recorded in the sheltered waters of the inner bay. Grey seals and the smaller common seal were also frequent but despite the long history of fishing there was practically nothing known about life beneath the waves.[11] The Cape Clear Bird observatory was established in 1959 and had been documenting the wealth of birdlife on and around that island. The range of coastal habitats, from cliffs and caves to rocky shores and shallow inlets promote a wealth of species, while the incessant currents bring with them nutrients from the deep. Locals were no doubt aware of the presence of reefs lest they damaged fishing gear but it was not until very recently, with the demands of European legislation, that the true biological riches of the bay were uncovered.

Surveying marine life presents a number of challenges. You need to be able to scuba dive for a start, but you will also need to operate a boat to navigate the reefs and skerries, be able to identify the myriad variety of fish and invertebrate species, and add to that the discipline and tenacity to adhere to, and document the results of, a methodological scientific procedure. And all this in cold waters that are subject to frequent storms and swells. It makes surveys on dry land seem like a piece of cake. The first survey came from research carried out in 1975.[12] It recorded 142 species of algae and 187 species of fauna including the presence of beds of the seagrass also known by the scientific name *Zostera*. Seagrass looks like a seaweed but is, in fact, a flowering plant. There are two species in Irish waters and their flowers are similar to terrestrial grasses, but they use currents for pollination rather than wind. Throughout the world, seagrass beds are known for the huge variety of life they support. They provide shelter in shallow waters for numerous invertebrates and act as nursery grounds for fish, many of which are of economic value, e.g. plaice or turbot. They grow and develop through rhizomes, a concealed mesh of roots and interconnections that stabilise the sand and mud and trap sediment. These hotbeds of diversity provide vital habitats for maintaining healthy marine ecosystems and are of direct benefit to local fishing communities.[13] A further study in 1980 examined reefs of maerl, commonly referred to as coral where it occurs in Ireland. While superficially similar to real corals (which are animals), maerl actually derives from a family of seaweeds that exude a hard shell of calcite, a bit like the stalagmites found in a cave. The dead remains can build up into substantial structures, many metres deep, and pieces also wash ashore to create the incredible tropical-looking beaches, such as Trá an Dóilín at Carraroe in County Galway. Maerl is slow growing but like the beds of seagrass is an incredibly rich and colourful habitat. Every nook and cranny becomes occupied with anemones and sponges, sea fans and brittle stars, grazed by sea slugs, patrolled by larger fish and kept clean by scavenging crabs and other crustaceans. The maerl reef at Roaringwater Bay is considered the largest in Ireland and covers nearly 4km^2. Maerl is high in calcium and has been extracted historically for use as fertiliser on agricultural land.

In addition to the seagrass and maerl reefs there are also kelp forests to be found in Roaringwater Bay. Kelp is another family of seaweed that is readily identifiable by its long, brown strap-like fronds and sturdy stipe (analogous to a stem in a plant) that ends in a fist-like 'holdfast'. The holdfast is used to secure the structure to a boulder or the rocky seabed. Kelp can grow to lengths of up to 4m and makes a wonderful place in which to snorkel at low tide. The shelter of the kelp forest provides a home to dogfish and the colourful cuckoo wrasse, as well as blennies, spider crabs, seahorses, pipefish and lots more. As the light diminishes so too the kelp thins and gives way to deeper-water habitats of soft corals, sponges and plant-like animals rooted in the mud, known as hydroids. One particularly interesting species is the peacock worm, worms of the sea being an entirely different order of beauty to those that burrow through the earth on land. This worm lives in a tube of its own construction, up to 25cm long, and held fast to a rock or stone in the sand or mud. When feeding, a set of red-striped tentacles resembling the finest emu feathers fan out from the tube. They use these to filter particles from the passing water and will retract suddenly, and in unison, if approached without care.

As well as these rich and diverse habitats studies have also noted that large areas of the bay are composed of sandy mud, cobble and gravel that can be home to burrowing shellfish.[14] Otters are still known from its shoreline and a regular occurrence of harbour porpoise, common dolphin and grey seals has been recorded, even the odd sunfish – a giant of the marine world.[15]

Clearly there is lots of wildlife under the sea in Roaringwater Bay and special habitats worth preserving. But does this colour and diversity represent a healthy marine ecosystem? This question has never been asked. The reason the answer may be 'no' is that amid all this diversity part of the picture is missing. We know that pilchards have long deserted the bay in its once innumerable shoals, as have the mackerel and the herring in their former abundance. Angel sharks that once scoured the sandy bottom are no longer to be found while the common skate, each of which could grow to a metre or more across, is not so common any more. Enormous cod and ling, up to 30kg in weight, are no longer caught. In the later eighteenth century there were reported to be thirty to forty boats fishing out of Cape Clear 'which took large quantities of fish, including cod, ling [and] hake'.[16] Most of these large fish are predators and dominate the ecosystems in which they are present. Records abound of their presence in these waters yet they have quietly disappeared from many of their former haunts.

The original attempt to provide legal protection to Roaringwater Bay as an area of scientific interest foundered in the Supreme Court in 1989 but the amendment of the Wildlife Act in 2000 allowed the government to designate so-called Natural Heritage Areas for conservation purposes. Many of the sites previously identified now fell into this category as *proposed* Natural Heritage Areas. It is in this limbo that most of these sites remain, unless, like Roaringwater Bay, they were deemed worthy to be of European value under the strict specifications of the EU's Habitats Directive, which came into being in 1992. On the face of it, the Special Area of Conservation (SAC) that now extends from Baltimore to beyond Cape Clear, north to Castle Point and encompassing the many offshore islands provides a strict protection for nature in the bay, enforced by the European Commission and the European Court of Justice. The Habitats Directive is, as described by its designers, 'the cornerstone of Europe's nature conservation policy'. The innovation in this legislation lay in the fact that people could carry on doing what they were doing in these areas so long as the key features that made them important were not harmed. An SAC for otters, for instance, simply meant that whatever the activity, agriculture, fishing or development, was fine so long as it safeguarded the otters. The special features were to be decided on strictly scientific criteria based on their uniqueness or level of threat across Europe. The features in Roaringwater Bay that marked it for this esteemed accolade are: otters, harbour porpoises, sea caves, grey seals, reefs, shallow sea inlets as well as cliffs and coastal heath habitats. Local people were to be consulted about the designation (although they couldn't oppose it unless on scientific grounds) and the authorities would draw up plans for the wise management of the area taking into account the special features and the multiple human-based activities. In 2013, over forty years after the bay was first recognised as being of scientific importance, Roaringwater Bay become the first such coastal site in Ireland to be subject to an analysis of its competing uses and conservation aims. The report, published by the Marine Institute, provided a fascinating insight into the range of commercial activities that rely on the sea.[17] So numerous and diverse are they that it is worth giving each of them a mention. According to the Marine Institute:

- There is an 'intensive' autumn pot fishery for shrimp
- There is also a year-round trap fishery for crab and lobster
- Crayfish and fish that stay close to the sea floor are targeted with tangle nets and gill nets
- Scallops are 'fished' in the upper part of the bay in winter and spring
- Bottom trawling occurs in the outer part of the bay throughout the year

- Mid-water trawling (i.e. for fish that swim in the middle of the water column or near the surface) is 'sporadic'
- Line fishing for mackerel and pollack is common in summer
- Mussels are cultivated using ropes
- Pacific oysters are cultivated on trestles in the intertidal zone
- Hand gathering of periwinkles is allowed
- Seaweed aquaculture is also encouraged

On the face of it, there is a thriving seafood and aquaculture industry based out of Baltimore that exploits a range of opportunities and employs a lot of people. So what explains the tweet of marine journalist Tom McSweeney in February 2014? 'Last fisherman on Cape Clear Island leaves the industry ... fishermen are an endangered species'.

It's worth going through the list again.

Pot fishing is ostensibly a very sustainable form of fishing that is selective (i.e. no unwanted by-catch) and leaves little footprint on the sea floor if the pots are dropped on sand or mud. The shrimp industry in Ireland dates from the 1970s and today nationally there are believed to be 117,000 traps in operation. There is a closed season but there are no regulations on the number of traps that can be set or the number of shrimp that can be taken from the sea. It is hard to know how many traps there are operating out of Roaringwater Bay and Baltimore but BIM gives a figure for 2006 of 22,000 in west Cork (which includes Kenmare Bay and Bantry Bay but based on further data in this report the vast majority of effort is based around Roaringwater Bay).[18] This is an enormous number given the unknown state of the shrimp population. The cylindrical metal traps are sizable, but portable, and if dropped on sensitive habitats, such as sea-grass beds or living maerl, and left to soak for two days or more, they can cause substantial damage.

Traps for crabs and lobsters are smaller and lighter than those for shrimp and so are likely to be less damaging. Away from sensitive habitats this fishing causes little damage to the sea floor and results in no by-catch. However, again, there are no controls on the number of pots that can be set and so there is a risk of overfishing. A system for lobsters has been established whereby females laden with eggs (known as 'berried' females) are marked with a 'V' in their tails and released. A lobster with a 'V' cannot subsequently be landed and only becomes a target again when the 'V' grows out, thus securing the next generation of lobsters. Although such a scheme seems logical and in the best interests of the fishermen, public funds are nevertheless provided for fishing communities to participate and this supplies them with training and equipment. Not all communities participate, however, and a lobster I bought in Howth, near Dublin, in 2013 was laden with eggs.

A tangle net is basically a mesh of monofilament line. It is left to soak in the water to entangle anything that strays into it. A gill net is similar but slightly more selective in that smaller organisms can escape (depending on the mesh size). In theory, these nets can be made to be selective as fish are not killed and can be released. However, the soak time, i.e. the length of time the net is left in the water, is crucial. The longer the soak time the more the net will catch and fish can die or become spoiled if left too long. The presence of fish, meanwhile, will attract predators, such as otters, seals and harbour porpoise which, if trapped and unable to breathe, will die in a short time. Tangle nets can get dislodged in heavy seas and cast adrift, destined to continue 'ghost' fishing indefinitely. The degree to which this is a problem in Roaringwater Bay is not known because there have been no studies to look at it; however, the Marine Institute identified the practice as 'high risk'. Bad weather, for instance, can extend soak times for days.

Scallops are an instantly recognisable shellfish. They live on the sea floor, often burying themselves slightly in mud or sediment. There are king scallops and smaller queen scallops, or queenies, and both are popular with diners. Because scallops live partly within the sea floor itself, to remove them involves ploughing the bottom with a row of metal teeth. The scallops then fall into a chain net to the rear of the teeth. Scallop dredging is among the most damaging fishing method in the world as it is indiscriminate and destroys everything in its path, frequently for very little return. There are parts of the world, such as in Norway, where scallops are collected through hand-picking by scuba divers and, if controlled, this would be a sustainable way for people to continue enjoying this delicious seafood. This has been particularly controversial in the UK where extensive scientific studies (e.g. in Wales' Cardigan Bay) have tried to find an accommodation between dredging and conservation. I am not aware of any such studies being undertaken in Irish waters.

Bottom trawling is slightly less damaging than dredging for scallops because it doesn't involve metal teeth ploughing up the bottom. However, with its heavy rollers it is also indiscriminate and will destroy anything that protrudes more than a few centimetres off the sea floor. Trawling is the principal method of fishing for bottom-dwelling creatures such as Dublin Bay prawns, monkfish and plaice but other fish, frequently juveniles, as well as any other marine organism in its path, will end up as 'by-catch', i.e. unwanted. Up to 90 per cent of what is caught in a trawl can be this unwanted by-catch, most of which is dead or dying and dumped overboard for the birds. There are major efforts under way to reduce this by-catch, as the practice known as 'discarding' is being gradually outlawed under the EU's Common Fishery Policy. This development is welcome but will not improve the damaged sea-floor habitats.

Mid-water trawling is similar to bottom trawling except it involves the net being dragged through the central water column instead of along the bottom. It therefore does not harm sea-floor habitats and in theory targets only shoaling fish, such as herring or sprat, resulting in little by-catch when compared with its bottom-based cousin. However, mid-water trawling will catch anything that also swims in the water column, such as whales, dolphins, turtles and seals (admittedly there are not so many turtles in Irish waters but one, the leatherback turtle, is found regularly near our shore as it migrates between the Tropics and the North Atlantic).[19] Globally, scientists have estimated that 650,000 whales, dolphins and seals were killed each year through the 1990s as a result of by-catch.[20]

Ireland's waters are important for marine mammals and in theory are a sanctuary for them, established by Charles Haughey in 1992. It is fair to say that a global problem on this scale will have its fair share around our coast; however, the only study I could find from Irish waters found no by-catch of whales or dolphins, although a small number of seals were trapped in this way.[21]

Line fishing is one of the more sustainable forms of fishing because it does not damage the sea floor and, in this part of the world, probably does not result in much by-catch. Elsewhere longlines can be many kilometres long and while they target tuna and swordfish they will also ensnare ocean-going birds, sharks and turtles. In inshore Irish waters they are much shorter and target edible fish so that there is likely to be little waste. Catches are in line with available quotas and so regulations are in place to ensure the stocks are managed.

Mussels are a cheap and popular shellfish that have been enjoyed by people for millennia. They are a native species and spend the early part of their life cycle as 'spat', floating as part of the plankton in open water before settling on a hard surface to grow their shell. In Roaringwater Bay the spat naturally settle on ropes that have been set and at this point they are referred to as 'seed'. Seed mussels are removed from the ropes, separated out and replaced onto new ropes. The new ropes are placed back in the water and the mussels allowed to mature. It is a simple technique that uses natural processes and requires no artificial input of food or nutrients for the mussels as all they require is filtered from the sea. Mussels stay on the ropes for up to a year and a half while they grow to a decent size. As filter feeders they will extract sediment and fine matter from the water and deposit it as faeces. If the water is very turbid (say, for instance, because of scallop dredging and bottom trawling) they will produce more faeces, the total volume of which is also influenced by the density of mussels and the currents which either disperse the solid matter or allow it to settle on the seabed beneath. Where it settles it can block light and provide nutrients upon which algae and other organisms will grow. While this has little impact where the bottom is already muddy, if it is on top of a maerl reef it will essentially smother the habitat. This is exactly what was observed by divers as part of a study to assess the condition of the maerl reef in Roaringwater Bay.[22]

Oysters are similar to mussels but are a much higher value product. Native oysters once carpeted the shallow seas and estuaries around Ireland's coast but were overexploited so that today the native oyster is found in only a few, small and isolated populations. Pacific oysters, meanwhile, are native to the East Asian coast where they have been cultivated for centuries. The ease and speed with which they can be grown to market size has made them attractive to the aquaculture industry across the world. They were first brought to Ireland in the 1970s and are grown in a number of shallow estuaries where once there were reefs of native oysters. While native oyster reefs are considered to be rich habitats for marine life, the Pacific oyster is classified as an alien invasive species in the UK and on mainland Europe (although mysteriously not in Ireland), where they can outcompete native species and smother reefs full of marine life. It was originally believed that Pacific oysters could not breed in Irish waters; however, this has been disproved and there are now a number of self-sustaining colonies.[23] There is no evidence yet that this is happening in Roaringwater Bay but given the risk, should there be cultivation of this species at all? Were Pacific oysters to become widely established they can grow in such densities that their commercial value diminishes. Dr Tasman Crowe of University College Dublin authored a report highlighting these concerns and says: 'experimental tests show that if it does become more widely established, the Pacific oyster would affect native biodiversity in intertidal habitats. The oysters also alter several biogeochemical properties and processes, potentially reducing the capacity of estuaries to support aquaculture and fisheries.'[24]

In other words, not only does the presence of Pacific oysters threaten the natural diversity, it also threatens the commercial rationale for which it was established in the first place. Pacific oyster aquaculture is already a risky business as it is susceptible to a herpes virus that can wipe out up to 80 per cent of a stock in a short time (something that happened in Carlingford Lough, County Louth, in

2013). Cultivating it at all seems like a bad idea, particularly when we have a native oyster, but bringing them into an area of high conservation value, like Roaringwater bay, is crazy.

Handpicking periwinkles and growing seaweed on lines, although not without potential negative effects (e.g. from overharvesting or seaweed blocking out the light at lower levels), are candidates for highly sustainable activities. With the proper controls they should be encouraged.

When you take in all of this information you are left with a glaring question: if Roaringwater Bay has such strict legal protection for the conservation of its natural heritage, the strictest in the land in fact, what exactly is being done to protect it? The answer: not much. This is not because fishing is unregulated or somehow a free-for-all. On the contrary, fishermen and women work in an environment that is highly regulated, and are burdened with an ever-increasing volume of inspections and rules coming from Brussels and Dublin. Nevertheless, these regulations do not prioritise the health of the environment upon which the livelihoods depend. Following the work of the Marine Institute in 2013 regulations were introduced which prohibit the use of trawls and dredges in sensitive areas of seagrass and maerl reefs (basically maintaining the status quo). There is also a requirement to notify the authorities on a weekly basis where fixed nets are to be used. It is hard to see how these measures will help marine life in the bay recover. For a start, it increases the administrative burden on both fishers and the authorities while doing little to address overharvesting, invasive species or other potentially harmful activities (in any case, when I rang the Sea Fisheries Protection Authority in late 2016 I was told that no fisherman had notified them about the use of fixed nets despite the regulation being in force since May 2015). It fails to establish a framework in which nature can recover and, in doing so, enhance the value of the local catch. Not that such an approach is necessarily easy. Short-term pain, in reducing fishing pressure or closing the bay off to particular activities, will be required and this, in turn, will affect people. But the history of successful conservation demonstrates that it only works through engagement with the local community, providing the necessary supports and developing sustainable alternatives. Despite all the activities in the bay, clearly there is still a lot worth saving. Recreational angling, whale watching and scuba diving provide local incomes and opportunities to enjoy nature without harming it. Yet what exists is only a fraction of its potential.

I chose to look at Roaringwater Bay because so much is now known about the extent of different fishing activities there and its special natural features. As similar studies become available about other coastal areas, a similar pattern emerges. I also chose it because these waters close to shore are the exclusive domain of the Irish authorities so we do not need the agreement of the EU to implement changes. Beyond the six-mile limit Irish boats have to share waters with other EU nations that wish to fish and have the required quota. So, for places like Roaringwater Bay, its future is entirely in our hands.

The history of fishing on the high seas is also one of lost opportunity. The assessment of fish stocks is supported in Ireland by the Marine Institute. It presents its data every year in its *Stock Book*, which, despite the density of information, is admirably readable. Recommendations are made on how much of each type of fish can be caught in particular parts of the sea. Their data is based on surveys carried out aboard a dedicated research vessel, which is then used to calculate the amount of catch that can be taken from the water without damaging the overall health of the stock, something called Maximum Sustainable Yield (MSY). The use of MSY is controversial because technically a fish stock can be decimated but fished 'sustainably' at this artificially low population. We see this with plaice, a flatfish which is caught at historically low levels but is nevertheless considered to be 'at MSY' and so being exploited 'sustainably'. In *The Stock Book* every year the state of play is laid out in a table that shows green or red ticks and crosses which indicate whether it is believed stocks are being exploited sustainably or not. The science is not perfect and for many stocks there are question marks where we would like to see ticks or crosses. But the problems really arise when the politicians get their hands on the numbers. Every December the Council of Ministers meet in Brussels and haggle – behind closed doors – with the science. According to the New Economics Foundation (NEF), between 2001 and 2015 EU politicians set fishing quotas beyond what the scientists advised by an average of 20 per cent.[25] In 2016, Ireland was singled out as the member state with the worst record of overfishing, coming away with quotas 26 per cent above the scientific advice.[26] This is the worst kind of false economy, what the NEF describes as 'standing in the way of more fish, more profits, and more jobs'.

I started this chapter with a quote from a popular fish cookbook and, in fairness to the author, she highlights the need to protect our seas. However, her exhortation for us to consume the 'fantastic, sustainable fish stocks' merely serves to highlight the extent to which things have gone wrong. Eight species of fish are listed: mackerel, pollack, monkfish, skate, flounder, bass, tuna and Arctic charr. When this statement was written in 2009 mackerel was certified by the Marine Stewardship Council as sustainable, the only Irish fishery at that time with such an accolade. This was stripped in 2012 due to overfishing and has yet to be restored. Pollack is caught in the North Atlantic and in the Celtic Sea, off our south coast. It is popular among recreational anglers as it inhabits shipwrecks and rocky areas close to the sea floor and for this reason it may escape the trawl. However, question marks stand in *The Stock Book*'s columns of data so stocks may, or may not, be healthy. We certainly don't know enough to assert that it is being managed sustainably. The same question marks can be seen beside the monkfish, or angler fish. These monstrous-looking fish live on the sea floor where they lure their prey with a 'fishing rod' which protrudes from their head. A fish coming close to investigate is quickly engulfed in the monkfish's oversized mouth and rows of pin-sharp, inward-pointing teeth. A recent specimen was caught off the south coast of Cork that measured 1.86m from its grizzly gape to the tip of its tail and weighed in at a whopping 68kg.[27] While specimens are uncommon greater than 1.2m this is still way above what we normally see at the fish market. Most of the delicate white tails that diners enjoy are from juveniles who may not even have had a chance to breed.

A skate is similar to ray and is a group of fish related to sharks as their bones are made of cartilage. There are no quotas for skates and they are caught in trawls that also catch lots of other things. A number of our skates and rays are in the Red list of the International Union for Conservation of Nature (IUCN) including white or bottlenose skate (endangered and declining), undulate ray (endangered and declining) and the misnamed common skate (critically endangered and, until recently, considered extinct from the Irish Sea).[28] Other skate and ray species are more numerous but the information is not gathered to tell how they are doing. The most popular flatfish on Irish plates is the plaice and it is fished in trawls all around our coast. Plaice in the Irish Sea is considered to be fished within sustainable limits; however, those in the Celtic Sea either are not or we have more question marks. In the Irish Sea up to 80 per cent of the fish caught were thrown away while discarding of other marine creatures is done on a huge scale.[29] Flounder is a relative of the plaice, being in the flatfish family. There is no directed fishery for it and it is not mentioned in *The Stock Book*.

Sea bass is popular as a sport fish and appears routinely on our menus. I have been in restaurants that boast how their sea bass was landed on the pier that morning but in fact this practice has been illegal since 1990. All sea bass in Irish restaurants are most likely to have been farmed in Greece following a collapse in the bass population here in the 1960s. Sea bass can still be caught by licensed anglers with strict bag limits, size limits and a closed season, and these cannot be sold for commercial consumption. Over 26,000 tonnes of albacore tuna were caught in Irish waters in 2012 and the vast majority of this was taken by Irish boats. Until 2016, the population of this tuna was assessed as being below its 'safe biological limit' –

not a good place to be. Due to conservation measures it is only now considered a healthy stock.

Finally, the Arctic charr is a species in the salmon family that is a relict from the ice age. Its status in Ireland is considered to be 'vulnerable' and it has disappeared from a number of lakes because of pollution.[30] There is no wild fishery in this country although it is farmed in County Sligo. Fish farming can have negative effects on the environment although aquaculture in tanks and with freshwater fish that eat only plant matter can be sustainable in theory. Arctic charr are carnivores (like their cousins the Atlantic salmon) and so the food they are fed must also be from a sustainable source.

The truth is that we don't have fantastic, sustainable fish stocks off our own shores. But it wasn't always like this. The richness of our seas has been alluded to for centuries and the cycle of over-exploitation, only to move on to the next species, has continued right up to the present day. The term 'shifting baselines' is attributed to fisheries biologist Daniel Pauley, in describing the long-term declines that stretch beyond memory, so that 'normal' is seen as an ever-deteriorating point of reference, with fewer fish and fewer species. It is hard to believe today the incredible abundance and diversity that once abounded around all our coasts. As far back as 1673 Sir W. Temple was quoted in a letter to Lord Essex as saying 'the fishing of Ireland might prove a mine underwater as rich as any under ground'. In 1847, at the height of the Famine years, Richard Valpy expressed his astonishment to the Statistical Society of London at the inability of the Irish to feed themselves:

> Few of the resources of Ireland are perhaps more capable of affording extensive and speedy relief than the sea fisheries, as in such industry no delay occurs in the return for the capital and skill required, and the yield is almost miraculous ... Cod, ling, haddock, hake, mackerel, herrings, whiting, conger, turbot, brill, bream, soles, plaice, dories and salmon, are the sorts most frequently met with; but several others are by no means uncommon, as gurnet, pollack, skate, glassen [saithe], sprats etc.[31]

At this time fishing was restricted to areas close to shore as boats were mostly small and powered by hand. The cod off the south coast were considered 'preferable to those caught in the American seas' and 'were much esteemed on account of their quality and the good condition in which they appeared'. Cod of up to 13kg and haddock up to half that were 'both being as good and cheap food as need be desired'. In 1829, 455 tonnes of cod were dried and salted along with 1,600 tonnes of hake but even by then Valpy tells us that 'the herring fishery of Ireland appears to have suffered a more extensive decline than any other branch of the fisheries'. English boats were engaged in an annual fishery in the North Irish Sea which engaged 100 boats while between 1783 and 1790 the fishing for herring off Donegal was 'so extensive that 500 vessels were generally loaded every year'. An eyewitness account from 1866 along the shores of Derrynane, County Kerry, describes the spawning of the herring, which was an event to say the least: 'The harbour of Sneem, Kilmachalogue, and Argroom were alive with fish. Several fine nights I visited the little herring fleet, which, inside the harbour, with their shallow nets, were taking fish in abundance ... the fish were seen rushing in bodies in every part of the harbour, and so numerously, that from the rocky ledges and at the shores they were taken by hand nets and baskets.'

Herring no longer frequent this area.[32] In one week in 1876 fishermen from Arklow, County Wicklow, hauled in £7,000 worth of herring (worth in today's money an astonishing €830,000) while in 1881 in one month, 1,000 tonnes of mackerel were taken off the Shannon Estuary.[33] In fact, the seas were considered inexhaustible. In a famous and catastrophically wrong analysis, Thomas Huxley (previously feted for being Charles Darwin's 'bulldog') predicted: 'I believe ... that the cod fishery, the herring fishery, the pilchard fishery, the mackerel fishery, and probably all the great sea fisheries, are inexhaustible; that is to say that nothing we do seriously affects the number of fish. And any attempt to regulate these fisheries seems consequently, from the nature of the case, to be useless.'[34]

By the turn of the twentieth century, despite overfishing in places, there were still plenty of fish in the sea and the Donegal herring fishery was described as 'unprecedentedly successful'. In reality, though, the cracks were beginning to show. Between 1921 and 1923 the herring and mackerel catch fell by nearly 50 per cent in quantity and over 50 per cent in value.[35] Through the post-war years total catches went up as well as down as new grounds were fished and technology improved. The marine historian John de Courcy Ireland grappled with the cause of the decline of the fishing industry from historical times to the early 1980s: [Regarding the state of the fishing industry in 1966] 'what critics ... failed to see was that the falloff in landings of pelagic fish were, at that stage, attributable mainly to inadequate research – insufficient knowledge of the habits of the erratic herring – insufficient training, and neglect of the once much sought mackerel'.[36]

There was no mention that mismanagement of a renewable, but finite, resource was at the heart of this decline. Even fifteen years on he pointed out that: 'Today [1980] the problems of the industry are not so much overwhelming penetration of EEC trawlers into our coastal waters, which has been largely avoided, but problems of diminution of certain species, lack of a coherent EEC fishery policy aimed at genuinely helping the weaker members, and the perennial absence of sufficient research to guide the next step forward.'

A 'diminution of certain species' is the nearest we get to an admission of massive overfishing that had at that stage been going on for 200 years. In the 1970s the herring off the south coast went into dramatic decline with attendant loss of livelihoods and ensuing conflict[37] while in 1975 serious concerns were being expressed about the depletion of salmon – an issue that would require twenty years of deliberation before concrete action was taken. In de Courcy Ireland's concluding remarks he proclaims that: 'The future production of fish for eating and eventually perhaps also for fertilizer and cattle-feeding, depends on rapid progress in fish-farming or mariculture; a real breakthrough in this field seems fairly close in Ireland'. The bonanza is just around the corner, without ever looking back and wondering how it all went so badly wrong.

Since 1982 the fish resource around Ireland has been managed by the EU's Common Fishery Policy. It is roundly acknowledged that in this time it has failed miserably in its stated aim to manage fisheries in an economically, socially or environmentally sustainable manner. The reasons for this include high levels of unwanted catch, throwing away fish in favour of larger specimens in order to stay within quota, repeated failure by politicians to follow scientific advice, and too many boats chasing too few fish – fuelled by an elaborate and generous system of public subsidies.[38]

Estimating the impact of commercial fishing on an ecosystem has always been hampered by a number of factors. Firstly, what is caught is not necessarily a reflection of the quantity of fish that exists. For this reason catches can be seen to be improving while in actuality the fish stock is being depleted. This can be done through more targeted fishing or improvements in technology (more efficient engines, radar, etc. As some populations get smaller they can also aggregate into larger shoals, creating the appearance of abundance). Secondly, unwanted by-catch is rarely, if ever, recorded so that landings may only be a fraction of what was originally brought on board in the net. Thirdly, there are illegal catches, about which, by their very nature, little is known. And finally there is legal, but unreported catches. In Ireland 80 per cent of the fishing fleet operates from small boats, less than 12m in length, and close to the shore (within 19.2km). Those boats are not required to keep logbooks and until recently many were unlicensed and unregistered.[39]

In a groundbreaking global initiative Daniel Pauley from the University of British Colombia in Canada has been looking at this problem in an attempt to reconstruct historical catches. The Seas Around Us project has scoured data from all manner of sources, from anglers' records and government research, to estimate the total quantity of marine life that has been extracted from the sea. It is an imperfect exercise as many assumptions need to be made but it is a fascinating study in its breadth and novelty. The Irish chapter of the research found that from 1950 to 2010 nearly 8.7 million tonnes of fish and marine invertebrates were removed from Ireland's Exclusive Economic Zone by Irish boats. This figure is 20 per cent higher than what had heretofore been officially calculated. It shows how the total catch ballooned by over twenty times the weight from 1950 to a peak in the late 1990s at 420,000 tonnes.

While industrial fishing has left nowhere untouched, in an Irish context the greatest damage has occurred in the shallow seas, and the Irish Sea in particular. Today the Irish Sea is an ecological wreck, a shadow of what it once was. Once there were extensive beds of oysters and mussels, great shoals of herring and mackerel and large predatory fish such as bass, cod, monkfish, sharks and rays. These are all now gone, leaving a fishing industry that is dependent, ironically, in preserving a permanent state of ecological damage and ill health. Officially, the fisheries for cod, sole and whiting in the Irish Sea have collapsed.[40] Bottom trawling takes place to such an extent that many parts of the sea floor are hauled over an incredible seven times a year. This results in a system that has not only destroyed the sea floor but produces a mind-boggling level of waste. The practice of 'discarding', i.e. throwing unwanted or unlandable catch overboard, has been widespread and generally affects juvenile fish. In 2012 all the whiting that was caught was thrown overboard.[41] Since the beginning of 2016 new regulations are phasing this practice out although serious questions remain as to how effective it will be. The quantity of spawning Irish Sea cod has declined ten-fold since the 1980s – hardly a time when the sea was virgin territory. Catches of haddock are about a fifth of what they were in 1998 and astonishingly over half of what was caught in 2012 was discarded. Plaice is considered not to be overexploited but catches in 2010 were the lowest they had been since at least 1964 (because that's as far back as the data goes). There is a herring fishery in the Irish Sea amounting to about 5,000 tonnes in 2014 and while this is considered to be fished 'sustainably', i.e. so as not to reduce the population further, this is a false hope as the stock is well below what it should be. In the 1920s and 1930s the catch was twice that.[42] The herring industry in Britain reached its peak before the First

World War with a catch of 600,000 tonnes only to collapse in 1955 after an orgy of overfishing, although much of this came from the North Sea.[43]

Reconstructing the ecosystem of the Irish Sea to a time before the advent of industrial steam trawling is no easy task. Historical records, eyewitness accounts of fishing, government reports and Commissions of Enquiry provide a basis for imagining just how many fish there used to be. In 1847 it was said that 'besides the large supplies of excellent white fish and herrings of which the Irish Sea can boast, many kinds of flat fish, of an equally fine and large description, are to be found there also. And lobsters and crabs are by no means strangers to several parts of the coast.'[44]

In Dundrum Bay, County Down, turbot, a predatory flatfish that can grow to a metre in length, were 'so abundant ... that they are speared close to dry strand'.[45] Oyster beds were found throughout. At the end of the eighteenth century there were oysters 'as large as horseshoes' off Howth, Malahide and Ireland's Eye, near Dublin.[46] In 1836 Carlingford Lough, between Counties Louth and Down, was described as being 'occupied by an immense bed of oysters, of which vast quantities are taken to Dublin and other towns'.[47]

The fortunes of Arklow in County Wicklow were built on a seasonal herring fishery but also on the beds of oysters, which were expansive. The oyster beds provided a hard surface upon which many other species settled while their incessant filtering of the water above probably meant that the Irish Sea was clear, and not dull and muddy as it is today. The oyster fishery was the chief earner in Arklow in the 1830s but as early as 1806 fears were expressed that overharvesting would exhaust the beds. By the 1860s boats from England, Wales, Holland and France, having fished out their own oyster reefs, were descending on Arklow, some of whom 'dredged at all times of day and night'. By the end of the century predictions of its demise came to pass and the Arklow oyster fishery was no more.[48]

Not only were there vast beds of oysters in estuaries along the coast but, incredibly, there is even evidence that a cold-water coral reef was once a feature of the Irish Sea, several lumps of which were kept as souvenirs by a trawlerman, and were discovered by Johnny Woodlock of the Irish Seal Sanctuary.[49] Until relatively recently there were beds of mussels, more resilient perhaps than oysters but also fetching a lower price and so historically in less demand. The juveniles are scraped off the sea floor for rearing on ropes but in recent years the traditional mussel grounds have been all but worked out.

The work of Callum Roberts and his team at the University of York in the UK has been illuminating the historical fishing effort in the Irish Sea.[50] He explored the problem of 'shifting baselines' and the forgotten wealth of the sea in his compelling book, *The Unnatural History of the Sea*. His work is now showing how the quantity of fish in the Irish Sea is a tiny fraction of what it was in 1800. Successive waves of technology from hook, line and nets, through to sail trawling, steam trawling, diesel engines and scallop dredging have progressively worked out the sea's productivity. Using data from international trawl surveys since the early 1980s he has found that while some species have benefited over this time, such as blonde rays, dab and gurnard, others have gone into a steep and steady decline – hake, spiny dogfish, cod, monkfish, whiting and common sole. Yet others are stable, albeit at historically low levels, and are showing no signs of recovery – lemon sole, brill, conger eel, haddock, cuckoo ray and thornback ray. The overall picture is one of a tattered ecosystem, the shattered remains of a submarine empire that has literally been dragged through the mud.

So where did it all go wrong? For some species the answer is simple. The herring and mackerel were lost as major sources of income because measures were never taken to protect them, or when they did, it was too little and too late. Mistakes were made; wise management remained elusive. Proper regulation and enforcement were absent. There are few documents in the Irish literature that capture the cycle of depleting one species after another without an attempt to reverse previous mistakes. In this regard Ed Fahy's *Overkill! – the euphoric rush to industrialise Ireland's sea fisheries and its unravelling sequel* is a revelation. Fahy is a pugnacious character, fisheries scientist and non-conformist. He has a way of communicating that I could only describe as tragic humour. Within the wider vista of a declining industry one of his parables is instructive. It regards the tale of a little-known shark found throughout the coastal waters of Europe variously known as the spurdog or piked dogfish. It's a small shark, males growing no more than a metre or so, grey and slender with large eyes and vaguely striped or spotted flanks. Hardly a shark that will feature in a Hollywood film but one that shares the typical shark features of longevity, slow growth and giving birth to live young. Unlike the bony fish that lay millions of eggs, sharks tend to give birth to only a handful of young at a time; each is given a head start in life and so have a greater chance of surviving to sexual maturity. In fact, the females have one of the longest gestation periods in the animal kingdom (up to two years), while the pups, as the young are called, enter the sea with a yolk sac that nourishes them through their first days. This strategy, shared with humans, of slow growth, late maturity and investment in young, contrasts with species which produce innumerable young with little or no investment of time in their rearing. Each strategy has its advantages; however, the downside of the spurdog's is that it leaves it vulnerable to overfishing as it doesn't have the capacity to rebuild its numbers quickly if depleted.

In 1920s Ireland spurdog was considered an ingredient for fish fertiliser but in the UK had been marketed as 'rock salmon' in fish-and-chip shops. They were not highly regarded by fishermen who saw them as a 'perpetual menace' and so were happy to find a use for them. Despite there being targeted fisheries by a number of European countries throughout the twentieth century, it was not until the 1970s that it drew the attention of the industry in Ireland. Spurdog began to be targeted with leftover gear from the declining salmon drift-net fishery and, in 1985, 8,800 tonnes were landed, mostly in County Donegal. Despite widespread knowledge of the vulnerable biology of the spurdog there was no effort to limit the impacts of the fishery (i.e. the targeted effort towards a particular stock) by releasing pregnant females or closing off the known spawning grounds. By the early 1980s the larger sharks, targeted by anglers, had disappeared while by the early 2000s catches had dropped to less than 1,500 tonnes annually. Only in 2008 were catch restrictions imposed and the North Atlantic population of spurdog is now classified as endangered in Europe by the IUCN. It was eventually acknowledged that 'the stock suffered a high fishing mortality for more than four decades, and was not managed during this time'.[51] Today, scientists recommend that there should be no targeted fishery and that recovery will be slow because of the shark's biology.

The spurdog is not alone. The following table details a number of species, some familiar, some now little known, which have been overfished in Irish waters in recent decades. In each case, intervention – if it came at all – was insufficient to prevent drastic declines in abundance.

Overexploited marine life from Irish waters. Each species represents lost economic potential and unknown repercussions for the ecosystem.

Species	What it is	Status
Palourde	Tasty shellfish gathered by hand at low tide at a small number of west coast locations	Intense period of exploitation during the 1970s referred to as the 'clam boom' as crowds descended upon beaches in Galway to dig them from the sand. Stocks collapsed by 1977.[52]
Whiting	Once-plentiful white fish	Landings peaked in 1981 at 15,600t. Stock collapsed in the Irish Sea since 2004.[53]
Cod	Well-known favourite with batter and side of chips	Irish Sea populations have collapsed and scientists have recommended no catch in this area since 2004 – advice which is routinely ignored.[54] There are currently no stocks in Irish waters believed to be in a healthy state.

Species (continued)	What it is	Status
Plaice & Haddock	Well-known dinner favourites	Fisheries scientist Ed Fahy calculated that the decline in the populations of these fish in the Irish Sea since the early 1920s is 97 per cent and 58 per cent respectively.[55]
Sole	Well-known dinner favourite	Irish Sea stock beyond safe biological limits. Zero catch recommended since 2008.[56]
Spurdog	A type of shark, marketed as 'rock salmon'. An important predator.	Commercial exploitation commenced in the late 1970s. Landings peaked in 1985 at 8,800t. By 2000 this had declined to 1,500t. Scientists currently recommend no targeted exploitation.
Sea bass	A favourite among sea anglers	Commercial fishing by Irish boats banned since 1990 and sale of sea bass is illegal. Stocks showing worryingly little sign of recovery despite conservation measures.
Lemon sole	Once common flatfish	Landings valued at €2.1 million in 1999 more than halved to €0.83 million in 2000.[57] Disappears from The Stock Book assessments after 2002.
Lesser argentine, deep water sharks, grenadier, black scabbard and orange roughy	Deep-water fish	Exploitation started around the turn of the century when little was known of the biology of these species, which live for a long time and reproduce slowly. 1,300t of orange roughy were landed in 1998. In 2000, 6,000t were landed, creating such a glut in the market that they were used as fishmeal. Marine Institute advice in 2012 stated: 'high levels of exploitation have led to the depletion of many deep water species' with a commensurate 'drastic reduction in fishing opportunities'.

Species (continued)	What it is	Status
Lesser argentine, deep water sharks, grenadier, black scabbard and orange roughy (continued)	Deep-water fish (continued)	In 2014 they said: 'zero catch advice for deep-water sharks is in place since 2006, but there have been no concrete measures to reduce the catch of these seriously depleted species. The zero Total Allowable Catches in place [i.e. recommendation for no targeted catch] are not effective because sharks continue to be caught and are discarded.'[58]
Purple sea urchin	Spiny echinoderm once widespread and common on rocky shores	350t landed in 1976 48t landed in 1986 6t landed in 1997 Not listed after 2003 Before exploitation it was estimated that one inlet in Galway Bay had 1,600 individuals per m^2. Today they are very hard to find.[59]
Turbot	Large flatfish once common in sheltered inland waters	Around 250t landed in 2000 worth €2.3 million. No longer commercially exploited but caught as by-catch in other fisheries. Now listed as 'vulnerable' in European Red list.
Crawfish	Large crustacean similar to a lobster	Landings declined from 300t in 1971 to 34t in 2006 due to use of tangle nets.[60] Now an endangered species.
Atlantic salmon	The king of fish	Commercial fishing effectively banned in 2007. Now listed as 'vulnerable' on the Irish Red list.
Whelk	Snail-like shellfish desired in the Far East	Landings peaked in 1981 at 10,000t. This had reduced to 3,000t by 2010. Biomass considered to be one third of its peak, currently the minimum landing size is less than its size at maturity, meaning overfishing continues.[61]

Species (continued)	What it is	Status
Razor clams	A bivalve shellfish that lives within sandy sea floors	'Significant' expansion between 2011 and 2014. Fished using hydraulic dredging which flushes out marine life living in the sand. Quantities considered to be half what they were in the late 1990s. Marine Institute said in 2014 that 'all indicators show a significant and persistent decline over time indicating that the north Irish Sea stock has been and continues to be fished down'.[62]
Mussels	Common bivalve	Natural reefs of juvenile 'seed' are harvested by dredging. Alarm bells ringing in the early 2000s as 60 per cent of all seed mussel culture was sourced in the Irish Sea.[63] Distress in 2015 as seed stocks virtually dry up.
European eel	Migrates from rivers to spawn in the Sargasso Sea	Commercial fishing closed down in 2009 following a 90–95 per cent reduction in the number of juvenile fish returning from the sea.
Boarfish	Small red fish heralded as a new bonanza in 2010	With no catch limits set, landings plummeted from 144t at peak to 17t in 2015.[64]

Marine creatures have to contend with a myriad of problems including a changing climate, pollution, acidification of the water and noise from shipping and fossil fuel exploration, all of which hamper efforts to rebuild fish populations.[65] But first and foremost too many fish were taken from the sea, spawning grounds were not protected, or migrating shoals were targeted along every stage of their journey. Usually these fish populations can be saved by implementing protection measures but it is too early to tell whether our herring or salmon will return in the same numbers as before.

Simple overfishing is only part of the problem. Remarkably, the more intractable element in this still-unfolding calamity was identified as a disaster when it first appeared. And it wasn't scientists or environmentalists who sounded the alarm. It was fishermen themselves.

'About 1852 trawling on a large scale was introduced. It was opposed, including with violence but persisted.'[66] Trawling is a method of fishing that involves towing a weighted net along the sea floor and dates to the Middle Ages. It had been trialled as an innovation, under the power of sails, in Galway Bay in 1820 but the catches from it were so great that despite the major reservations of fishermen, 'progress', it seemed, could not be resisted. The fishermen of the Claddagh quarter in Galway city relied on the abundant mixed fishing in the bay onto which they faced, and opposed trawling by all means available to them. Their prescient warnings of the damage trawling was doing was dismissed as indicative of old-fashioned ways: '... though they sometimes exhibit a great shew of industry, they are still so wedded to old customs, that they invariably reject, with the most inveterate prejudice, any new improvement in their fishing apparatus, which is consequently now very superior to that used centuries ago by their ancestors.'[67]

In 1848 the wonderfully named Wallop Brabazon, in his book *The Deep Sea & Coast Fisheries of Ireland*, in relation to the Irish Sea highlighted his concerns about overfishing, and trawling in particular: '... the distress amongst the Fishermen on the East Coast is caused by the spawn on the fishing banks being destroyed, along with the small fish, by the Trawlers, which has made a great scarcity of fish, and its effects have been severely felt by the poor fishermen who supported themselves by line fishing out of the small shore boats.'[68]

Fishermen in Arklow in the 1830s saw their summer herring fishery disappear and attributed the loss to large trawlers from Skerries, Howth and Baldoyle depleting the stock.[69] By-laws were introduced, such as one prohibiting trawling in Bantry Bay in County Cork in 1894. But these regulations provided no match for fast, steam-powered boats, sometimes from other jurisdictions, and patrols proved ineffective.

A few years ago I was walking along the beautiful broad strand at Bettystown, County Meath. Looking out at the sea it looked probably much the same as it did a thousand years previously, except for that muddy brown colour that the Irish Sea tends to be. It is highly likely that before the advent of trawling, which disturbs the sediments on the sea floor, and with millions of particle-filtering mussels and oysters, the waters here were clear. As I approached the shore there was a growing stench of stale fish. Looking down at the tideline I was shocked to sea a flotsam of thousands of dead and dying marine creatures. Starfish, urchins, whelks and brittle stars, these were creatures that do not normally inhabit the very shallow waters of the high-tide line although they do live on, and in, sandy bottoms. The beach was literally strewn with these small animals all the way from the town to where the beach meets the estuary of the River Boyne. Everything else seemed so normal – people walking their dogs, children playing in the early spring sunshine – that I thought at first this was some natural phenomenon. Perhaps rough seas had driven these creatures onto dry land. Except there had been no storms and in all my experience post-storm beachcombing I had never encountered anything like this. It was only later that I heard about the fishery for razor clams in the Irish Sea off Louth and Meath. Razor clams are those long, half-pipe-shaped shells that are common around our coastline. They are about 20cm long and when alive they bury themselves quickly, descending vertically into the soft sand or mud. They have been found to move their shells in such a way that they turn the surrounding solid sand into a liquid – something that is being exploited for new drilling technology.[70] When disturbed they can in this way quickly dig themselves deeper to evade would-be predators. To extract them on a commercial basis a device called a hydraulic dredge is used. This involves pumping water through tubes of metal that penetrate the sand and blow the razor clams up into the awaiting metal box. Of course, it doesn't just expel the clams but also anything else that might be living in the top few centimetres of the sea floor. The animals, which are not adapted for life in the turbulent water column, are thrown in all directions, delicate shells are broken and heavier creatures that are surplus to requirement are cast overboard. Hydraulic trawling for razor clams, which began only in 1997, expanded in the north Irish Sea so that in 2013 it was worth €2.1 million and supported sixty-four jobs. However, the absence of restrictions, including trawling inside the Dundalk Bay Special Protection Area, saw the fishermen themselves stepping in to ask for regulation as catches started to dwindle: 'The fishery is currently profitable but is unlikely to be sustainable given that there is no input (effort) or output (catch) control in place.'[71]

Today, bottom trawling is responsible for the catches of scallops, cod, Dublin Bay prawns, various flatfish, monkfish, haddock and ray. While most trawling is not of the hydraulic kind it nevertheless involves relentlessly scraping a weighted net over and back across the ocean floor. The Marine Institute tells us that most of the seabed around Ireland is trawled at least once a year and some regions are trawled more than ten times a year.[72]

Trawling has a number of effects. It disturbs the bottom, sending sediment into the water column. This in turn increases the turbidity of the water, reducing the amount of light that can penetrate through it while the sediment can go on to settle on sensitive organisms, such as corals. The trawl itself will take anything that is in its path. The indiscriminate nature of the trawl means that even if a fisherman is only looking for one or two types of fish, lots of organisms for which there is no available quota, or which are simply not wanted, get caught and are thrown overboard. Many fish and invertebrates (although not all) do not survive this 'discarding' and either sink to the bottom and rot, or are taken by seabirds near the surface (hence the flocks of seagulls that follow trawlers). Some populations of bottom-dwelling species can recover from this onslaught but slow-growing sea fans and cold-water corals have little resilience. Continued trawling will eventually smooth out the natural unevenness of the sea floor thus homogenising the habitat and reducing the overall diversity of species. Repeatedly trawling areas has been shown to cut biodiversity by up to half, reduce the animal life in the sediment by up to 80 per cent and presents a major threat to sea-floor ecosystems on a global scale.[73]

Perhaps most seriously, trawling can contribute to alterations in the entire food web. Fishermen selectively target larger fish because they are easier to transport and get a better price at market. However, fish take time to grow to their full size and eventually all the large fish are removed, leaving smaller, less fertile specimens in their wake. Large fish are generally carnivorous and so play a vital role in keeping other species further down the food web in check. Cod, haddock, monkfish, whiting and ling are all species that fall into this category. Their removal leads to what is known as a 'trophic cascade' (trophic being a level of hierarchy in the food chain) in that it results in multiple and frequently unexpected consequences that affect nearly every other organism. This was documented in the Canadian North-West Atlantic after the cod fishery there collapsed and it is likely to have also happened in the Irish Sea.[74] One of the consequences of this is a much-simplified ecosystem where crustaceans, jellyfish and a few other creatures dominate. Structure and diversity give way to a banal monoculture and is reflected in the target species caught by the fishing industry. Where once a range of large, valuable fish were available throughout the seasons, now there are only one or two invertebrates that may not be available year round. This is a global phenomenon, revealed by Daniel Pauley and referred to as 'fishing down the food web'.[75] It explains why, in the absence of healthy populations of cod, haddock, skates and other top carnivores in the Irish Sea, most of what is left of commercial value is prawns, given that there are few predators left to eat them. Indeed, the 2016 *Stock Book* from the Marine Institute indicated that there are fears that even the prawns are being overfished.

Of great concern is whether the ecosystem now has the ability to recover. Small fish, even predatory ones, are themselves prey for jellyfish and other invertebrates when they are in their larval stage.

The Grand Banks off Newfoundland in Canada were closed after the epic collapse of the cod fishery there in the early 1990s and have begun to show some signs of recovery. But the loss of livelihoods has been massive and scientists say it will be ten years at least until fishing reopens.[76] In the Irish Sea we have yet to see any effective effort to restore lost fish.

In March 2010, while I was chairman of the IWT, I introduced the documentary film *The End of the Line* to an audience gathered in Dublin's Lighthouse Cinema. Those assembled included representatives of fishing organisations, environmental groups and interested members of the public. A few months prior to that I had been approached by Mike Walker of the Pew Charitable Trusts (and member of the IWT) who asked us to host the screening and join his growing alliance of organisations like ours from across Europe, known as OCEAN2012. The catalyst was the looming reform of the EU's Common Fisheries Policy, which had widely been regarded as a disaster on all fronts. Irish environmental groups up to then had not been vocal in their cries for the protection of our seas but at the Lighthouse Cinema that night this began to change. Out of the 193 groups from all across Europe fully sixteen were from Ireland and I'd like to think our call to 'end overfishing' had a disproportionate weight as the final Common Fisheries Policy was negotiated in the summer of 2013 under the Irish Presidency of the EU. The final outcome was a firm commitment to end overfishing, to rebuild fish stocks to their former abundance and to end the wasteful practice of discarding – so loathed by fishers and consumers alike. It was a very positive outcome and one that gives hope for the future. Since then we have seen greater heed being paid to scientific advice and help directed towards fishermen in developing new gear which reduces unwanted by-catch. But will it be enough to save the Irish Sea and places like Roaringwater Bay? Can we protect the marine environment, indeed not just conserve what we have but start to rebuild it, while also reversing the decline of the fishing industry and providing a future for communities that depend on it? I believe we can.

Two thousand kilometres to the north-east of Donegal lie the cold and dramatic Arctic islands of the Lofoten archipelago, off the coast of Norway. On a March morning the air bites at your cheeks and ears and the sky is a deep, infinite blue. Sharp mountain peaks rise abruptly out of the sea with only the precipitous cliffs dark against the brilliant mantle of snow. Wooden cabins painted red with white trim cluster in the sheltered bays, some almost in the sea itself, connected by wooden boardwalks perched above gentle waves that slap against the rocks. At night, the fortunate witness the spectacle of the aurora borealis as its fiery curtains whip across the Arctic sky. This community faces the sea, as it always has. The wooden drying racks that prepared the cod before being shipped off to Spain and Portugal as *bacalão* are scattered across the hinterland, and still very much in use. Lofoten's coastal communities have much the same hopes and concerns as those of any around the Irish seaboard. Their seas are dangerous, they live far from centres of power and gripe that politicians in far off Oslo neither know nor care for their plight. Yet there is one worry that an Irish trawler skipper can only dream about. I heard about the Norwegian system of fisheries management from Peter Gullestad, who is Specialist Director of the Norwegian Directorate of Fisheries in Bergen. He gave a short talk in Dublin in 2013 and, despite his matter-of-fact demeanour, five minutes in I was scraping my jaw off the floor. I emailed him much later to confirm that I hadn't just dreamt it all up, and this is what he wrote in reply:

> In recent years trawlers have experienced examples of the bursting of nets due to too large catches. In one extreme case,

an estimated 70 tonnes was caught after only three minutes of trawling, and 20 tonnes in three minutes has not been unusual. The trawl is filled so quickly that the 'damage' has happened before being registered by the sensors that normally tell the skipper that the trawl should be hauled. To solve this problem we are in the process of introducing a device (an escape hatch is maybe the right word) on top of the trawl, in front of the codend [the bit where the fish end up], which due to water pressure opens and releases the extra catch alive after the codend has been sufficiently filled.

Could this be true? Norwegian fishermen have problems catching too much fish? Norway is a part of the same Western European 'historical space' as Ireland so what have they being doing, or not doing, that has provided them with this glorious problem? The fishing industry would probably jump at the obvious fact that Norway has never been a part of the European Union and so has missed out on that whole disastrous saga. Their fish are not being systematically robbed by foreign vessels; they are full masters of their own seas. This is true, but in fact it explains more how they found the solution, than how they avoided the problem. Long before there was such a thing as the Common Fisheries Policy (which only came into being in the early 1980s) Norwegians were as enthusiastic fishermen as any in Europe. However, by the 1960s they were beginning to feel the sharp end of overfishing as their lucrative spring herring fishery collapsed, leading to economic hardship and the same problems we have become accustomed to here in Ireland. Increases in engine sizes and advances in technology were drawing down the sea's natural capital and by the early 1980s even their world-famous cod stocks were beginning to show the strain. So far, so familiar.

However, here our stories diverge. Where the Irish fishing industry got caught in a vortex of wrongheaded policies, misdirected subsidies and official apathy, the Norwegians took decisive action. They began to implement a system that strictly regulated access to fishing grounds that would turn their declining, subsidy-dependent industry into a profit-making export powerhouse with practically no public subvention and thriving local communities. They banned the throwing overboard of all unwanted catches, imposed real-time closures of fishing grounds to protect juvenile fish, set conservative limits on how much fish could be caught and backed all of this up with a strict regime of enforcement and surveillance. Fishermen have active and meaningful participation in the setting of regulations which has gradually turned them from sceptics to supporters of the new system. They took the short-term pain to tie up their boats to allow stocks to recover, heeding scientific advice that told them catching cod which were nine years old would double their yield compared to catching cod that were only three years old. This alone was estimated to have increased the value of the cod catch by US$100 million per annum. Sustainable fishing has been at the heart of their approach because they recognised that without the fish there simply would be no fishing industry.

In fact, since implementing the new regime, the average spawning stock of their economically most valuable fish has tripled since the 1980s.[77] In 2008 the Norwegian government passed the Marine Living Resources Act which recognised the intrinsic value of marine biodiversity and the integral value of its conservation to sustainable fisheries management. This could not contrast more with the Wildlife Act in Ireland which goes out of its way to exclude marine fish and invertebrates from even being classified as 'wildlife'.[78] Conservation here is seen as a threat, not as an opportunity. Consequently, as the Norwegians have seen the value of their fish catches grow, they have simultaneously seen the number of marine endangered species in their waters fall. In Ireland there is such a reluctance even to speak about marine conservation that we don't know how many of our sea creatures are threatened (but trust me, it's a high number – see chapter 4). The Norwegian experience shows us that it is possible to turn around damaged ecosystems in a relatively short space of time and protect the environment while also increasing incomes for fisherman and social stability in towns and villages that have traditionally depended on fishing for a living. The corollary is just as clear: without ecological sustainability, everything else is doomed. The recently reformed Common Fisheries Policy gives cause for hope, but the Norwegian experience also shows that without the buy-in from the fishing community the problems are likely to persist.

The new regulatory regime is committed to ending overfishing and rebuilding historic levels of fish in our seas, but will this new dispensation improve the prospects for communities like Baltimore in County Cork? In my view the answer will be yes, but it won't be enough. However, there is a solution that could transform the fortunes of these communities. It is an innovation that has been known for some time that has been driven not in the Western world but in the developing nations across the tropics, and is supported by a growing mountain of scientific research. The problem is that the central idea is totally counter-intuitive: fish less and you will catch more fish. Marine Protected Areas (MPAs) seem to promise the impossible and yet they are working across the world as I write these words, to the enormous benefit of wildlife and people who depend on the sea for a living.

On land about 12 per cent of Ireland's surface is protected in some way or another. On paper, at least, a significant portion of our seas are now also strictly protected under the EU's Habitats Directive. In theory we have a substantial network of MPAs, of which Roaringwater Bay is one. In reality, as we've seen, there is no protection and the future looks like yet another conflict between local interests and EU law that will further alienate people and provide no incentive for protecting the environment. Only a tiny fraction of our seas, to be found in the Lough Hyne Marine Nature Reserve not far from Baltimore in County Cork, enjoys any actual protection (and that's because it is not accessible to most boats). So what would a real MPA in Roaringwater Bay look like? What would it do for people and wildlife? And ultimately who would enforce any regulations so that the protection is real and lasting? These are questions that have never been asked in Ireland but which have been answered in communities from Indonesia to Mexico and the Pacific South Sea Islands.

Mackerel is among our most economically important fish but is not being sustainably managed.

Firstly, the conservation of marine life is the top priority. This means that fishing techniques that damage the environment would stop. In Roaringwater Bay this would mean no more dredging for scallops, trawling for prawns, tangle netting or piling in pots to catch too many shrimp. Only low-impact fishing would be allowed such as hand picking, using rod and line, potting or mid-water netting, all of which would be monitored to ensure that the level of fishing effort is sustainable. There is no doubt that this is a hard proposition for many in the local fishing industry to swallow as it would put some operators out of business. This is why the very idea of MPAs is being ignored by both government and industry alike.

What would happen if this was actually implemented? Marine life would start to recover. Individual fish would be allowed to grow bigger, fully to maturity instead of being caught as immature juveniles. Bigger, older fish are more fertile and so the total population of fish begins to recover. With the end of bottom scraping, the sea floor would begin to be recolonised with sea fans and sea pens and the myriad other creatures that add structure and diversity to the otherwise impoverished stone and mud. Larger predators, such as small sharks and rays, would return to feed on this new abundance – free from the risk of being snagged and drowned in a tangle net. The food chain would gradually be restored. The sustainable fishing methods are gradually resulting in more fish to be caught, and these are larger, more valuable fish. Local restaurants feature fat, Baltimore-caught cod and mackerel – delicious and naturally organic. The abundance of giant fish draws the attention of sea anglers (strictly practising catch and release), divers and snorkelers who stay in local hostelries and eat in the local restaurants. Eco-tour companies providing kayaking or dolphin-watching trips can genuinely proclaim that the Roaringwater Bay marine reserve is the real deal, and worth travelling to experience. And what if some sneak wants to take out his dredger to fill a net full of scallops? Well, he'll have the local community to answer to because by then it will be obvious that to take a dredger to the precious marine reserve is an act of vandalism.

I am oversimplifying to some degree but local governance and management of MPAs is working elsewhere. To those to say it won't work locally I direct them to evidence, this time along the south-east coast of Norway, which created an MPA for cod of a mere 1km^2. This wasn't even a full 'no take' zone and some fishing was allowed. Nevertheless, the cod in the area lived longer, grew bigger, and spilled over into the surrounding sea to non-protected areas where they could be caught.[79] The researchers said that had their experimental area been fully 'no take' then the effects would have been even greater. In this scenario fishermen have a direct say in how the resource is managed. They not only work with scientists but they are gathering much of the data needed to carry out the scientific assessments, because they are the ones at sea and with the greatest knowledge of what is going on. The great joy about this approach is not only that it provides a genuinely sustainable way of protecting incomes and the environment but it cuts the government out almost completely. Government agencies are there to assist in scientific analysis or the enforcer of last resort but are not the main instigators.

In a quote attributed to Albert Einstein, insanity is defined as doing the same thing over and over and expecting a different outcome. We've been doing the same thing to our seas over and over and each time there are expectations that this time the bonanza is about to happen. If we really want healthy seas and a sustainable future for fishing and coastal communities, we will have to have the courage to do something different.

3

The whittling-away of our iconic landscapes

> 'Protected areas are essential for biodiversity conservation. They are the cornerstones of virtually all national and international conservation strategies, set aside to maintain functioning natural ecosystems, to act as refuges for species and to maintain ecological processes that cannot survive in most intensely managed landscapes and seascapes. Protected areas act as benchmarks against which we understand human interactions with the natural world. Today they are often the only hope we have of stopping many threatened or endemic species from becoming extinct.'
>
> International Union for the Conservation of Nature (IUCN)[1]

NATIONAL PARKS ARE THE ICONIC landscapes of the world. Whether it's the Grand Canyon or Yosemite in the United States, the Serengeti or Kruger in Africa, Iguazú Falls or Yasuní in South America, Ranthambore or Chitwan in South Asia, or Kakadu and the Great Barrier Reef in Australia, they define not only whole countries but whole continents. In Europe, national parks such as England's Lake District preserve a rural idyll while in Poland the primeval forests of Białowieża National Park preserves a living glimpse of our deep past. The term 'national park' today is as much a logo as any marketing tool and draws tourists in their millions who want, and expect, to experience a country's quintessence.

Credit goes to Yosemite in California which first held the national park designation in 1890 through a United States Act of Congress. It was Scottish-born naturalist John Muir whose exhortations on the spiritual qualities of landscape and wilderness would form the *zeitgeist* of post-Civil War America and its increasingly self-confident government. It was the creation of an idea that was infused with God and conquest as much as any notions of conservation of nature for its own sake. For Muir: 'everyone needs beauty as well as bread, places to play in and pray in, where nature may heal, and give strength to body and soul'.

He practically invented the concept of 'wilderness', which even today is contentious. Implicit in his idea of wilderness was the absence of people, and little regard was given to the fact that the Yosemite valley had been inhabited by the Ahwahnechee tribe and their ancestors for over 3,000 years. With the signing over of the land from the state of California to the Federal Government by President Roosevelt, they were promptly evicted. The US went on to create a series of national parks that would ingrain themselves in the psyche of Americans (arguably the white, Protestant type) and define for them their country – from sea to shining sea. This was cemented by the use of the emerging technology of photography, and in particular by Ansel Adams who brought the drama of the Yosemite peaks to people who had little opportunity to see it for themselves. Today over 3.5 million people make the pilgrimage annually. Adams defined wilderness as 'a mystique. A religion, an intense philosophy, a dream of an ideal society'. The national park encapsulates that dream, packages it and sells it to the world.

The national parks of America were protected to safeguard the dream that was fast vanishing beneath the rapid expansion of farming, railroads and sprawling cities. However, by the early twentieth century it was becoming clear that they could also be useful in protecting vanishing species. It was at the first International Conference for the Protection of Flora and Fauna in 1933 that an attempt was made to define the national park and promote its use in the protection of nature which also included other types of protected area such as nature reserves or wilderness areas. By 2014 the UN estimated that 209,429 protected areas covered over 32.8 million km^2, equivalent to 14 per cent of the Earth's terrestrial surface and 3.4 per cent of its seas.[2] In 2003 these figures included 3,881 national parks, 273 of which are in Europe.[3]

The national park is no longer the chief tool for the protection of nature and the exponential growth of protected areas since the 1940s is attributable to a range of legal designations that provide for different levels of protection or with very specific aims. The EU, for instance, has rolled out an extensive network of protected areas under its Birds and Habitats Directives which in Ireland currently covers 594 sites spreading across 13 per cent of the land area. These are very far removed from national parks, however, with very targeted aims to protect defined lists of species or rare habitat types. They are vital for the protection of our biodiversity but contain none of Ansel Adams' mystique, and certainly when it came to protecting our disappearing peatland habitats, have been more of a nightmare than a dream.

More often than not they are in private hands, some are large, like the combined lengths of the Rivers Nore and Barrow, while some are tiny, limited to an old shed or church that provides a refuge for roosting bats. In the west of Ireland they cover whole mountainsides or expanses of blanket bogs. While these European sites are bringing species like the freshwater pearl mussel back from the brink of extinction, in the public mind they are a failure. If anyone has heard of a Special Area of Conservation (SAC, for habitats and species that are not birds) or a Special Protection Area (SPA, for birds) it is usually in the context of derision, such as when then Minister for the Environment, John Gormley, paid out €125,000 to count frogs during the depths of the recession, or outright scorn, as the government enforces bans on turf-cutting on the tiny number of raised bogs that are still worth saving. The legislation enacting the Habitats Directive was signed into law by then Minister for Arts and Heritage, Michael D. Higgins (later Ireland's president), in 1997. Shortly after, an explanatory note from Dúchas (the predecessor organisation to the NPWS) stated: 'Dúchas will produce a draft conservation plan for each SAC, SPA and NHA[4]. A plan will list the resources of the area, the current human uses, any conflicts between the two, and strategies for retaining the conservation value. This draft document will be given to the liaison committee and other interested parties for discussion and consultation. Dúchas will then prepare a final version of the conservation plan. Consultation on plans began in 2001. Plans will be reviewed on a 5-year cycle.'[5]

The plans never materialised and the lack of consultation with landowners has left a bad feeling across the countryside. Even today, the only plans that exist for these areas have emerged after the rancour of protecting our raised bogs (see chapter 6). After over twenty years of these directives it is debatable as to whether they have made matters better or worse for our struggling wildlife. For many people, they are part of the interfering reflex of distant bureaucrats who have nothing better to do than tell people in rural Ireland to stop doing what they've been doing all their lives. It has allowed native politicians to sign up to laws only to outsource the protection of the environment to Brussels and then feebly explain to an angry electorate that if we don't do what we're told we'll be subjected to punitive fines. The barrel of acronyms and technical guidance notes has served to divorce people from the will to protect what is best of our heritage. It couldn't be more different from the love that people feel for national parks.

To add to the problems it has tied up the state agency responsible for the protection of our wildlife, the underfunded and politically neglected NPWS, in meeting the stringent data gathering and reporting requirements of the directives, to the detriment of our wildlife in the 87 per cent of land (not to mention our sea) that is not in a protected area. And while data is needed to understand the condition of our rarest habitats and species, there are precious few resources left to actually address the underlying problems that have brought them to the brink in the first place.

In 2013 the NPWS reported that of fifty-eight types of protected habitat only five are considered to be in good shape – two types of caves, bog woodland (because it is regenerating on worked-out bogs that have been abandoned), marine sandbanks, and something called 'chenopodion rubri', an obscure community of vegetation found along lakeshores in a handful of counties. Of the sixty-nine species listed on the directive that are found in Ireland a much better thirty-four are of 'good' status, but it is still less than half.[6] Defenders will say that things would be much worse without the Habitats Directive and certainly these areas are much better protected from a planning point of view. But it's hard to argue that the protection of our wildlife is on the right course. In 2016 the European Court of Justice issued legal proceedings (yet again) against Ireland for not doing enough to protect those areas which have been designated.

Beyond these European sites legal protection for nature falls away steeply. In 1981, before there was such a thing as the Habitats Directive, the state's scientific body proposed a list of over 1,000 Areas of Scientific Interest (ASI) which promised to be the basis of a protected area network for the country.[7] These had been meticulously surveyed and mapped through the 1970s. One of these in County Galway was to protect intact areas of blanket bog. A report at the time stated that: 'the Roundstone Region is renowned for its magnificent undisturbed expanse of blanket bog, so well seen from the summit of Errisbeg. The complex of scientific sites in this region stretches to include the Ballyconneely Peninsula, well known for its wildfowl habitat and dune grassland. Taken in conjunction with the Twelve Pins nearby, with their Arctic-Alpine flora, the whole region should be considered as a National Park.'[8]

In 1989, when a consortium of Clifden-based businessmen proposed a 1,200m runway that would have eaten into the bog, a group called Save Roundstone Bog was mobilised in opposition to the plan. Galway County Council had refused permission, citing the negative effect the development would have on the ASI. The developers sued and brought the case as far as the Supreme Court, where the ASI designation was found to be in breach of the constitution. Some years later, then Minister for the Environment, Noel Dempsey, told the Dáil that the legality of designating ASIs was being examined and that in the interim the most important areas were to be designated as NHAs, to be enforced by statute via the Wildlife Act. According to Mr Dempsey:

Government approval was given in December last year [1992] to designate 1.25 million acres of land as natural heritage areas. These are the most important areas for conservation of Ireland's native flora and fauna. As a first step Areas of Scientific Interest will be resurveyed and designated as natural heritage areas where they meet the necessary criteria. I intend that this shall be completed by the end of 1993. Legislation has been drafted which will provide protection for natural heritage areas. This should be before the Oireachtas [parliament] later this year.⁹

It never happened and today over 1,000 NHAs have yet to receive official legal protection. Many of these sites have been lost or damaged in the intervening twenty years although the full scale of neglect is not known because no one has been back to look at them. Even during the boom years, when there was money for everything, the NPWS never had the funding to advance surveys or conservation measures that would protect NHAs. In the meantime, they were built on, in-filled, used as dumping grounds or allowed to be taken over by alien plant species like rhododendron or cherry laurel.

In many European countries there is a third tier of protection for areas in need of safeguarding at a local, or county level. In Northern Ireland for instance, local authorities can define a local woodland or wetland as a Local Nature Reserve. Perhaps they don't have the scientific value to meet the criteria for some of the other designations but many of these places are loved all the same. After all, to most people, a beautiful woodland where they walk their dog or bring their children at the weekend is no less beautiful because the trees are not native enough or the species around them are common. In Northern Ireland the law does not need to be invoked to define these areas but they are clearly identified in the various statutory development plans. This gives them a certain official status so they can be protected when lands are being zoned or plans for transport corridors are being made, for instance.

The value of a local nature reserve system is not hard to see. For the past five years a local Cork community group has been trying to save one of the few areas in Cork city that hasn't been poured with concrete or intensively manicured. It is the site of an abandoned quarry at Ballintemple. Even though it had been used by local people, especially children who were attracted to its state of semi-wilderness, it was only ever seen as a blank space on the map. In the early 2000s plans were drawn up to turn it into a municipal dump. Fortunately, the economic collapse intervened and in the hiatus the volunteers moved in. They studied the cave systems in the cliffs and found three species of bat using the site. Their dedicated surveyors recorded 134 species of plants and have clocked up thirty-three species of birds. Through dogged perseverance and lobbying of local politicians, by 2013 the plans for the dump were quietly removed. The volunteers have been busy removing the tons of rubbish that have accumulated since the quarry was closed in the 1960s and have plans to install interpretive signage and lead nature walks. Their work has paid off so that the Cork City Development Plan 2015 now contains official recognition that Beaumont Quarry is a 'non-designated area of natural heritage importance'. Gradually, there is an acknowledgement across Irish development plans that such areas have value, but it needs to go further so that protecting these areas relies less on the struggles of dedicated volunteers.

In 2014 when councillors in Galway City Council planned to drive a bus corridor through a small ancient woodland known as Merlin Woods, again there was no official recognition that this was a valued local amenity with ancient oak woods, red squirrels and more right in the heart of the city. By default this lack of recognition meant it was useless and available for development. Only a vocal campaign, again led by local volunteers, managed to bring these plans to the attention of the national media and were eventually successful in convincing the city council to abandon the plans.

Why does it have to be so hard? It seems that the way we protect our natural areas is just all wrong. The really valuable places are protected reluctantly at the threat of fines from the EU while a national campaign is needed to protect a well-loved local woodland. Why does the active protection of natural areas leave some feeling threatened? Is it the idea that some areas might be 'sterilised' in terms of their development potential? Or do we place individual rights of land ownership above those of the community at large?

This chapter is not about the thousands of places that don't have the status they deserve. We'd need a separate book for that. Nor is it about some predetermined, terminal decline of nature – after all we have plenty of beautiful and wildlife-rich areas without any form of formal guardianship. Just take a walk along one of the canal towpaths or the north shore of Galway Bay. Nature has evolved to survive but yet the human influence is so pervasive in Ireland, and the economic pressures so fierce, that without an active system of protecting and managing areas for wildlife the decline in our natural heritage will continue.

And so to test our ability and willingness to protect nature it is worthwhile to look at our national parks. We have six in total: the Burren in County Clare, Killarney in County Kerry, the Wicklow Mountains, Glenveagh in Donegal, Connemara in County Galway and, most recently, Ballycroy in County Mayo. They are all in state ownership, 'sterilised' from development, and so the state (i.e. the people of this country) has full control of what goes on inside their boundaries. They are all in areas of the highest scenic value in Ireland and they attract a lot of tourists – in other words, it is not difficult to draw a direct line between conservation and money in the local economy. Connemara National Park alone received over 190,000 visitors in 2015. Fáilte Ireland reported that over a fifth of the 8 million overseas visitors that year identified national parks as one of the activities engaged in during their stay.[10] Not only that but these are the places we really love. They are the places that make us who we are – our history and self-identity enmeshed with millions of years of geological upheaval and thousands of years of growth and decay. They have driven our finest bards to wax lyrical; to capture in words their 'savage beauty', as Oscar Wilde referred to Connemara. In the introduction to our first National Landscape Strategy, then Minister for Heritage, Heather Humphreys, opens by stating that: 'Our landscape is our ultimate resource.'[11]

These are our crown jewels. They are the outdoor equivalent of the Ardagh Chalice or the Proclamation of Independence – treasured and minded. The public should expect them to be managed to the highest of standards. What hope for the future of our landscape if we can't preserve these places, relatively small and few in number, but outsized in the public imagination?

To find out how we are getting on in preserving each of our national parks I visited all of them, sometimes more than once, between 2013 and 2016. I visited them at high season, when I thought tourists would be crawling all over them and they'd be looking their best. But I also tried to visit them off-season, in the depths of winter, when the savage can sometimes outweigh the beauty in the western mountains. Since the national parks themselves are all quite small, I also tried to look at the surrounding countryside – as a sort of 'compare and contrast'. Is there a defining line between the National Park boundary and those areas beyond these imaginary lines. After all, they each act as a nucleus for a wider cultural area that retains high landscape value despite being in private ownership. The one national park omitted from this chapter is Ballycroy in Mayo. It is at the centre of something all together different and new and its story is in the last chapter about how our landscape could look into the future.

Killarney National Park

> 'Now, however, so glowing has become the love of nature, and so ardent the desire for travelling and wandering about the country, and hunting after interesting and pleasing scenes, that it has also taken its natural beauties under its especial protection, and adorned them with the most charming songs, poems and colours. Thus it is that certain spots have obtained so great a celebrity that it is regarded as little less than barbarism to have been in the country and not to have seen them. To these places belong the Lakes of Killarney.'
>
> Kohl's *Travels in Ireland*, 1844

Killarney National Park justly deserves its unofficial title of 'jewel in the crown' of the Irish National Park system. Long before there was such a thing as a national park, it was regarded as a 'must see' destination. The quote above comes from Johann Georg Kohl, native of Bremen in Germany, and widely regarded as an astute and seasoned observer of the lives and ways of Europeans of his time. He travelled extensively throughout Europe in the 1840s and wrote prolifically as he went. Strolling today beneath the gnarled boughs of hefty oaks, preferably on a 'soft' day when the rain varnishes the leaves, the forest seems timeless. Much, perhaps, as Kohl would have seen it 170 years ago.

Although it is small, little over 100km^2, it seems perfectly formed. A rich embroidery of lake, river, mountain and forest, it is possible to find quiet solitude beneath the branches even in summer when the hordes descend on Muckross House, the park's HQ. And it is full of wildlife. It is not just that it is the greatest expanse of native oak forest in Ireland, or that it is situated in such a scenic spot, it is also home to practically every species of terrestrial animal we have. Not only that, but there are creatures here that are not found (or at least scarcely found) elsewhere, such as the spotted slug and strawberry tree. The red deer are not native in the sense that they have been here since before humans but they are relics of the first herds brought to Ireland by Neolithic people. The lakes are home to the Killarney shad, a fish stranded by the receding glaciers and to be found nowhere else in the world. Thanks to a reintroduction programme white-tailed eagles once again soar above the woods and, as of 2015, are once again successfully rearing chicks. Its magic extends through the year, with the passing of the seasons and ever-changing light as rain turns to shine, and back to rain again. Fáilte Ireland estimates that nearly 130,000 tourists visited in 2015 (a very conservative figure as it counts only those who paid into Muckross House) making it one of the most popular places in Ireland to visit. Perhaps better than most places then, Killarney demonstrates our conflicting attitudes towards the environment: on the one hand you won't find a person in the land who thinks that it should not be cherished and protected; on the other, the park is under serious threat and next to nothing is being done about it.

It is possible to invert the commonly held sense of awe and wonder that surrounds Killarney. In other words, instead of asking 'why is this place so special?' perhaps we should be asking 'why doesn't the rest of Ireland look like this?' The fact is it did, at least enormous swathes of it did, up until relatively recent times. Although unlikely ever to have been a uniform blanket of dense woodland, what with bogs and other wetlands breaking it up, there were significant patches of woodland right up to the sixteenth century.[12] Although rebuilding the landscape at this time is no easy task it is possible to imagine that Killarney-type forests cloaked the western fringes from Cork to Donegal. Massive woodland clearance continued right up the mid-twentieth century with even unique and ancient woods being swept aside for development, such as the flooded forest of the Gearagh along the River Lee in Cork (cleared for a hydroelectric scheme in the 1950s). The woods at Killarney were lucky to escape the axe at all, and even though they cannot be considered 'primeval' (in the sense that they have never been subjected to human influence) they are, to all intents and purposes, ancient.

There must have been a time when the forest was what modern ecologists would refer to as a 'functioning ecosystem'. Even those with limited knowledge of ecology will probably recall the food web from primary school science. Although in reality they are impossibly complicated, there is still use in the idea of a small number of large predators at the top, preying on the larger number of deer in the middle which, in turn, feed upon the innumerable plants at the bottom. In a standard European ecosystem it is the wolf that lords it over the beasts beneath, while various species of deer and smaller mammals occupy their individual niches below. The complex interactions between all organisms (don't forget the parasites, bacteria and viruses) is difficult if not impossible to grasp fully but one element of the puzzle is becoming increasingly clear and that's the crucial role that top predators play in shaping how an ecosystem works.

In North America the wolf had been all but exterminated from the 'lower 48' states by the turn of the twentieth century. In 1995 wolves were reintroduced to Yellowstone National Park and few foresaw at the time the effect this would have, not only on the other creatures of the park, but the entire landscape. Very quickly, wolves changed the behaviour of the deer. Reared without the need for vigilance against predators the deer soon learned to be scared again. So not only did the wolves reduce the number of deer by eating them, it also made them more stressed and less likely to linger in the same area for long. Released from intensive grazing, more trees started to appear on the plains and along the rivers. Trees stabilised the riverbanks, which had become eroded, and so the water started to flow differently. The carcasses of wolf kills attracted eagles and bears, the numbers of which flourished. Coyotes, which had boomed in the absence of wolves, went into steep decline as wolves either hunted them to eat or chased them off. In the absence of coyotes, the numbers of smaller mammals and reptiles bounced back. In short, the presence of the wolf underpins the entire ecosystem and engineers the landscape.[13]

The wolves of Kerry are gone but were considered common in this area right up to the 1600s. The last wolf in Kerry is believed to have been shot around 1710.[14] Gone too are the wild boar which rooted the soil and helped to spread seeds. Before that again bears, wildcats and lynx were to be found. The red deer, now such a familiar sight to visitors of Killarney, were the last of their kind in Ireland in the wake of the Great Famine. They had been protected within the Muckross Estate for their value as quarry but increasingly unregulated hunting continued up to the early 1970s. At this stage a mere 110 deer remained in the upland areas.[15] With the prohibition of shooting and the establishment of the National Park in 1985 the deer population recovered and the current population estimate is 500–600, although a formal census has never been carried out.

In losing its wolves Killarney National Park has lost one layer of the ecosystem. With the loss of other species, key links in this system have also been lost. The extinction of deer would have been disastrous although sheep and goats have been grazing the uplands and lowlands of Kerry for as long as there have been people. The effect this ecological diminishment has had on the forest ecosystem is hard to know because there is little data on the vegetation of the forest before these changes took place. Killarney, however, has ceased to be a 'functioning' ecosystem. Now, as deer numbers increase, the effect of too much grazing is pretty plain to see. The red deer, goats and sheep have been joined by the introduced sika deer, a separate species but genetically closely related and capable of interbreeding with the red deer. All the munching is taking its toll. A healthy forest will have three layers to it: the plants that grow on the forest floor, the canopy of mature trees high over head, and what's known as the 'shrub layer' in the middle. This middle layer provides shelter for many nesting birds and a nursery zone for the new trees that will eventually replace the old ones upon their inevitable death. Walking through the forest the visitor today will notice that this middle layer is all but absent. While pleasing to the eye to have open vistas of mossy rocks under the forest canopy this lack of a shrub layer means that new, juvenile trees are not present to ensure the ongoing health of the forest. Together the deer and sheep are killing off the woods. In effect, unless something is done, this sylvan idyll is doomed.

What to do is a vexed question. The importance of some grazing was highlighted by a long-term study that looked at the effect of fencing-off small patches of the woodland in Killarney. These fences are designed to keep out deer and sheep and can be seen by walkers adjacent to paths throughout the park. The study found that eliminating grazing altogether resulted in an abundance of brambles, ivy or holly. In fact, the diversity of plants in areas with too much grazing ended up being *higher* than in places where there was too little.[16]

In Killarney, what was once a finely balanced ecosystem has been tipped over and the pieces have spilled onto the floor. The crenulated bark of the ancient trees gives the illusion of stability but the ecosystem of the Killarney forests is lurching this way and that, albeit at a timescale unobservable to the human eye. If we want this magical place to retain its allure it can only be with the active and constant intervention of people. This means one of two things: either carefully control the numbers of deer and sheep in the park at any one time; or bring wolves back to do the job. Neither of these things is happening. Sadly for Killarney National Park it would be doing well if it only faced this single existential problem. Alas there lurks another.

Rhododendron is a family of plants related to the heathers. There are hundreds of species, mostly found in Asia, and reaching their zenith through the foothills of the Himalaya. Their lustrous flowers made them an instant hit with the landowning classes and they were enthusiastically imported to decorate estates. Most never went any further than the high stone walls of Victorian-era gardens and continue to delight visitors today. However, one, the *Rhododendron ponticum,* transferred from its home around the shores of the Black Sea, found Irish conditions to its liking. In particular, the peaty soils, mild temperatures and high rainfall of the west coast seem ideally suited to this tall shrub, which soon set seed and spread out across the landscape.[17] I remember as a child discovering rhododendron thickets in forest parks when on holiday and being transfixed with a sense of magic and mystery as I clambered through its tangled branches and dense undergrowth. At the time I was not appalled at the absence of native flora or the lack of nesting birds. Alas the burden of knowledge has set me against the malign beauty of the rhododendron and when I see the dense walls of the stuff as I climb above the Torc waterfall in Killarney National Park all I can think about is the absence of native flora and lack of nesting birds.

In 1977 Bill Quirke, then a student of zoology in University College Dublin, decided, after a visit to Killarney, to do something about the rhododendron problem. In 1981 he partnered with the Office of Public Works (OPW) and the IWT in setting up Groundwork, an organisation with the aim of clearing the forest of the invasive species. It was based entirely on voluntary support and attracted people from all over the world to participate in summer camps and days in the woods spent sawing and hacking. It is slow, tedious work. Rhododendron seems evolved to defy any attempt to remove it – Hydra-like, for every branch that is sawn off another ten seem to spring up. However, behind the hard labour lay a deep understanding of the ecology of the plant and the perseverance for repeat visits to stamp out re-sprouting stumps or newly germinating shoots.

By 2000, volunteers had contributed 4,000 weeks of their time and had succeeded in clearing 40 per cent of the affected areas.[18] While the management of Killarney National Park and the NPWS gave Groundwork every encouragement it has to be emphasised again that this work was done for free. This included a voluntary committee that was run professionally and single-mindedly. Rhododendron, it seemed, could, and would, be eradicated. The NPWS acknowledged the contribution of Groundwork in the management plan that was produced for the park in 2005: 'Groundwork (a voluntary conservation organisation) work camps have been very successful in dealing with rhododendron. Their clearly defined strategies and extremely well co-ordinated work programme has resulted in a significant proportion of the formerly infested oakwoods in the park being maintained free of rhododendron.'

However, in 2005 there was a shift in attitudes that led to disagreements between Groundwork and park staff. The relative autonomy that Groundwork had enjoyed was being eroded and new experimental methods were being forced through despite the proven effectiveness of the tried-and-tested techniques. The parties fell out altogether and in 2009 there was no Groundwork camp in Killarney. And there has been none since. Trained and dedicated volunteers were left to watch as their Herculean achievements began to unravel. Since 2009 the NPWS has borne sole responsibility for the maintenance of the woods and the rhododendron eradication programme.

In 2013 the Groundwork team returned to the forest to resurvey areas in which they had spent nearly twenty years trying to restore the native oak woods.¹⁹ Three areas were systematically surveyed for signs of regenerating rhododendron. In particular if rhododendron were to be flowering (and hence producing seed) this would signal the ultimate reversion of the forest to dense thickets of the invasive weed. Of the three areas surveyed none was free of flowering rhododendron. Their report had harsh words for the NPWS: 'Whereas the areas visited were relatively small, the results of these visits clearly indicate that at least in these limited areas, the new programme and new methodologics introduced by NPWS have already failed, and that exponential re-infestation of these woods is now imminent unless systematic maintenance clearance of each of these woods is carried out immediately.'

The management plan for Killarney National Park is prefaced by the then Minister for the Environment, Dick Roche, who said: 'It is remarkable that such a popular site can remain, at the same time, a place for relaxation, reflection and where nature can still be appreciated in its pristine state.'

If it was 'pristine' in 2005 (and it wasn't) then it is far from pristine today. At least it has a management plan, out of date perhaps, but unlike every other National Park bar Wicklow, and that indicates at least some intent.

The way in which we tend to heap too much praise on ourselves was exposed in the review of proposed UNESCO sites in Ireland. UNESCO world heritage sites are those of 'outstanding universal value' and there are over 1,000 of these very special places around the world. A mere three of these are on the island of Ireland: the Giant's Causeway in County Antrim, Skellig Michael in County Kerry and the Brú na Boinne complex in County Meath (not to be confused with UNESCO biosphere reserves, of which Killarney is one). UNESCO status comes with prestige and tourist cachet as well as international responsibilities for the management and maintenance of the sites. It has long been considered that two sites in the Republic was not a fair reflection on the wealth of our heritage given, for instance, that Croatia has seven and Norway eight. For years Killarney National Park was on a list of UNESCO hopefuls, referred to by the government as a 'tentative list' and was alongside the Céide Fields in County Mayo, Georgian Dublin and the Neolithic complex of Lough Gur in County Limerick.

When the Green Party joined the government in 2007 the then Minister for the Environment, John Gormley, set up a review group to progress the tentative list towards actual UNESCO status for at least some sites. The review group consisted mostly of government agency representatives and experts from within Ireland but had, crucially, two international experts in the field. The minutes of these meetings, which were released to me under a Freedom of Information request, show how the government struggled to explain why Killarney would be of 'outstanding universal value'. The fact that it had ancient Atlantic oak woodlands, perhaps the best example of their kind in the world, in a unique geological setting was not enough, especially considering the long-term management problems which were tacitly acknowledged by the review group. Killarney was dropped. By international standards there simply was no evidence that the park was ever cared for as an outstanding place.

The problems facing Killarney National Park are not easy to solve. Is it feasible to reintroduce wolves, boar, lynx and wildcat in an attempt to rebuild the ecological order of things? I think it is and genuinely hope and believe that this will happen in time, but am realistic enough to accept that there are major hurdles to achieving this. In the shorter term, I expect that management plans are implemented by the people who drew them up. This would see, at a minimum, the active exclusion of sheep from the park and the reduction of grazing pressure so that new forest can regenerate. The feasibility of reintroducing lost species should also be examined. I would also expect to see the threat of rhododendron being tackled aggressively and in an inclusive manner – i.e. to allow Groundwork back in to do what they had been doing best. We will have to wait and see if this will happen.

Wicklow Mountains National Park

'Though I could see for many miles, apart from distant plantations of Sitka spruce and an occasional scrubby hawthorn or oak clinging to a steep valley, across the whole, huge view, there were no trees. The land had been flayed. The fur had been peeled off, and every contoured muscle and nub of bone was exposed. Some people claim to love this landscape. I find it dismal, dismaying. I spun around, trying to find a place that would draw me, feeling as a cat would feel here, exposed, sat upon by wind and sky, craving a sheltered spot. I began to walk towards the only features on the map that might punctuate the scene: a cluster of reservoirs and plantations.'

From *Feral* by George Monbiot (Allen Lane, 2014)

George Monbiot was not talking about the Wicklow uplands, but he might have been. He was actually referring to his dispiriting view of the Welsh uplands, although anyone who has driven from the southern Dublin suburbs up towards the Sally Gap will know exactly what he's talking about. In my youth it was a place where the bodies were buried or where you might have been brought in the boot of a car by gangsters. It is a disorienting place, being so close to the capital city and yet having a Wild West feel: Utah, but with mist and rain. The roads up here are mostly occupied by commuters taking a short cut to avoid city traffic – determined types who know the twists and turns – and tourists who have come up in their rented cars on their way to Glendalough. They seem wide-eyed and a little unnerved. I imagine them to be keeping one eye on the petrol gauge and quietly locking the doors. Sometimes they pull in and you can see them struggle to decide what they should be photographing before hurrying back to their cars. On foot, the landscape is scarcely more welcoming. The featureless contours are, up close, a monoculture of knee-high heather. A colleague and I spent whole days walking across War Hill looking for red grouse for BirdWatch Ireland in 2007 and without the GPS it would have been impossible to navigate. The weather can close in but more than that, it's the lack of defining features that makes this whole expanse so monotonous. Like Monbiot's view of the Welsh hills, it is only the distant blocks of dark plantations that anchor the landscape. The Wicklow uplands comprise the largest continuous expanse of land over 300m in elevation to be found on this island. The vast majority of this space is composed of either heather or blocks of non-native evergreens.

The Wicklow Mountains are in a bad way. Although you will hear that the habitats to be found throughout these uplands are of 'international value', listed for extra special protection under European law, they are, in fact, degraded, neglected and increasingly abandoned. Peat, the characteristic soil of the uplands, has, in most places, stopped growing and is eroding; a consequence of a changing climate, burning, turf-cutting, recreational pressure and over-grazing.[20] The scale of this erosion can be seen from high up on the peak of Lugnaquilla, Wicklow's tallest mountain at 925m. The top is less of a peak and more a surprisingly flat hump that drops sharply into cliffs on one side. The ground here is grassy and stony, not boggy and peaty as you might expect. On the other hand, there are these incongruous pillars of peat, some 2–3m high, but isolated from the surrounding land. From a distance these features have the appearance of a crust that has dried and cracked, rather like those pictures of scorched mud used by newspapers to depict drought in Africa. Up close these 'peat hags' (a Scottish term that seems to derive from Scandinavian words for a cliff or overhang) present a brown wall of flaking soil with an overhang of heathery scrub. There's a dramatic difference between the plants that grow on top of the hag – heather, tormentil, bog-cotton and mosses – and those that grow on the stony ground beneath – bent grasses mostly and not much else. The tops of the hags are remnants of the blanket bog that once clothed the hilltop here and stand witness to the massive amount of erosion that has taken place since they started to break up. Literally tonnes of peat have been washed down the sides of this mountain, so much so that what remains looks perfectly normal and it's the hags that look out of place. This landscape-scale erosion is continuing to this day as sheep, wind and rain take their toll. Soon even the peat hags will be gone and with it any trace that a cloak of blanket bog once spread out across this now denuded vista. This colossal erosion is not just a problem for the 'internationally important' habitat that now isn't there any more, but all that sediment goes on to clog up riverbeds and muddying once-clear lakes. Then there's the release of carbon from the eroded soils, estimated at 1 tonne per hectare per year.[21]

It is small wonder that these supposedly high-value habitats were all evaluated as 'bad' in the report provided by the government to the European Commission in 2013 as per its requirement under the Habitats Directive. These include blanket bog, alpine heath, wet heath and dry heath.[22] Upland habitats in general cover approximately 19 per cent of Ireland's land surface and a disproportionate 40 per cent of the areas designated for conservation under EU law. The extent of damage to the uplands was outlined at a conference on agriculture and biodiversity in 2011, the proceedings of which stated that

> loss and degradation of extensive areas of upland habitats increased since the introduction of forestry grants and ewe headage payments in the 1980s and encroachment or intensification of other human activities including wind energy developments. Negative impacts include changes in plant species composition, habitat fragmentation, drainage, soil erosion and, in some areas, landslides. Upland habitats may also be especially vulnerable to climate change, including extreme weather events ... The most serious impacts recorded to date are overgrazing by sheep and peat erosion.[23]

Vast tracts of the Wicklow uplands have historically been important for birds of European conservation value: the peregrine falcon, the fastest bird in the world, its diminutive cousin the merlin and the ring ouzel (similar in appearance to a blackbird but with a distinctive white bib). The ring ouzel now is of critical conservation concern in Ireland.[24] Once widespread in Wicklow it is now likely to be extinct as a breeding bird there (there was one 'possible' record during the mammoth Bird Atlas project from 2007 to 2011).[25] Habitat loss is no doubt the driver of this decline but increased recreational access to the hills also seems to be a factor. Merlin is an upland bird which has taken to nesting in the coniferous plantations in Wicklow but even so the Atlas project shows that while its distribution remains stable in the core upland areas it has declined across the periphery to the point where it is now absent from Dublin or the higher parts of Carlow and Wexford. The peregrine falcon has a happier story to tell, having rebounded substantially since a ban on organochlorine pesticides in the 1960s. Peregrines exclusively nest on cliff faces (although tall buildings are also known to do the trick) so they are less susceptible to habitat loss from erosion or burning.

The story for all upland birds is a dismal one. Red grouse numbers, for instance, have collapsed by 66 per cent since the late 1960s, although Wicklow remains one of their strongholds. Even the doughty meadow pipit, whose high-pitched chirps are familiar to any hillwalker, have fallen off in number by nearly 40 per cent since the late 1990s. It, too, is now on BirdWatch Ireland's critical Red list of species. Hen harriers, another threatened raptor, have not nested in Wicklow in years, despite the heather habitat being suitable for them.

The playwright John Millington Synge once wrote a short book about his travels and encounters with locals in Wicklow. In a chapter entitled 'The Oppression of the Hills' he recounts an evening stroll, now over 100 years ago, 'where the nightjars were clapping their wings in the moonlight'. Nightjars are scarcely recorded in Ireland any more and are likely to be extinct as a breeding bird in Wicklow.

The plants of the Wicklow mountains have suffered a similar fate. For Robert Lloyd Praeger, who enthusiastically chronicled Ireland's botanical diversity in the decades around the turn of the twentieth century, in Wicklow 'great tracts of featureless peat-bog, with only a few plants of interest, cover the higher ground'.[26] He nevertheless went on to list two specialist alpine plants, the alpine lady's-mantle and the alpine saw-wort, found only at Tonelagee during his extensive searches in the early part of the twentieth century. Thankfully they're still hanging on in this single outpost. Their peers have not been so lucky. I count fully thirteen species – marsh clubmoss, parsley fern, tunbridge filmy-fern, ivy-leaved bell flower, narrow-leaved helleborine, variegated horsetail, great sundew, intermediate bladderwort, broad-leaved cottongrass, roseroot, mossy saxifrage, mountain everlasing and rock whitebeam – all presumed lost, that is, locally extinct, from Wicklow.[27] It is hard to be definitive about extinction because these small plants, growing in difficult to access places, can be overlooked. Surveys in 2008 found a number of plants that had not been recorded in decades.[28] Nevertheless, it is hard to ignore the fate of many of these species, given the extent of the pressures in this area. The loss of biodiversity across these hills has been enormous. Uncontrolled and illegal burning, in particular, is having a devastating and possibly irreversible impact.

While the plantations of evergreen trees are a relatively recent addition to the Wicklow landscape (most were planted after the Second World War), the carpets of heather have an older provenance. Landscape paintings of the area from the early 1800s show even fewer trees than today. Yet it was not always like this. Archaeological remains of ironworks dot the hills. The remnants of the charcoal used to fire the furnaces date from the seventeenth century.[29] Charcoal is the result of burning wood in a low-oxygen environment to yield a fuel predominantly composed of carbon. It burns hotter and steadier than 'raw' wood, thereby producing the temperatures required for the furnaces. The presence of charcoal is a sure indicator of forests and the ironworks were likely to have been driven onto the hills following the clearance of woods in the lowlands. These high-elevation woodlands were no doubt expansive but probably did not blanket the hills completely. Higher up, blanket bog formation kicks in. This is generally wet, with poor soil and high winds, conditions that promote peat formation. Good-quality blanket bog can still be found at the source of the Liffey. These bogs are not suitable for grazing of any kind and don't respond well to fires. It is an environment of which trees are not fond and it is likely that in the days before deforestation the hills were a variety of different habitats providing opportunities for a range of species. Once these upland trees were stripped, sheep farming, deer grazing and burning have combined to keep any regeneration of the woodlands at bay.

That upland areas should be forested landscapes should not come as a surprise when you think about it. Irish hills are not high and anyone who has been on a skiing holiday will know that mountainous areas, in the Alps or the Pyrenees for instance, can be densely wooded. In northern Scandinavia the treeline continues right to the 1,000m contour – about the same height as Carrauntoohil. Because of our climate, the treeline here is around 450–500m. Here in Ireland generations have become so used to looking at brown, desolate expanses of hillside that it is assumed this is their natural state. It isn't. What grows in any particular place is a product of many factors, such as the climate, the land use, the soil and the biological history of the land. John Cross is a vegetation scientist with the NPWS and has worked for decades researching the native forests of Ireland. In 2006 he wrote a paper on the 'Potential Natural Vegetation of Ireland' which was the result of a thought experiment into what would happen if humans and their animals ceased to exist.[30] The detailed map published with the paper shows the Wicklow uplands as a core area of blanket bog on the highest ground, ringed by an extensive band of 'montane birch forest'. This research stated clearly that 'extensive areas of upland heath are a direct result of grazing and associated burning' while 'overgrazing is ... a cause of degeneration of upland vegetation'. As George Monbiot would have said, our hills are 'sheepwrecked'. Get rid of the sheep (and maybe the deer too) and the missing montane birch forests would be a feature not only of the Wicklow hills but also the Mourne Mountains in County Down (which are particularly denuded), the Cooley hills in Louth, the Slieve Blooms (now mostly carpeted with conifer plantations), the Cork and Kerry Mountains, the Comeraghs and the Galtees, as well as mountains in Donegal, Mayo and Galway. Despite their scarcity, oak woodlands are familiar to most people but few in Ireland would recognise a birch forest as being a native woodland type. This is because few people have ever seen one and I know of none that could be described as easily accessible. Yet they do exist and it was only when I found myself wandering in one that it dawned on me just how degraded the heather deserts beyond really are.

On a clear winter's day I had ascended Luggala from the west, a particularly barren expanse of Wicklow hillside, with the intention of snapping some pictures of Lough Tay on the far side of the ridge. Scooped out of the mountainside by long-gone glaciers, Lough Tay and Lough Dan are particularly pretty lakes adorned with oddly-out-of-place golden sand beaches set against precipitous cliffs and dark, icy water. A herd of sika deer caught my eye and I followed them as they bolted into the lower ground. I soon found myself in a world scarcely recognisable as an Irish landscape. Ivory white boughs of silver birch lay twisted and contorted over boulders on the steep slope. Mosses as dense and green as any to be seen in Killarney embraced the rocks while ferns were draped on the boughs of the trees themselves. The birch were interspersed with holly and mountain ash, along with a few outposts of Scots pine. A troupe of long-tailed tits scoured the branches for food while other birds, such as treecreepers and blue tits, flitted through the trees. What a contrast from the wastes beyond where birdlife is thin and made up of only a handful of species. Even in winter it was a lively, vibrant scene that buzzed with wildlife. Here it was, proof positive that forests are meant to be up here. So why are they so rare? And are there any attempts to restore this, one of the rarest of Irish habitats?

The Wicklow Mountains National Park does not encompass all of the Wicklow Mountains but does stretch over much of it, albeit in an erratic and fragmented way. The vast majority of visitors head straight for the Glendalough valley and this is wise as it is by far the most attractive and accessible spot. Glendalough combines an outstanding natural and cultural history with patches of old oak woodland (albeit also with conifer plantations which have found their way even to this idyllic spot). In 2005 a management plan for the national park was drawn up by the principal interest bodies ('stakeholders' as they're known in the jargon). One of its main aims was to 'conserve nature' within its boundaries and yet, despite the acknowledgement of historic deforestation, there is a subtle but tangible reluctance to confront the issue of intensive sheep grazing. 'Sheep grazing has been practised on WMNP [Wicklow Mountains National Park] lands for at least 200 years. This traditional management has maintained the open upland landscapes seen today. The national park policy is to maintain sustainable levels of grazing on such habitats.'

Yet this is not for the conservation of nature. This high priority for 'sustainable levels of grazing' is designed to keep the landscape looking as it has done since the eighteenth century. It is a refusal to acknowledge that the landscape could or should be anything other than a cultural relic and is far from having nature conservation at its heart.

Sheep farming brings with it a dark side which devastates the uplands even more than that constant munching. In order to encourage fresh growth of heather and grass, and be rid of the woody stuff that even sheep turn their nose up at, it has been common practice to set fire to the heath in the early spring. Controlled burning is a familiar practice in the Scottish Highlands where red grouse are effectively farmed so they can be shot by the wealthy. Red grouse like to feed on the tender shoots which sprout in the wake of the flames. Yet even for this type of single-species management the idea is that burning is controlled, i.e. closely monitored by trained moor managers and confined to defined strips, ensuring that the flames don't get out of control. Burning in this kind of scenario is usually done on a ten-year rotation so that the heather has a chance to recover. Yet in Ireland burning is of the uncontrolled variety. Burning during the bird-nesting season (March to August) is illegal and yet it happens annually, usually in April after a week of dry weather.

The short-term impact is a bonfire of our wildlife with the curtains of dark smoke visible from many parts of Dublin city and beyond. In the longer term even the heather cannot withstand year after year of incineration. Where there are few sheep it encourages the growth of bracken, a native fern that invades the hillsides to the exclusion of all else. Being poisonous to livestock it is of use to neither man nor beast. Where sheep, goats and deer are at work, the heath is taken over by more resilient grasses so that specialist species like red grouse no longer have anything to eat.

The burning of hillsides is not driven by pyromaniac tendencies but by the insistence that for farmers to be eligible for the 'single farm payment' (the standard subsidy given to farmers by taxpayers) the land must be 'grazable'. The urge to burn is a direct result of the struggle among sheep farmers to make a living.

The management plan for the WMNP highlights the problem of uncontrolled burning and yet the authorities seem powerless to deal with it. The fire services are reluctant to call out for what are generically referred to in the media as 'gorse fires' unless property or lives are at risk. The Gardaí don't have a wildlife crime unit so illegal fires are rarely investigated. While NPWS staff have powers under the Wildlife Act, the burden of evidence is so high that perpetrators have to be caught red-handed for a conviction to be won in court. It is left to NPWS staff to tackle the flames, a dangerous job for which they are grossly under-resourced. Frequently all they can do is record what happens and where, something they have been doing since the late 1990s. These records show that there is scarcely a part of the Wicklow Mountains that has escaped burning in the last fifteen years and many areas have been burned more than once. The dates of fires extend well into the bird-nesting season, frequently up to May, three months into the prohibited period.

At least there is concern in Wicklow for the uplands and there are people who are willing to take action. An extensive report for the non-governmental Wicklow Uplands Council details in an honest manner the many challenges faced by landowners and others who have an interest in the hills.[31] Disappointing, though, is the total absence of proposals to increase the cover of native woodland despite the obvious advantages (reducing erosion, recreation, carbon storage, etc.). There is unfortunately little tradition of forest management in Ireland, as there is in many other European countries and we lack the research or the incentives that would encourage native woodland restoration alongside hill farming in these areas.

Yet the hills are in crisis. We need new ways to bring life back to the uplands. Some habitats are not suitable for grazing sheep as they are too sensitive to the negative effects (blanket bog, for instance). In some parts of Ireland innovative farmers are turning to traditional breeds of cattle, which were once the mainstay of upland farming. Ultimately, farmers have an essential role to play if the landscape is going to receive the maintenance it needs. The path to recovery means reversing the decline in upland farming as well as the depletion of habitats and species. With the right leadership in place, preferably by bodies like the Wicklow Uplands Council, and with a government that listens to genuine concerns, I see no reason why upland birch forest and habitat restoration cannot go hand in hand with vibrant upland farming.

Connemara

> 'In Connemara, when salmon and sea-trout fishing is quite hopeless, an expedition to some little lake tucked away in the mountains, after "brownies", is always a pleasant way of spending a day. The climb is often stiff enough, but there is much to interest one by the way – on a clear day, views of mountain stretching away on every side, of moors and lakes, and of many a river winding down to the Atlantic. The absence of trees is a sad feature of a Connemara landscape. Seen from a distance the very bareness of mountain slopes makes them look savage, and indeed almost repellent in a hard light.'
>
> From *Irish Bogs* by J. W. Seigne, 1928

Loving Connemara is a bit like being in a difficult marriage. You want it to work, you remember the times when you first fell in love, there are still flashes of that first romance, but you can't ignore the stale banality, the ugly neglect or the resentment that only comes from opportunity bitterly lost. Like most people the first time I saw Connemara, I was impressed. Heading west from the lush green fields and tree-lined lake margins which lead to the gateway town of Oughterard, so civilised by the hand of man, one is suddenly plunged into an altogether different landscape. The field boundaries, such definitive features of the pastoral countryside, are gone. Of trees there is none. The land spreads out before you as far as the eye can see – off to the hills, the Twelve Bens and the Maumturks, so often shrouded in mist or revealing only shadowy glimpses between showers. On and on it goes. Green has now turned to brown, except for the few short weeks when the purple moorgrass comes to life – not purple but green – and splashes of orange, magenta and red, the flowers of montbretia, rhododendron and fuchsia. Occasionally a lake appears, now and again with a pretty island. Sometimes sheep are on the road and tourists slow down to take their pictures. It is wild countryside supposedly, unleashed upon an unsuspecting public, like it somehow broke its chains and is now roaming free like some latter day Hound of the Baskervilles. It is, as I still hear quite a lot, 'unspoiled'.

At its heart lies Connemara National Park, one of the most popular tourist attractions in the county. Not long ago the visitors thronging to the park were unwittingly causing great damage to the vegetation along the route leading to the top of Diamond Hill, one of the peaks in the Twelve Bens. To halt the erosion the park authorities had to close the path, which was a shame, since it was pretty much the only thing visitors can do inside the park once they've seen the visitor centre and had a cup of tea in the café. Happily, construction was carried out on a sturdy path with steps and boardwalks so that for a number of years now visitors have once again been able ascend the summit of Diamond Hill, this time with few effects on the surrounding habitats. On a clear day the view from the top is stunning – on one side the offshore islands cast into the North Atlantic, on the other, the bunched-up pinnacles of the Twelve Bens looking like an inverted cow's udder. From here the mountains look smooth, 'skinny' as a German visitor told me once, and she didn't mean it as a good thing. 'Where have all the trees gone?' she lamented, with a sad look in her eyes. She was puzzled as to why people thought this landscape was wild when there was scarcely a tree to be seen. In Germany wild places are mostly wooded, she explained, and there are wild things in them. Here it seems the landscape had been stripped naked, leaving, well, very little. 'Look', I pointed down at the imperial Kylemore Abbey, nestled in woods that clamber high up the side of the hills, 'trees, lots of them, oak and holly and, er, rhododendron.' 'Yes,' she replied, sweeping her arm in a wide arch like she was casting seeds at her feet, 'very pretty, but where are the rest of them?' At the time I was caught on the back foot. Everyone knows that trees either can't or won't grow in Connemara – or anywhere in the west of Ireland for that matter. It's the wind ... or maybe the unceasing rain ... or probably the fact that the soil is so peaty. That's it: trees don't grow on bogs. A nearby sheep bleated before hopping down to a lower ledge to resume munching.

The dioramas in the visitor centre explain vividly the story of humans in Connemara, from their first arrival to a densely wooded landscape which was swiftly felled and burned, to the denuded plains we see today. Charcoal from fires left a layer on the soil that rainwater found hard to permeate so the ground became waterlogged, promoting the growth of bogs and, presumably, preventing the re-emergence of trees. Having cut most of the peat away for fuel over the centuries, the ghostly stumps and roots of these pines emerge from the bogs, zombie-like, to remind us of this sylvan past.

The treeless Connemara mindset, however, has one flaw: there are trees everywhere, and not just cloaked on the hillside around Kylemore Abbey. Take a look at those islands in the lakes: they're covered with a mixture of pine, oak, rowan and holly. On Lough Inagh there is the Derryclare Nature Reserve, a remnant oak woodland that would not be out of place in Killarney. Granted it is tiny, but the fact that it is there at all points to the fact that trees grow well in Connemara. And the most glaring evidence of all? It's the blocks of non-native pine that have been plonked onto the landscape since the 1950s and seem to be able to grow perfectly well on the peaty soil. They are everywhere, and far more widespread than our native woods, even in this area that we're told is of such outstanding beauty. In fact, once you get your eye in, trees are to be found everywhere around the Twelve Bens and Maumturks. There are mountain ash up on the rocky slopes, holly clinging onto the edges of rivers and hawthorns looking like a vegetative version of Munch's 'The Scream', being dragged from the earth against their will.

The real reason there are so few trees is not down to elemental wildness, but quite the opposite: sheep. That symbol of rugged pastoralism and west of Ireland snapshot favourite. Like everywhere in Ireland Connemara is, and pretty much always has been, a farmed landscape. So the fact that this continues, by itself, should come as neither a surprise nor a cause for particular alarm. What *is* alarming is how economic goals, whether farming or forestry or quarrying, have been allowed to squeeze the life from Connemara with little regard for the landscape or its natural ecosystems; and how trees – the native type at least – are seen as worthless and pointless. Tim Robinson, a great chronicler of life in this corner of Ireland, put it well: 'In certain lights, certain moods, Connemara looks sick. Here, a quarry is spreading like a festering sore; there, another hillside has come out in a rash of deep ploughing for forestry; a favourite boreen has its hedgerows ripped out and replaced by barbed wire and a calf-shed the size of a small factory. A village development committee proposes that a road be widened, a verge sprayed with weedkiller, a green field turned into a car park; everywhere, nature stamped down in favour of convenience.'[32]

Connemara is far from all bad. Tourists, myself among them, still flock there, after all. But there seems to be a blithe negligence, taking its attractions for granted in the pursuit of short-term gain and the convenience that Tim Robinson decried. Even in this age of cycling greenways and outdoor adventure, there is little evidence that Connemara is participating in the boom in off-road walking and hiking. The Western Way, one of the first marked walking trails in the country, sends its few pilgrims on a route that is still largely through grim plantation monocultures and along busy roads. This dramatic part of Ireland could be exceptional but the vision is lacking. Why, for instance, is there no place for native trees? To look at it from a reverse perspective, Connemara should look more like Killarney with its damp forests and abundance of wildlife. The lush growths of mountain ash, holly and Scots pine that are given life on the lake islands, spared from the sheep, give the tiniest glimpse of what an amazing place Connemara could be. Nearly all the species that are found in Killarney are also here, including red deer, red squirrel and pine marten. A pair of white-tailed eagles had returned after an absence of more than 100 years. Tragically, a female about to lay eggs was poisoned in the spring of 2015, setting the recovery of the birds in Connemara back perhaps a decade.

Yet there is no room here for romanticising: 150 years ago there was probably less wildlife than there is today. Connemara was densely populated and while today it might look treeless, back then it was a bedsheet of a landscape. Certainly by then, red deer had been driven out and the lack of any woodland meant there was no foothold for red squirrels or pine martens. So I do not see this place as a paradise lost, more of a paradise waiting to happen. Sheep farming has a long pedigree on the hillsides of the west of Ireland. Cattle, too, were once on the hills, before rules and regulations from the EU made them too much hassle. Before there was such a thing as subsidies, no doubt farmers knew how many sheep could be grazed without damaging the vegetation upon which they fed. This was mostly heather, a collective term for the clutch of shrubby species that only reveal their differences in close-up. In the 1950s, local author Seán MacGiollarnáth described how 'the ling and heather give the prevailing colour to the moors, high and low, but it is the bonny common heather that fills autumn with its restrained brightness, so beloved of our artists'.[33]

Subsequently, disastrous policies from the EU in the 1980s and 1990s, where payments were made to farmers based on the number of animals, effectively destroyed the habitats of the hills. The sheep population of Galway and Mayo doubled between 1980 and 1989, to over half a million.[34] Animals were stuffed onto the hillsides and the result was vegetation nibbled to the roots and bare peat washing off the slopes into the rivers and lakes below. This came at a price. The heathery hills had traditionally been home to red grouse, one of our few upland birds which both feeds on and shelters in heather. Today the red grouse is on BirdWatch Ireland's Red list and a national survey from 2006 to 2008 found a paltry one record from the Twelve Bens and none at all from the Maumturks.[35]

The Twelve Bens were known as one of the best places in the world to see a very special community of mosses and liverworts known as the North Atlantic hepatic mat. It is a collection of non-flowering plants that exists under very unique climatic conditions at high altitude and where rainfall is not only high but near incessant. Not only are these plant communities very beautiful but they are highly significant from a scientific perspective. One species, Lindenberg's Featherwort *Adelanthus lindenbergianus* (sadly many of these so-called lower plants lack easy recognition or even common English names), in Europe is found only in the west of Ireland and one location in Scotland. Elsewhere it is found only in the mountain ranges of Africa's Rift Valley! Ireland therefore has sites of international value but even these only run to about seven locations. It is devastating that one of these areas, the Twelve Bens, has been obliterated by overgrazing from sheep. An account from the British Bryological Society's meeting in Connemara from 1994 (and reproduced here from the account in the *Rare and Threatened Bryophytes of Ireland*[36]) gives a good first-hand account of the scale of the damage: 'The problems of overgrazing were evident during our ascent [of Benbreen] ... We were scarcely prepared, however, for the devastation which awaited us in the northern corrie. Here, on the stony north-facing slope, a few broken fragments of heather and the dead remains of large *Herbertus* tussocks [a type of mountain moss] bore sombre witness to the destruction of the dwarf shrub heath which once clothed these slopes.'

A. lindenbergianus, which 'must have been plentiful here only a few years ago', had been reduced to 'small and sorry pieces'. Although a small colony was recently rediscovered on Benbaun the species has been wiped out and the unique hepatic mat community with it. Studies in the 1990s, when the damage was largely done, confirmed that overgrazing on the Connemara hills was devastating the vegetation and soil structure. One article stated that 'the bare appearance of the Connemara hills is obvious to even the most casual observer; peat erosion, in particular, poses a very serious threat to fish life in nearby rivers and streams, owing to a considerable increase in suspended solids'.[37]

Long after the horse had bolted, the EU got around to mending the error of its ways and decoupled subsidy payments from sheep numbers. By then the Twelve Bens had been trashed. The lightly grazed heather that had sheltered the red grouse, mosses and other plants was gone, replaced by mud and green algae. In 2000 the government commissioned studies of the commonage areas across the country to determine what the right sheep stocking levels should be. Reductions in the Twelve Bens area ranged from 12 per cent to 50 per cent with the aim of restoring vegetation and plant diversity. By this time the EU's Habitats Directive was being implemented and all these mountain ranges fell under Special Area of Conservation designations. The government now had a duty to restore to 'favourable conservation status' the bog and heath habitats.

Fifteen years on from these first framework plans and the designation of much of Connemara as of outstanding wildlife value, it would be reasonable to assume that recovery is well under way. Yet optimists are still waiting on a management plan to rebuild the shattered mountain ecosystems. During the Easter break of 2014 I took advantage of the glorious spring sunshine to set off on a hike up Bencorr, in the heart of the Twelve Bens and overlooking the glassy Loughs Inagh and Derryclare.

It was one of those bracing days with a clear blue sky that you're lucky to get anywhere in Ireland but when you have planned a day on the hills the feeling is akin to winning a moderate lottery prize. Crossing the small stream that bisects the two lakes it was not long before we entered the band of dreary conifers that coils its way around the base of these mountains. In the distance I could see the lofty oaks from the tiny Derryclare Nature Reserve on the far bank of the eponymous lake, now swamped by plantation forestry, and wondered what this vista would have been like had a mix of native trees been planted rather than an alien monoculture. Through the forest tracks we went until it opened out to a scene of utter devastation. It seems like the only thing worse than plantation conifers in scenic areas is the post-nuclear-style 'clear felling' that follows when swathes of trees are chopped down at once. Sawn stumps and piles of dead branches had cut a swathe through what life had developed in the stands of Sitka spruce.

Onwards and upwards and, having left the plantation, it is a scramble up a rocky ridge all the way to the summit. The view of the Bencorr horseshoe is stunning in every direction and takes in all the Atlantic islands, south to the village of Roundstone and the entrance to Galway Bay. On the way back down I passed through classic sheep farming hills. I was watching out for the effects of overgrazing and in particular was anxious to see if heather was making a comeback after the reduction in sheep numbers. I was sorely disappointed. Miniature cliffs of bare peat spoke of the soil that had already been lost and the erosion that is continuing. The ground was slippery with the green alga that colonises bare peat giving the ground a gone-off appearance. Of heather there was none, only the telltale signs of roots nearly pulled out of the ground by hungry sheep. More green alga in the stream flowing through the valley pointed at pollution while a single holly tree poked up out of a pair of boulders, like a middle finger telling us what it thought of our commonage framework plans and Special Areas of Conservation. After six hours I returned to the car, high on mountain air, but depressed at the damage done to the land.

Sheep on their own are not the problem, rather it is too many sheep and sheep in the wrong places. Grazing is a natural and vital part of many ecosystems and, in the absence of large herds of deer, sheep do a capable job. However, some of the delicate moss-based habitats are best not grazed at all. In the state-owned national park, sheep numbers have been greatly reduced but this has given way to excessive growths of purple moor-grass. This leaves a dense, and flammable, mat of dead leaves through which other species cannot penetrate. This is not good for wildlife, either. Apart from the grazing issue Connemara faces threats from invasive species (rhododendron and various knotweed species), continued plantation forestry and the spread of bracken – a native invasive species that thrives after fire. Clearly, Connemara needs major surgery if its ecological decline is to be reversed.

Inspired by the success of 'High Nature Value' farming schemes in the Burren, the state-run Heritage Council commissioned a study into how these problems could be addressed.[38] It highlighted how the grazing issue is thorny and complex. How would sheep be kept off the most sensitive habitats without erecting fences, or stationing shepherds on the hills? How can stocking levels be determined without putting farmers out of business? And, in such a degraded landscape how can species for which there is little seed source (especially the heathers) regain their foothold? Clearly the economics must make sense. A healthy landscape must include people working the hills and they must earn a decent wage. Given viable alternatives farmers have shown great flexibility. The fact is that sheep farming is only viable with state subsidies but government funding for environmental measures is erratic and unreliable. They were the first to be axed during the age of austerity. The Heritage Council's worthy report, which included well-thought-out financing proposals, sits on a (virtual) shelf and so the neglect moves up the political hierarchy. To see how state ownership makes a difference, we just have to look next door to the Connemara National Park (most of Connemara is not in the park) where, incredibly, there has never been a management plan for what is supposed to be a national treasure. The hills here are more heathery and perhaps a pair of red grouse or two hang on. But the park, at around 2,000 hectares, and taking in only four of the peaks of the Twelve Bens, is too small to make much of a difference to the wider landscape. Certainly it's a far cry from what was originally envisaged.

Connemara badly needs a new vision. This must include farmers who are farming (the right number) of sheep and cattle, eradicating alien invasive species and restoring the upland habitats by reseeding heathers or planting native trees. It also must include a review by the state-owned forestry company, Coillte, to rethink its strategy in scenic areas like the Twelve Bens. They have a stunning opportunity to replace the plantations around Derryclare Nature Reserve, and indeed elsewhere in Connemara, with oak and Scots pine and the myriad other species that make a real forest. Imagine all that magical woodland providing habitat for our native plants and animals, maybe even seeing the reintroduction of long-lost birds such as the capercaillie.

Seán MacGiollarnáth talks of an emigrant leaving Ireland, and her beloved Connemara, for the last time. On the night before her departure she is found staring pensively into a lake, with its wooded islands, and tells the author *'tá mé a'tabhairt lán mo shúl liom'* – 'I am taking with me the full of my eyes.' Would that Connemara were a feast for the eyes once more.

Glenveagh National Park

'A National Park exists to conserve interesting plant and animal communities and associated landscapes in their natural state and, under conditions compatible with that purpose, to provide appreciation of them by the visiting public.'

From www.glenveaghnationalpark.ie

After a busy summer it was some time towards the end of September 2013 that I had a day with no deadlines or places to be. The email account was due a clear-out so with one finger on the delete button I started trawling through the months' worth of debris. Among the detritus there was a brief but pointed message. There was a photo of a hillside that could have been anywhere in Ireland and rows of turf that had been sucked out of the ground with a machine that extrudes it in sausage-like rows. It came with a message claiming that the photo was from Glenveagh National Park in Donegal where, the sender maintained, 'illegal turf-cutting is now rampant'. It was the fact that the email had so little detail, with more than an edge of drama and not signed off by anyone, that I thought it was probably a crank and had moved on. Now, months later I belatedly felt that it might be worth following up.

There is always a moderate stream of enquiries in the IWT from people who sprout a sudden concern for wildlife when they think it might be a useful stick to wield in planning objections or festering gripes with state agencies. Turf-cutting was one particular running sore in rural Ireland throughout 2013 and while the IWT was vocal at a national level we simply didn't have the resources to get involved with every angry standoff. Yet here was an email claiming that mechanised turf-cutting was going on illegally, inside a national park. It seemed a bit far out but I was feeling guilty that I had done nothing and feared that we might have lost an opportunity. I eventually tracked down the sender of the email and when we spoke he didn't seem too put out by my inaction – the problem, he maintained, had been going on for years. The authorities, he asserted, were well informed but had no interest or will to address it. He had that slightly edgy tone of someone who had been fighting too many battles but he spoke with authority and offered to show me what was happening on the ground. I like to think that as an organisation the IWT has always had the courage to speak out without fear of reprisal. It is one of the advantages of being a small and poor organisation – there is very little to lose and we are beholden to no one apart from our members. On the other hand, we only get stuck in when we feel we have a full grasp of the facts and where we can stand over our claims. This was a case in point. Turf-cutting goes on across Ireland all the time. Contrary to a commonly held belief, there was never any push to ban turf-cutting away from designated conservation areas from any quarter, not the EU nor the NPWS nor environmental groups like the IWT. So my main concern was to

The whittling-away of our iconic landscapes

Broken picnic seating in Glenveagh National Park.

verify whether the turf-cutting was going on inside or outside the national park.

I set off to Letterkenny on a grim October morning with incessant rain and slow-moving traffic. As I left the town, with the drone of the window wipers grating in my ears, the Donegal uplands emerged reluctantly from the mist. The dreary, brown landscape merged with the monochrome sky lending a more desolate aspect than usual. Like Connemara and other upland areas across Ireland, the only trees that seem to be permitted are those of a non-native, plantation variety. Although there are substantial remnants of oak woodland along the sides of the Glenveagh valley, elsewhere saplings are kept in check by munching sheep and myopic government policies. I was brought to four areas along but inside the northern boundary of the national park. For anyone familiar with the traditional view of turf-cutting with basket-laden donkeys, children with freckles and red, curly hair eating sandwiches on turf banks, or the gentle rows of carefully stacked slices of turf, the vista that awaited would have come as something of a surprise. For hundreds of square metres the vegetation had been stripped away, leaving bare earth and muddy pools. The telltale parallel lines of extruded peat could only be left in the wake of the 'sausage machines' that can devastate a bog in a matter of hours. In one location the destruction was embellished with a liberal sprinkling of waste tyres and other debris – as if to emphasise how this place is seen as little better than a landfill.

A gushing river marks the boundary of the national park and the (pretty identical-looking) land beyond. It cuts a deep gully through the land and provides sheep-proof refuge for holly and mountain ash, as well as flourishes of ferns and mosses. The stream was sadly full of household debris including plasterboard and an old radiator. I checked my map, now damp, with ink running. Yes, I was standing inside one of only six national parks on this island, on land with the strictest designation for nature conservation in Europe, and one of the main tourist draws in the county of Donegal. Is this really conservation of the landscape's 'natural state'?

I catalogued what I could, taking photos in the dim autumn fog and carefully noting the locations with a GPS and crosses on maps. It was late when I got home and I was tired. The next day I wrote to the then Minister for Arts, Heritage and the Gaeltacht, Jimmy Deenihan, the head of the national parks division of the NPWS and the manager of the national park itself. Presenting the evidence that I had gathered I wanted to give them a chance to respond directly to our claim, i.e.: Glenveagh was being illegally trashed under their noses and there seemed to be little being done about it.

Some weeks later, having received no response from any government quarter, the IWT sent out a press release detailing our findings. I was encouraged that we were greeted with a sense of alarm among the media and a degree of interest in our story. I appeared on RTÉ Radio 1's *Morning Ireland* and RTÉ dispatched a crew to Glenveagh to film a piece for the evening news. The next day there were features in most of the national newspapers and I did interviews with local radio stations. I was thrilled that it gained such traction and felt that these headwinds would drive the issue of lack of management right into the heart of the NPWS, and the authorities in Donegal in particular. My hopes were to prove naïve and misguided. Throughout the media interest the NPWS refused to comment. They put forward no spokesperson and issued no statement other than to say that traditional 'turbary' rights (small-scale turf-cutting with hand tools) existed in certain parts of the park. There was to be no defence, no admission that problems existed, no claim that they would even look into the evidence that we provided. It was a deeply frustrating experience.

Early in 2014 I re-sent my letter to all parties but again there was no response. I wrote to all the elected representatives in Donegal but even accepting that concern for environmental issues runs shallow in the body politic, I was disappointed that there was no meaningful response from any of them. My hopes that someone in authority might see the vandalism of the county's greatest tourism asset as something worthy of urgent action were dashed.

In July 2014 I drove along the Doochary road, along the valley of the Barra river, to the head of the Glenveagh valley which overlooks the popularly visited parts of the national park. This long, long road brings the visitor to areas seldom seen by tourists and stopping by the scenic Lough Barra provides a welcome opportunity to take in the landscape. The lay-by has room for a few cars. A faded and broken interpretive panel tells us that the lake is a national nature reserve and is important for merlin (a tiny bird of prey) and meadow pipits (a brown bird). I would have sat down but the concrete seat had been smashed, leaving the pre-poured pillars at odd angles and with protruding iron rods. A smattering of litter completed the scene but what depressed me most was the sight of a large area just off the road that had been worked over with turf-extracting sausage machines. The desolation of Glenveagh for me was complete.

The irony is that the first time I visited Glenveagh I was impressed. Its ancient oak woodlands cloaked the valley near the visitor centre and there was clearly evidence that the invasive rhododendron was being controlled. The long walk from the visitor centre to the waterfall is a dramatic combination of woodland, mountain, lake and river as heady as any in Ireland. Unlike in Killarney, or the rest of Donegal for that matter (where the invasive Japanese knotweed festoons the roadsides), there were few obvious problems. It had a 'native' red deer population up to the nineteenth century which was wiped out but a mixed-breed deer had been subsequently reintroduced behind a deer-proof fence to provide game for the hunting lodge (Glenveagh remains possibly the only national park in the world with a fence around it). The deer had badly damaged the woodlands through overgrazing up to the time the lands were donated to the state by Henry McIlhenny of Philadelphia in the 1980s. In the absence of natural predators the deer numbers are kept in check by the NPWS through annual culling; illegal poaching also does its bit.

Prior to my turf-cutting experience the problems of the deterioration of the upland environment in Donegal were to be brought into relief by the launch of the most ambitious nature conservation project this country has ever seen. In 2001 the park was the centre of a radical initiative to reintroduce golden eagles.[39] Golden eagles are enormous birds, with a wingspan up to 2.3m. Found throughout the northern hemisphere they are a bird of mountains and out of the way places. Shy of people, they are the quintessence of wildness and their formidable silhouette, should you be lucky enough to see it, is a transfixing sight. Although common in Ireland for centuries they were systematically shot out of the sky and their nests raided until they vanished altogether around 1910. Possibly the last of these was in County Donegal, which had until then been a stronghold. The reintroduction was a remarkably proactive manoeuvre on the part of the authorities here and hinged on taking hatchlings from their nests in Scotland and releasing them in Donegal. This is a gross oversimplification, of course, as the programme was based on meticulous planning and a significant logistical operation which adhered to international standards for the reintroduction of a species. The pressures under which they were originally extirpated must have been lifted, there must be sufficient resources in terms of food and habitat in their new home, and the numbers involved must be of such a magnitude to provide a sustainable population in the long term without the risk of inbreeding. I remember the enormous public interest at the time, the extensive media coverage and sense of possibility that wildlife issues in Ireland were not doomed to development quarrels over what Taoiseach Bertie Ahern dismissed as 'swans and snails' holding up everything.

The Golden Eagle Trust deftly managed the project from the outset and has been dogged in its mission to raise awareness among landowners and ensure the acceptance of the project among local people. The initial omens were good: by early 2003, 100 per cent of the birds that had been released up to then were alive and well. They were being fed with deer carcasses and were enjoying the relative safety of the national park confines. Throughout the summer of 2003 they began exploring their new country and there were sightings from Counties Fermanagh and Tyrone. By the end of that year at least one was known to have died but new birds were still arriving from Scotland and there were twenty-four believed to be still out there, although not all accounted for. A certain degree of mortality from natural causes must always be expected in a project such as this and during 2004 a number of birds were found dead. The Golden Eagle Trust speculated that it was a result of malnutrition or disease. Hopes remained high as a nest was found in early 2005 near where a pair was spotted circling. After over 100 years' absence, golden eagles were again breeding in Ireland.

The efforts of this pair were ultimately unsuccessful but this is normally expected for a first attempt at egg laying. There would no doubt be many more breeding seasons, with greater success as the parents grew in confidence. At the beginning of the following year there were a total of five occupied territories with four paired couples. Two of these couples laid eggs but sadly none hatched. A total of forty-six birds had been introduced since 2001 but nearly half of these were known to have died. At this stage, the Trust had set a target to release sixty to seventy-five birds from donor sites in Scotland but there were problems with this as persecution was suspected to be impacting on birds in the east of that country while elsewhere birds were laying only a single egg (birds could only be removed from a nest where two had hatched). Finally, in 2007, the first Irish-born golden eagle was successfully reared by its parents in Donegal, to much jubilation and media excitement, and was christened Conall. It was a huge achievement after seven years of expectation and hard work. In its news bulletin the Trust expressed its hope that in the future we would look back at 2007 as a turning point – the moment when golden eagles resumed their rightful place in the uplands and crags along the west and north of Ireland. The Trust was feeling confident that despite the challenges the future was bright. 'County Donegal was chosen,' their blog proclaimed, 'as the best golden eagle release site in Ireland due to the quality of its mountain environments. Traditional hill farming methods in Donegal have helped retain these features.'

It was always expected that perhaps only one third of the released birds would survive to maturity and the hope was that by 2010 there would be six to eight established pairs. Eagles live long and breed slowly so it was always going to be a marathon rather than a sprint. The problem of persecution was never far off and at least two birds are known to have been shot or poisoned. This also affected the reintroduction of white-tailed eagles in County Kerry, and red kites, another bird of prey, in County Wicklow. Many farmers were *legally* laying poison on their land to control crows and foxes but this indiscriminately affects any other carnivore that stumbles across the inoculated carcass. There was also an unfortunate illegal element which was prepared to target the eagles despite the groundswell of popular feeling behind the reintroduction project.

In 2008 concern was rising for one pair of birds which yet again failed to hatch young despite four consecutive attempts. Fears were raised that the parents, particularly the female, were somehow not in sufficient condition for successful breeding, malnourished in other words. The volume of prey in the territory was low, it seemed. Golden eagles in Scotland feed on hares and red grouse and while they were turning up as prey items in Donegal the Irish birds were also found to be feeding on badger and fox cubs, as well as hooded crows. Was this a sign that their traditional prey was in short supply?

There was devastating disappointment in early 2009 when a dead bird found within Glenveagh National Park was confirmed by the state laboratory to have been poisoned. It was not the first bird to die in this area and the Golden Eagle Trust believed that a pair that had established a territory nearby had also perished in this way. It was most disheartening after the efforts that had gone into engagement with the local communities and farming organisations and it is inconceivable that the person responsible did not know the potential consequences of their actions. 'The loss of several golden eagles, probably due to human persecution, on the edge or within Glenveagh National Park represents a real threat to the entirety of the golden eagle project' was the grim summation by Lorcon O'Toole, the Trust's manager.

Eight years into the project the government had yet to ban the laying of poisoned meat baits and this was undermining the viability of the entire reintroduction project across Ireland.

In May 2009, spirits lifted as two new chicks were hatched from a single nest. Because the second chick generally has poor prospects the decision was made to remove the younger chick and rear it in captivity. Although breeding success was low there were now a possible eight occupied territories and finally the government of the day recognised the severe threat that poisoning was posing to the project. Legislation to outlaw the practice was prepared by Green Party Environment Minister John Gormley. By November the fostered chick had been returned to the nest and both it and its sibling were gracing the skies above Glenveagh. It was only the second successful breeding attempt for the project.

There was another setback when, in February 2010, Conall, the first of the three Irish-bred chicks, was found poisoned. Conall had gained a sort of celebrity status among birdwatchers with his aerial antics and long-distance travels. He had turned up outside Sligo town just before Christmas 2009 and seemed to be getting fond of the Sligo/Leitrim border area. What a boon it would have been for local tourism interests had these magnificent hills become the new home for a golden eagle! It was not to be.

The knowledge that the next poisoning was waiting to happen tempered the excitement surrounding the arrival of a new chick from the now experienced pair in Glenveagh. This was to be the fourth chick to fledge successfully. Elsewhere the news was not so good as another Donegal pair yet again failed to hatch any egg they had laid. Yet there was joy when a third nesting pair in the county succeeded in fledging two young in an area of sheep commonage. The Golden Eagle Trust had received great support locally from hill farmers and, interestingly, there was growing anecdotal evidence that the presence of eagles was reducing the abundance of hooded crows. It has traditionally been for crows and foxes that farmers have laid poisoned bait so it was now being demonstrated that the golden eagles could act as effective pest control. That year saw the release of five new eagle chicks from Scotland, bringing the total to fifty-eight out of the target of seventy-five.

Through 2011 and 2012 the eagles struggled through horrendous weather that brought near-incessant rain to most parts of Ireland. They were not helped by the scale and frequency of fire in the uplands, something that has become more common (despite being illegal) and which is dramatically changing vegetation and ecosystems. Fires leave the landscape bare, wiping out nesting birds, and leaving the eagles with nothing to hunt. Thanks to the perseverance of the Golden Eagle Trust and local volunteers, poisoning of land is now on the decrease. Initial fears among some farmers that large birds of prey would threaten their livestock have proven unfounded. The reintroduction has overwhelming support among the farming community and the public at large.

However, the uplands of Donegal still present problems, not least of which is the ongoing neglect of upland and marginal farmers though the EU's Common Agricultural Policy. The spending of vast quantities of public money on the agricultural sector has yet to recognise public benefits to landscape and nature from low-intensity farming. Instead, it remains production-based and the greatest sums are still diverted to large intensive farms in the east and south. Abandonment of land can lead to hillsides blanketed with bracken or rhododendron, of use to neither man nor beast. Misguided policies still promote the removal of scrub, further depleting the opportunities for wildlife. Meanwhile, lack of economic opportunity creates pressures from wind-farm or plantation-forestry interests, resulting in greater conflict with wildlife and the continued degradation of habitats and landscape.

By the time of writing in 2016 the golden eagle population remains on a knife edge. A Facebook post from the Golden Eagle Trust in November of 2015 brought grim tidings:

> The small golden eagle breeding population struggled to cope with the conditions in the Donegal mountains in 2015 ... In summary four pairs built nests, three of which laid eggs and two of these pairs hatched young. But all three breeding pairs failed and no new golden eagles were added to the population in 2015. The Glenveagh pair laid eggs in the same nest they have used in recent years. But unfortunately the chick that hatched died after a week, during a spell of poor weather.
>
> The other established pair also had young but their breeding attempt failed when the chick was only three weeks old. And though the weather was wet at the time, it was quite disappointing to lose a chick at this age. A shortage of food availability locally may have been exacerbated by the poor weather and resulted in a weak and vulnerable chick ...
>
> We have come to the opinion that it is primarily the Department of Agriculture who hold the key to improving the lot of the upland eagles. Farm policies regarding upland vegetation are the key ingredient ... There is very little habitat reflecting the true potential of Irish upland wildlife.[40]

The national park website says that 'such a great wilderness is the haunt of many interesting plants and animals'.[41] It remains to be seen whether these magnificent birds will remain one of them. Was it naïve to think that national park status would mean the conservation of the eagles would be a priority? Glenveagh National Park remains a nice place to visit but its edges continue to be eaten away. There is no management plan, never has been, and as far as this author is aware, no plans to produce one any time soon.

The Burren

> 'At a distance, these bare rocky hills seem thoroughly devoid of vegetation, and a desert-like aspect thus imparted to the landscape has been compared to that of parts of Arabia. But on a closer inspection, it will be found that all the chinks and crevices, caused by the above mentioned joints, and the action of rain, are the nurseries of plants innumerable, the disintegration of the rock producing a soil, than which none is more productive.'[42]
>
> M. J. Foot, 1871

The mountain avens is a small plant with white flowers and, like all small plants, is best appreciated up close. To get the best 'bee's-eye view' the viewer really needs to be flat out on the ground, face down, almost nose to petal. At this proximity the mountain aven's corona of overlapping ivory-white petals reveal their starry beauty. Its amber anthers huddle in its centre, each of its eight (although sometimes sixteen) marble-smooth petals stretching the short distance to the cushion of crenulated leaves, pine green, pressed against the meagre soil. Such is the expanse of the flower, carpeted across the stony grey hillsides, that, at a distance, it could pass for a fall of snow. But this is not winter, it is spring in the Burren.

Dryas octopetala is a plant of the Arctic and Alpine mountain ranges, its habitat stretches across North America from the Rocky Mountains to the treeless plains of Iceland and throughout the European peaks as far east as the Caucasus Mountains. But only in the Burren does it attain its luxuriant abundance in an area that is neither Arctic nor Alpine. This corner of Atlantic-salted limestone has a life force that is all its own. You don't need to be a botanist to appreciate the otherworldliness of the Burren. Its 'fields' frequently don't have any grass, or even soil for that matter. Where these are present its grass is noticeably less green than the pastures from elsewhere in Ireland, and distinctly less grassy too. These fields (and visitors have ample opportunity to amble through them on colour-coded trails) are teeming with life. From twenty types of orchid to the flamboyant bloody crane's-bill, the straw-yellow carline thistle and delicate-flowered quaking grass, the variety and number of plants will astound anyone familiar with the banal plains of ryegrass found in most parts of Ireland. On a summer's day the air hums with movement – butterflies and day-flying moths flit, bees and dragonflies duck and weave. While there are some rarities, most of the plants and animals of the Burren can be found elsewhere. What makes this place so special is the abundance in which they are found. It's of Disney-movie proportions and while this is no doubt a product of the unique location and history of the Burren, it is also a pointer to the diminution of wildlife across the rest of the countryside. This biological cornucopia at first seems incongruous for it is enmeshed with a human presence as old as the landscape itself. The appearance of all sorts of human relics, from the tombs and forts of the ancients to the boundary walls of farmsteads, blur the lines between bedrock and architecture. The textured limestone of which each is made makes the geology organic, mutable, plastic. This indelible but unmistakably human influence continues right to this day. Understanding why people, landscape and wildlife not only coexist but nurture each other is vital if we are ever going to get our heads around the idea that humans are not separate from the natural world, but an integral component of it. Here in the Burren a living laboratory marks the way. It is, as the BurrenBeo Trust (an organisation dedicated to protecting the Burren) puts it, Ireland's 'premier learning landscape'.[43]

The Burren is defined mostly, but not entirely, by its karstic limestone geology. The universal laws of water chemistry – weak acids dissolving the relatively soft rock – have produced some of the most stunning landscapes in the world, including the fairy-tale pinnacles of Yangshou in south-west China, the jungle-clad caverns of Mexico's Yucatan Peninsula and the impenetrable stone 'forests' of Madagascar. In Ireland it is not confined to the Burren: there are areas of karst in Galway around Lough Corrib, on top of Ben Bulben in Sligo and in the hills of Cavan, but there is no doubt that the Burren is one of the most stunning areas of its kind anywhere. Here it is confined to the northern shoulder of County Clare but, after stretching under the waves, resurfaces along each of the Aran Islands, although the islands are rarely considered to be within the Burren.

There is a Burren National Park but it is an odd entity. At 1,500 hectares it covers barely 3 per cent of the Burren landscape and visitors would scarcely notice its existence. Throughout the 1990s, however, it was the epicentre of a protracted and divisive dispute which encapsulated the divergent views regarding conservation versus 'economic progress' in Ireland.[44]

The row revolved around the construction of a visitor centre at Mullaghmore, at the centre of the recently created national park. Mullaghmore is a remarkable-looking hill that reminds me of a fossilised heap of pancakes drooping under the weight of too much syrup. Frozen in geological time it looks set to topple entirely into the mystical, topaz-hued waters of Lough Gealáin beneath. In the early 1990s, with a newly forward-looking economy and injections of European financial aid there was certainly merit, as well as obvious pitfalls, to the idea that a visitor centre at Mullaghmore would stimulate economic activity. Knowing that tourists like their epic scenery with coach parking, souvenir shops, toilet facilities, a café and photo opportunities that don't require breaking a sweat, the concept of the visitor centre is well established around the world and it has its pros and cons. On the one hand it allows an opportunity for tourists to experience the landscape in an easily digestible manner, a half a day would do it before trundling to the next attraction in the afternoon. There's no doubt that it creates economic activity, hiring local people, etc., and encouraging visitors to stay and eat nearby. The extent of such an economic kickback is debatable, however, if staff are hired seasonally and most of the souvenirs are made in China. It also calls into question the entire national park ethos which, according to guidelines from the IUCN (and which the Irish government professes to follow) must prioritise conservation, not economic exploitation.

From an environmental perspective the visitor centre tends to cluster activity towards one point in the national park. Here again there are pluses and minuses. In large national parks, such as Australia's Great Barrier Reef for instance, most areas are off limits to clumsy novice scuba divers who bash into coral formations and sneak off with shells. This means the vast majority of the park is protected from this kind of pressure. On the other hand the intense demand for 'consuming' nature brings with it intense pressures within a small area. Australia has, in effect, sacrificed small parts of the Barrier Reef so that the greater whole can be saved. Here in Ireland, isn't it better to have most of the visitors in one place where impacts can be controlled, as opposed to spreading the impact across the landscape, making it harder to deal with water pollution, litter and trampling?

The area around Mullaghmore seemed ideal for just such an enterprise since it was already government-owned land but its great flaw seems to have been what Liam Leonard, in his book *The Environmental Movement in Ireland*, referred to as a 'market analysis of EU or state bureaucrats'[45] which lent a totally different perspective to that of many of the local people or conservation groups at the time. The lines were quickly drawn when Minister of State Vincent Brady indicated the government's intention to develop the Mullaghmore site in 1991. Opposition coalesced into the Burren Action Group (BAG) which set out an alternative vision for development, one that favoured regional, community-led management over the state-imposed version. The prospect of employment, in an area and region blighted by 'brain drain' and underemployment was sufficient argument for many to support the project. BAG was to meet its match with BNPSA (the Burren National Park Support Association), which included the Irish Farmers' Association, the Gaelic Athletic Association and Fianna Fáil. It is interesting to note, however, that farmers were present on both sides of the argument. National politics intervened in 1992 when Labour TD Michael D. Higgins, subsequently President of Ireland, was appointed Minister for Arts and Heritage in the Labour/Fianna Fáil coalition. Well known for his sympathies, which leaned decidedly in favour of preserving our island's heritage, he was to become a decisive player in the visitor-centre saga. By this time, the EU was also getting weak-kneed about funding the project and the head of its Environmental Directorate announced that the enterprise might no longer qualify for European handouts.

In November 1992 BAG took the case to the High Court on the grounds that the OPW didn't have the authority to build the centre and should not be exempt from planning laws. Remarkably from today's perspective, not only did the state's engineers (OPW) not need planning permission for their projects but they were not even required to consult with local authorities on their implementation. Meanwhile the OPW went ahead and started building work, including a wastewater treatment plant, clearance for a car park and building foundations. Further legal wrangling, including the involvement of the UK chapter of the WWF and the European courts saw the imposition of an injunction on works at the site in early 1993. Ultimately the ball fell into Michael D. Higgins' court who was now minister in charge of heritage in the so-called 'rainbow coalition' (without Fianna Fáil), and he killed it off by prohibiting the OPW from responding to a planning request for further information from Clare County Council. By 1995 the plans had been scrapped altogether while funds originally set aside for building the centre now had to be used to demolish the half-built remains and restore the land. The Fianna Fáil TD Síle de Valera, who would herself go on to be minister for heritage, told the Dáil that 'The Minister's pre-emptive strike which was both insensitive and authoritarian has caused great hurt, anger and further division within the community of north Clare ... I was aghast at the suggestion by the Minister for Tourism and Trade, Deputy Enda Kenny, that the [existing] car park could be used for picnics.'[46]

Minister Higgins' response was that a new management plan for the Burren National Park would be prepared to guide future development and avoid the calamitous acrimony that left the community of North Clare scarred and divided. The management plan was still being discussed in the Dáil in June 1996 and thereafter all trace of it disappears. In an effort to track down this mysterious document I submitted a Freedom of Information request to the relevant department and was informed that: 'the record concerned does not exist ... while a commitment was given in 1995 by the then Minister, Michael D. Higgins, that such a plan would be prepared, due to pressure of other work commitments within the Department, such a plan was never finalised.'

It was a low ebb for the Burren. Ireland, modernising fast and with lax planning regulations has seen great damage to landscape and environment since the so-called Celtic Tiger began its roar. The Burren did not escape this and a mixture of one-off housing and agricultural 'reclamation' (i.e. where an ancient field is ploughed up, showered with artificial fertiliser and reseeded) began to erode its edges. In 1997 David Drew wrote of the Burren that 'the situation is now precarious; without coherent policies, landscape and general environmental quality will rapidly deteriorate. Remedial action requires a new harmony between farming, tourism and the landscape, involving the collaboration of a wide range of public and private interests.'[47]

And then along came a new vision for the Burren. In 2004 the Burren Farming for Conservation Programme was launched with significant grant aid from the EU. At its heart was a recognition that the Burren is a man-made landscape. Along with intensification of farming a significant threat has arisen in recent decades from land abandonment. With the removal of cattle, thickets of hazel develop in the dense shade of which the unique grassland community of plants and insects is lost. However, unlike the rhododendron in Killarney, hazel is not an alien invasive species. Hazel scrub is a type of native woodland that would probably develop into tall woodland, with ash and Scots pine, if left to its own devices. It would be rich in wildlife and, indeed, older areas of hazel in the Burren today are valued especially for their unique mosses and lichens. But it has to be recognised that the Burren as it has existed since the arrival of humans millennia ago would be lost in such a scenario. The cultural loss of a farming landscape with its ancient artefacts and living traditions would be bad enough but grasslands full of wildflowers and insects would pass into oblivion along with it.

Putting the cattle out to pasture in winter, something that has been done here for generations, prevents hazel scrub encroachment and makes way for new growth in spring. The new Farming for Conservation Programme was a radical departure in the way such issues are approached in Ireland. For a start it involved farmers from the beginning. It geared the farm-support payments (i.e. the subsidies paid to farmers with taxpayers' money) specifically to conditions in the Burren and the desired outcomes in terms of habitat conservation. It was complemented by the education charity BurrenBeo, which delivers a constant stream of walks and talks, helping visitors and locals alike to understand the treasures of the Burren. It is assiduous in supporting the sustainable development of the region through local businesses and tourism initiatives. By 2012 the programme had 143 farmers conserving 13,000 hectares of designated habitat. As one of BurrenBeo's founders, Dr Brendan Dunford, has said: 'the beauty of the approach is threefold: the farmer has great flexibility in managing the land; the taxpayer is guaranteed good value; and the policymaker has an in-built monitoring system as to the programme's effectiveness.'[48]

Its effectiveness has been marked by international acclaim including a 'Best of the Best' award from the European Commission and a short-listing for a 'Tourism Tomorrow' award from the World Travel and Tourism Council in 2014.

The Burren still faces issues from land abandonment and pressure from tourism. There is still no decent car park at Mullaghmore so people tend to park along the side of the road. But visiting in January, as I did early in 2015, these problems seem remote. Certainly compared to our other national park landscapes described in this chapter, the management of the Burren is exemplary. I watched the cattle graze on the ledges high on Mullaghmore and wondered how they managed it without breaking their ankles – the limestone pavement seems like a giant geological cattle grid. Their continued winter ruminations are crucial for the future of the Burren and for now this seems assured. I also watched a couple of farmers out early on that crisp, still morning, repairing a stone wall. It gave me hope that we're not doomed to see every traditional field boundary swept aside for a cheap barbed-wire fence. If the people of the Burren, the EU and the government can work together for the benefit of this amazing place then surely there's hope for the rest of the country.

Which leads swiftly on to my next thought which is precisely why, after demonstrating how to go about conserving nature and landscape hand in hand with farming in one place, are we still faced with appalling degradation in places like Donegal, Wicklow and Connemara? Why is there not a Farming for Conservation Programme right along the west coast from the peninsulas of Cork and Kerry right the way up to the Inishowen Peninsula in Donegal? What about the Wicklow uplands, Louth's Cooley Peninsula and Waterford's Comeragh Mountains? The places where traditional farming life is most under threat overlaps with most of the land areas designated for nature conservation, our national parks and the Wild Atlantic Way. Could it be that prohibitive cost stands in the way? Between 2004 and 2010 the Burren LIFE project (LIFE being the EU's funding mechanism for its biodiversity conservation policies) cost a total of €2.23 million. That's approximately €410,000 per annum. Expensive? It should be compared to the €3 *billion* of taxpayers' money that was spent on environmental farming subsidies in Ireland from 1994 to 2010 (approximately €429 million per annum) and which has shown little evidence of achieving any positive environmental outcomes.[49]

The good news is that progress is being made in what is now referred to as High Nature Value farming. A Burren-style initiative has been introduced to the nearby Aran Islands while detailed studies on the implementation of similar initiatives for Connemara and the Iveragh Peninsula in Kerry have been published by the state's Heritage Council. And yet progress is far too slow.

In this chapter I've looked at all our national parks and the landscapes in which they are nestled with the exception of Ballycroy in County Mayo. Ballycroy and its landscape represents something different and for this reason discussion of it is left until the last chapter. Of the five under scrutiny here I have found that only one, the Burren, comes close to what could be described as 'well managed'. Killarney still has a lot going for it but needs urgent intervention. In Connemara, Wicklow and Glenveagh, pretty places near the car parks are kept photogenic but beyond that things have mostly gone to rot. As for being managed for their 'characteristic species and ecosystems' (as per the IUCN definition), they don't even come close.

4

Extinct: Ireland's lost species

> Mr Power (Fianna Fáil): asked the Minister for Fisheries and Forestry if his Department are monitoring the various species of endangered flora and fauna; if he will list these species and the measures being taken to protect them.
>
> Minister for Fisheries and Forestry Mr Fitzpatrick (Fine Gael): No species of flora or fauna which occurs in Ireland could be said to be in danger of extinction.
>
> Dáil Exchange, 17 December 1981

THE BEAST BEFORE ME IS HUGE. I can see its armour-like plates run in rows down the full length of its body, which stretches over 2m from its nose to the tip of its tail. Its pointed snout seems to be all upper lip, and its sensitive feelers, known as barbels, have all the appearance of a moustache – giving it a slightly aristocratic, if comic, appearance. Its beady, unblinking eye stares back at me but its gaze passes right through me. It is a sturgeon, a giant bottom-dwelling fish that uses its barbels to detect prey in the muddy depths of estuaries and the shallow sea floor. This specimen once called the River Liffey home; the only problem is that it is long-dead, at one time a prize catch, now stuffed and on display in the Natural History Museum on Merrion Street Upper in Dublin. The museum card states: 'taken from the Poolbeg Salmon Fishery, River Liffey, Dublin, in June 1890. Purchased from Mr J. G. Powell (fishmonger) for ten shillings'. Large as it is, it is not enormous by sturgeon standards, specimens of which have been known to grow as long as 6m and weigh up to 400kg, equivalent to five men.

This sturgeon on display in the Natural History Museum in Dublin was caught in 1890. It is now extinct in Irish waters.

People today may be familiar with the sturgeon as the source of caviar, strictly speaking the eggs of the beluga, ossetra or sevruga varieties which are found in the Caspian and Black Seas. The Atlantic sturgeon has not been commercially valued in this way but it nevertheless always occupied a very special place on the tables of those fortunate enough to have it laid before them. In the thirteenth century King Edward II of England decreed that any sturgeon caught should be the property of the Crown and it has, since this time, been regarded as a 'royal' fish. Such was the value bestowed upon it that in the seventeenth century specimens were valued gifts or even peace offerings between rivals. Reports of such exchanges are recorded from Howth in Dublin, Cork and the Inishowen Peninsula in Donegal. In Charles I's charter to Waterford in 1631 he granted upon the mayor, sheriff and citizens the rights to fish salmon and 'other fish of every kind' – except, that was, for the sturgeon. In 1754 a specimen found off Dundalk was reportedly 3m long and weighed 136kg.[1]

Sturgeon are a large, slow-growing, long-lived and late-maturing species. They can live to over 100 years and don't become sexually mature until seven to nine years in the case of males and eight to fourteen years in the case of females.[2] Animals (or, indeed, plants) with this life strategy (which is shared by humans) are very vulnerable to changes in the environment or over-exploitation, whether through fishing or hunting. Arthur Went's study of the sturgeon in Ireland suggested that what was once a widespread and reasonably common fish was, by the eighteenth century, already becoming scarcer although still regularly occurring in river estuaries from Counties Down, Louth, Wexford, Waterford and Cork. Sturgeon breed in the shallow, sheltered waters of rivers and estuaries and it is clear from this evidence that they were once found throughout the large river systems of Ireland. Indeed, up to the nineteenth century it is possible they were commonplace across the waters of Britain and Ireland.[3] They were caught in drift nets set for salmon and, given how scarcer in number they would have been compared to salmon, it is easy to see how their numbers became quickly depleted.

During the twentieth century sporadic reports of sturgeons surfaced here and there. In 1950 Robert Lloyd Praeger wrote of the sturgeon that it was 'not infrequently taken'.[4] A 28kg specimen, a youngster given its size, was landed at Ardglass in County Down in 1966. Conservation not being part of the mindset of the time, it was dispatched there and then and a fibreglass cast is now on display in the Ulster Museum in Belfast.[5] Also in 1966 a small specimen was landed in Dingle, County Kerry, and caused a stir in then President Éamon de Valera's office and in far away Buckingham Palace where Queen Elisabeth politely declined delivery of what was still considered a 'royal' fish.[6] In 1967 another was hauled up in a trawler off County Cork weighing 29kg.[7] This is the last written record of a sturgeon I can find from Irish waters. What was once a reasonably common breeding species found around the Irish coast slipped noiselessly and without fanfare into extinction. It had most likely been overfished although changes to rivers from extensive drainage schemes may also have played a part. Long-lived and slow breeding is no match for regularly set drift nets.

The Atlantic sturgeon is not *extinct* extinct. Technically the IUCN, which monitors the status of many species through its Red lists, has declared *Acipenser sturio* to be critically endangered and decreasing. There are still breeding populations on the Garonne River in France and it has been reintroduced along the River Rhine in the Netherlands and the Elbe in Germany. Ireland was not the only place where overfishing took its toll while industrial pollution in the European heartland hastened its demise further. Although now present in only a tiny fragment of its once considerable range, for a species to be totally extinct there must be 'no reasonable doubt that the last individual has died'.[8] Clearly this is not the case and today considerable efforts are being made to ensure that sturgeons do not die off completely. But there is no question that the sturgeon is extinct as an Irish species. Its passing was not noticed, never mind lamented. It was the subject of no action plan and nobody today misses its medieval scales or aristocratic snout. It has passed not only out of the places where it once lived and breathed, but also out of memory, as if all trace of it has been erased.

In fact, the sturgeon is not the only species to become extinct as an Irish species. I have found confirmed records of over 115 species of plant and animal that have been lost since the arrival of people to this island. These range from the brown bear, an enormous animal which died out no later than 8,800 years ago, to the corn bunting, an inconspicuous brown bird which ceased breeding here altogether only in the 1990s. Some are familiar, such as the wolf and the wild boar. That we ever had lynx, wildcats and flocks of cranes will come as a surprise to many. Others are known to those with an interest in nature, such as the osprey (a bird of prey), the capercaillie (a giant grouse) and the bittern (a relative of the heron that lurks in reed swamps, revealing its location with its great booming call). But the vast majority are animals or plants that only the specialist has ever heard of. Who misses *Donacia semicuprea* (a water beetle) or remembers a time when *Gymnocarpium dryopteris* (the oak fern) was to be found in the mountains of Sligo and Leitrim? Some species that were wiped out have returned, either naturally, like the great-spotted woodpecker, or through concerted reintroduction programmes, such as the golden and white-tailed eagles and the red squirrel. Only one has gone completely extinct and that's the giant auk, the original penguin, that may have had a large breeding colony off the south-east coast. I was able to count over 100 species because there are dedicated naturalists who have spent their days knee-deep in mud and water, or reaching out on a mountain ledge counting and cataloguing the small things in nature. The nuts and bolts as it were, the cogs and washers that keep the wheels of life turning. But even at that there are whole groups of species about which practically nothing is known. So 115 is likely to be a considerable underestimate.

There are at least 5,500 species of fungi in Ireland. Fungi are essential organisms for the decay of waste and the recycling of nutrients. Many plants cannot function without them as dedicated fungi frequently cling to their roots, manufacturing the minerals and complex sugars needed for growth and development. Yet hardly anything is known about how widespread these organisms may be in Ireland, whether any have gone extinct, or whether any are on the brink of oblivion and in need of urgent action. This is partly because the number of people in Ireland capable of carrying out this highly specialised taxonomic work (taxonomy is the naming of life) would probably fit comfortably in a lift. There are an incredible 585 species of annelid worm rummaging through Irish soil (among which is the common earthworm of which we are all so fond). A square metre of healthy soil in a pasture can host between ten and fifteen species of these worms with up to 390 individuals! A 2008 study on the economic value of biodiversity to Ireland found that worms conservatively provide services to agriculture worth around €1 billion.[9] Yet there is precious little information on how our worms are doing; surely it would be beneficial to know whether some species are declining or not? There can be little doubt, therefore, that some species have gone extinct from this country entirely unnoticed. Part of the problem is knowing for sure whether something is really gone or if it is still out there but no one has gone to look for it.

The divided sedge is an inconspicuous grass-like plant which grows up to 75cm. It was historically recorded from its salt-marsh habitat at the mouth of the River Liffey in Dublin as well as along the Barrow estuary in County Wexford. In 1986, when botanists made the first conservation assessment of plants in Ireland, in what is referred to as a Red Data book, the divided sedge was declared extinct. 'Despite a number of searches this plant has not been seen at any of its five sites since 1930'.[10]

That was that, it seemed. Until it was re-found not long after the publication of the Red Data book; it is now considered 'locally frequent' along the Barrow, albeit still gone from the Liffey.[11] So naturalists are understandably hesitant to declare a species extinct. Some old records of obscure species can be unreliable and we are left to wonder if it was a case of mistaken identity or if maybe it really is still lurking out there somewhere. I've tried to find rare plants, knowing within a hundred metres or so where to look, and found it incredibly frustrating. And plants are stationary; imagine hoping to come across a hoverfly or ant! More often than not luck needs to align with incredible perseverance. The nightjar is a bird that was once widespread in Ireland. It has not been recorded breeding in years, but biologists won't say it is extinct because it is nocturnal and nests in out-of-the-way places. Maybe it is just that the right surveys have not been done. The angel shark was once found by sea anglers all around our coast. It too may now be extinct from our shores but, as they are not counted, we cannot be sure.

In some areas it is practically impossible to know whether a creature once called Ireland home. Even today very little data is gathered on the biodiversity of our seas. We know that the sturgeon was once widespread because it had significant social and economic value, and so there are numerous written records throughout the centuries. The North Atlantic right whale (it was the 'right' whale for hunting as it was slow moving and floated when dead), or Nordcapper, was once found throughout the coastal areas of Western Europe. They are well known to have bred in the shallow waters of the Bay of Biscay, where Basque hunters were skilled at their capture. James Fairley, in his account of the Irish whaling 'industry' (such as it was) recounts how Nordcappers were captured in Donegal Bay in the mid-1700s.[12] Five of the beasts were caught at the whaling station on Iniskea, County Mayo, in 1908 while a further five were harpooned in 1909. They probably did not breed in Irish waters but passed close to the shore in June, on their way north to feeding grounds off Norway and Iceland. No more of these behemoths were captured after 1910 and shortly after, it seems they were entirely extirpated from European waters. Even today, according to the WWF, only 300–350 individuals still live off the eastern coast of North America, where conservation measures came in the nick of time.[13] They nevertheless remain vulnerable to impacts from shipping and industrial fishing, and their numbers have sadly shown no sign of rebounding in recent decades. Whales, too, were valuable but what of all the myriad creatures of the deep about which little is known? If any have gone extinct from Irish waters we will probably never know.

The point is that my list of 115 or so species needs to be treated with a certain amount of flexibility. These 115 are predominantly records for which Red Data books have been presented by teams of experts in the relevant field. The scientists that produced them are reflexively cautious creatures and if anything 115 is much more likely to be an artificially low number, rather than artificially high. In the list of mammals, which I have shown below, I have gone beyond the published Red Data book and added some of my own species. This is because some, like the brown bear, lynx and wildcat, were excluded, perhaps because they vanished such a long time ago. There may also be an element of politics in this. The wild boar, even though there is ample evidence to show it is a native animal, doesn't get a mention in the Red Data book for mammals. The reason may be because wild boar is a troublesome species for arable farmers and foresters. The wolf, hunted to extinction by the end of the eighteenth century, does make it onto the list, perhaps because this is relatively recent and perhaps the authorities don't fear (yet!) that there will be serious calls for its reintroduction. So the tables below are the lists I have come up with:

Mammals known to have become extinct from Ireland since the arrival of humans and their last known record[14]

Species	Last-known record
Brown bear	8,880 years before present (bp) – a tooth from Derrykeel Bog, Co. Offaly[15]
Lynx	8,875 years bp – a femur from Kilgreany cave in Co. Waterford[16]
Wildcat	Unknown but bones were excavated from Edenvale, Co. Clare, and Newgrange in Co. Meath.[17]
Wild boar	Definitive evidence from a scapula found in Dalkey Island from 6,870 years bp but thereafter archaeological remains are confused with domestic pigs.[18] It may have gone extinct during the Neolithic period, 5,000 years ago but some think it may have persisted much longer.[19]
Wolf	Widespread and probably abundant in Ireland until a concerted extermination programme finished off the last one in the late 1700s.[20]
Red squirrel	Generally considered a native species. Went extinct, probably due to deforestation, although also exploited for its pelt, around the end of the eighteenth century. Reintroduced with stock from England between 1815 and 1876.[21]
North Atlantic right whale or Nordcapper	Regularly occurring up to 1910, a handful of records since then but none in recent decades.[22]

Breeding birds known to have become extinct from Ireland since the arrival of humans and their last known record

Species	Last-known record
Hawfinch	A woodland bird. Bones were positively identified from a cave in Co. Clare and are undated but possibly Bronze Age.[23]
Great-spotted woodpecker	Present from prehistoric times. Its demise seems to have coincided with the loss of woodland in the seventeenth and eighteenth centuries.[24] However, breeding through natural recolonisation was confirmed in 2006. It now breeds in a number of east-coast counties and its expansion continues.
Crane	A once widespread, if secretive, bird of fields and wetlands, it ceased breeding in Ireland probably in the mid-1700s. Breeding again in England after a reintroduction project there, Ireland still gets the odd flyover.
Capercaillie	An oversized grouse of woodlands, the last record of which was from Thomastown, Co. Kilkenny, in 1760 but which was already very rare for some time.[25]
Osprey	Breeding on inland lakes including Lough Key in Roscommon in 1779 but no records thereafter, save for the odd vagrant.[26] Given its relatively abundant status elsewhere it is somewhat of a mystery why ospreys have not bred in Ireland in recent times and many suitable nesting sites are keenly watched each summer for activity.
Great auk	Once nesting in enormous colonies in Iceland and Newfoundland, these flightless birds were hunted to extinction globally. Gordon D'Arcy speculated that there may have been a breeding colony on the Keeragh Islands off the Wexford coast but definitive evidence is lacking. Their bones do, however, appear in abundance in middens. A live bird was brought ashore in Waterford in 1834 and nursed for a few weeks before dying. The last two birds in Iceland were butchered in 1844. There was a plausible but unconfirmed report of two birds in Belfast Lough in 1845.[27]

Species (continued)	Last-known record
Goshawk	Last breeding record from Co. Derry at the beginning of the nineteenth century.[28] Recolonisation may be under way as breeding has been confirmed from a number of areas in recent times.[29]
Red kite	A scavenger of town and city it was rare by the 1820s.[30] Recently the subject of a reintroduction programme in Counties Wicklow, Dublin and Down. There is good reason to be confident that the kites are here to stay this time.
Redstart	It is hard to say that redstarts definitively went extinct although D'Arcy felt that its Irish name *Deargán alt* ('red winged'?) meant unequivocally that it was well known before the oak woods were cleared.[31] Breeding since then has been sporadic and very localised.
Spotted crake	Probably common up to the mid-nineteenth century, there was a record of them breeding on the once extensive marshes around Dundalk, Co. Louth, in 1892.[32]
Marsh harrier	Last known breeding at Lough Corrib in Co. Galway in 1897.[33] Marsh harriers are seen more and more frequently in Ireland, coinciding with a recovery in England, and breeding was confirmed in Northern Ireland in 2009 and 2011.
White-tailed eagle	Methodically shot out of the skies. Last nesting in 1898 probably in Mayo or Kerry.[34] A reintroduction programme with birds from Norway has seen much success and breeding now occurs in several counties.
Bittern	Eaten in medieval times, it nested in undisturbed wetlands until the mid-1800s. Winter visitors continued until 1900 when the only reliable site was in Durrow, Co. Laois. This wetland has since been drained. Bitterns are recovering in the UK and are strictly protected under EU law. They still turn up intermittently as winter vagrants.[35]
Golden eagle	Persecuted systematically. Last known nesting in Mayo around 1914.[36] Birds from Scotland were reintroduced to Donegal in the late 1990s but this shy bird has not seen the success of its white-tailed cousin.

Species (continued)	Last-known record
Woodlark	Formerly common along the eastern counties, it had disappeared as a breeding bird by the start of the twentieth century.[37]
Yellow wagtail	Nested on lake shores until the mid-twentieth century.[38]
Black-necked grebe	Bred at Lough Funshinagh, Co. Roscommon, until 1957.[39]
Corn bunting	A victim of changes in agricultural land use, breeding still occurred in remote outposts in west Galway and Mayo up to the early 1990s but there have been no records of breeding since the end of that decade.[40]

Insects known to have gone extinct from Ireland

Mountain ringlet butterfly[41]

The moths[42]
Barbary carpet
Blossom underwing
Common fan-foot
Cream-bordered cream pea
Hornet moth
Mullein
Pale shining brown
September thorn
Slender-striped rufous
Suspected
Sword-grass
Triple-spotted clay
V-moth
Wormwood

The bees[43]
Andrena rosae
Nomada sheppardana

The water beetles[44]
Bagous glabrirostris
Bidessus minutissimus
Donacia semicuprea
Helophorus alternans
Hydraena pulchella
Hydraena pygmaea
Hydrochus angustatus
Plateumaris rustica

Beetles dependent on dead or decaying wood[45]
Agathidium rotundatum
Amphicyllis globus
Anthocomus fasciatus
Calambus bipustulatus
Conopalpus testaceus
Cypha seminulum
Euplectus punctatus
Malthodes dispar
Nossidium pilosellum

Plants extinct from Ireland

Flowering plants[46]
Club sedge
Dense-flowered fumatory
Fen wood-rush
Holy-grass
Meadow saxifrage
Oak fern
Purple spurge
Rannoch rush
Rough poppy
Saw-wort
Scaly buckler fern
Sea stock
Sea knotgrass
Shepherd's needle
Stinking goosefoot

Mosses and liverworts[47]
Acaulon muticum
Aloina rigida
Anastrophyllum hellerianum
Atrichum angustatum
Barbilophozia kunzeana
Bartramia halleriana
Bryum turbinatum
Buxbaumia aphylla
Calypogeia suecica
Campylopus schimperi
Conardia compacta
Ctenidium molluscum var. robustum
Dicranum undulatum
Didymodon icmadophilus
Entosthodon muhlenbergii
Eurhynchiastrum pulchellum var. diversifolium
Fissidens curvatus
Grimmia crinita
Grimmia laevigata
Grimmia longirostris
Meesia triquetra
Microbryum curvicollum
Microbryum starckeanum
Myurium hochstetteri
Pohlia proligera
Pohlia wahlenbergii var. glacialis
Pterigynandrum filiforme
Pterygoneurum lamellatum
Pterygoneurum ovatum
Ptilidium pulcherrimum
Rhynchostegiella curviseta
Rhytidiadelphus subpinnatus
Seligeria calycina
Syntrichia princeps
Targionia hypophylla
Tayloria tenuis
Tortula protobryoides
Tortula vahliana
Tortula wilsonii
Ulota drummondii

In addition to those animals listed, we can add the sturgeon, and two land snails *Helicigona lapicida* and *Omphiscola glabra*.[48] In a study of Irish mayflies[49] remarkably no species was found to have gone extinct but in every other group of animal or plant that has been studied in this way it has been found that species have disappeared. It is certain, therefore, that in groups that have not been examined, such as ants, hoverflies, wasps, weevils, earwigs, grasshoppers, flies, spiders, lichens, fungi, etc. there are examples of creatures that have quietly vanished from our countryside. In England the Lost Life Project has documented 413 species that have disappeared from there in the last 200 years and many indeed are from these lesser-known groups of animals.[50] The recording of natural history is far more extensive in the UK than it is in Ireland (or anywhere else in the world, for that matter). In Ireland we find that some records from the 1800s or early 1900s are too sketchy to conclude for sure whether the organism in question has been correctly identified or was really established in this country. This has led to many species being assigned as 'data deficient' rather than the more conclusive 'regionally extinct'. Indeed, the Lost Life Project was able to assert that seventy species of fungus and thirteen species of lichen have been lost in England since the early nineteenth century but in Ireland, meanwhile, there is simply no data to show how these groups have fared.

There has never been any assessment of the biodiversity of our oceans apart from relatively large and easily identified species such as whales or seals. With the exception of the sturgeon and the North Atlantic right whale, however, it does seem that fewer species have gone extinct in the sea than on land. This may be because marine creatures have greater powers of dispersal than those on land or it may be because their ranges tend to be much larger. It may also be because data from the sea is so much scantier than those from terrestrial systems (and so extinctions are occurring without our knowledge). Certainly it is not because the pressure on marine habitats is any lower than those on land and there is evidence to believe that extinction rates are much higher than commonly assumed.[51] Indeed, research published in 2015 examined why there seemed to be a ninefold decrease in historic extinction at sea and a fourfold decrease in current extinction risk when compared to land-based ecosystems.[52] It found that once the figures were adjusted to take account of only those species for which full assessments had been carried out, then extinction rates were similar. The research cites the continued pressures from pollution, climate change and overfishing which are threatening the existence of up to a quarter of marine species.

We have come to live with the lexicon of ecological destruction. It is easy to forget that the whole concept of extinction has been with us only since Victorian times when the bones of long-gone creatures started turning up in numbers. The idea that a species could disappear entirely at the hand of humans came even later, but for my generation and the young people of today mass extinction is a part of the common dialogue of environmental pressure. Extinction is a part of nature and, remarkably, most creatures that have ever existed have disappeared, mostly during one of the five extinction events that are recorded by palaeontologists. Away from these massive die-offs, like the one that did away with the dinosaurs, the rate of 'natural' extinction is tiny – maybe one species every 400–500 years. It is commonly accepted that today's extinction rates are much, much higher, perhaps 1,000 times higher, than this natural rate and that we are in the midst of a sixth extinction event.[53] Yet it is not well understood that this problem is not confined to South American jungles or the plains of Africa; it is right here in Ireland, on our doorstep.

The loss of individual species can be acutely felt. With no wolves, Ireland is one of the few countries on Earth with no apex predator, although many people will see this as no great loss. But the rise in deer numbers across the country is now presenting a significant economic threat to foresters and other landowners. Deer themselves were once on the verge of extinction but numbers have rocketed with legal protection and the expansion of conifer plantations since the 1950s. In 2011 an adult sika deer was seen swimming in the sea near Greystones in Wicklow and it is believed that this was as a result of what has been described as a 'phenomenal increase in deer distribution, abundance and density in Ireland over the last three decades'.⁵⁴ Hunting can go some way to keeping deer numbers in check but ultimately it is no substitute for the natural order of things. Wolves have affected our countryside in other, less expected ways. It is estimated that the population of red foxes was in the region of 8,500 in the Mesolithic Era but has exploded to about 100,000 today.⁵⁵ This is probably the result of a phenomenon known as 'mesopredator release' and something that has been observed in the United States where coyote numbers boomed after wolves were exterminated there. While 'mesopredator release' is a relatively recent field of scientific research, its effects were summed up succinctly in 1879 by a Mrs Houston, when commenting on the continued persecution of golden eagles, ostensibly to boost the grouse population, in the west of Ireland:

> Both on our own moors, and on those of Achill [County Mayo], where the same murderous policy had prevailed, the grouse, instead of increasing in numbers with the slaughter of their foes, were found to sensibly diminish, the truth being that stoats and weasels, together with such like destroyers of infant bird life, gained immeasurably by the doing to death of these (to wit, the eagles), who had in the days of their strength made many a meal of the vermin class which did now unduly preponderate.⁵⁶

Along the River Shannon the seasonally flooded grasslands, known as callows, have traditionally been home to enormous flocks of wading birds, including curlews, redshanks and lapwings, which came to the callows each summer to breed and rear their young. These birds lay eggs on the ground, relying on long grass and camouflage to evade detection. In recent decades, the numbers have collapsed, by up to 88 per cent since the early 1990s, and a study by BirdWatch Ireland found that one of the principal causes for this was the increase in predation by foxes in particular.⁵⁷ However, this study never asked why fox numbers would have increased over this time. Since 1989 badgers have been culled from the region in their thousands in the name of controlling bovine tuberculosis in cattle. But could there be a link between the removal of badgers and the increase in foxes? A UK study suggests it might. The researchers found that culling badgers resulted in a doubling of fox populations in the affected areas.[58]

Recent developments in Irish woodlands have also shown the importance of species. Red squirrels have been under threat from the advance of their grey cousins as they cannot cope with the competition for food and space. Grey squirrels are native to the eastern half of the USA and Canada. Having been introduced deliberately to Britain, they were brought to Ireland in 1911 and released on the grounds of Castle Forbes in County Longford, apparently as part of a wedding gift (one wonders how the bride and groom felt about this!). It went on to colonise most of the counties east of the River Shannon and, wherever it went, the red squirrels soon disappeared. Greys arrived in Dublin in the 1970s and were first spotted in the Phoenix Park in 1979, where at the time there was a healthy population of reds. It took less than ten years for the reds to be replaced entirely with greys. Invasive Species Ireland, an organisation established to monitor new species arriving on our shores, labelled grey squirrels one of Ireland's 'most unwanted' alien invasive species. They are bad news for reds but they also do damage to forestry plantations where they strip bark and eat buds (something the reds were also guilty of when they were at their most numerous). Controlling grey squirrels, meanwhile, seems like an impossible task. In one study in County Meath, where two woodlands were cleared of greys it took a mere ten weeks in most cases for them to return to their former numbers.[59]

In Dublin's Phoenix Park, a study by University College Dublin in 2007 and 2008 estimated that there were 2,400 grey squirrels at one point (although it fell back considerably at a later stage).[60] The task of controlling the grey population seemed immense and there was even research into developing a contraceptive in an attempt to stop them breeding. Until recently there has been a depressing sense of resignation that, as sightings of greys west of the Shannon increased, it was only a matter of time before the reds would disappear entirely. It had long been thought that the broad River Shannon was too great a natural barrier for the greys to cross. The thought of grey squirrels displacing the reds in Killarney National Park or the Burren would be enough to bring a tear to any nature-lover's eye.

But then something entirely unexpected happened. Pine martens, carnivorous members of the badger family, and which not long ago were considered one of our rarest animals, began to increase in number. Like deer, they too seemed to benefit from a combination of legal protection and an expansion in conifer plantations. So as grey squirrels were moving west so too pine martens moved east and, unbeknownst to naturalists and researchers, the stage was set for an ecological clash as these two waves collided. It turns out that the collision of native pine martens and invasive grey squirrels has not only halted the advance of the greys into the west of Ireland but has resulted in a collapse in the squirrel population in those midland counties where pine martens are common. The red squirrel, absent not that long ago, is now considered common in counties Laois and Offaly.[61] This research from the National University of Ireland, Galway is interesting because they found that the crash of grey squirrel numbers could not be attributed to predation alone. In other words, it was not possible that the pine martens were simply eating all the grey squirrels. A telling finding was revealed in the comparison between greys in two woodlands, one (Charleville in County Offaly) where pine martens were common and one (Tomnafinnoge in County Wicklow) where they were present but at low densities. The researchers were live-trapping squirrels and weighing them as an indicator of their general health, or 'fitness'. They found a significant difference between grey squirrels at the two woodlands which otherwise are very similar habitats (oak dominated with lots of acorns and other squirrel food). The overall health and status of the grey squirrel population in Charleville was extremely poor, in stark contrast to the population in Tomnafinnoge who are thriving.[62]

Where there were lots of pine martens the few remaining grey squirrels were emaciated compared to their chubby compatriots with few pine martens on the block. Although the researchers don't draw the conclusion, it certainly seems that the grey squirrels in Charleville are being terrorised witless as their cosy predator-free life has come to an abrupt end. The fear has yet to reach their friends in Wicklow but it is only a matter of time.

This fear response was observed in Yellowstone National Park in the USA after the reintroduction of wolves there, as the behaviour of the deer changed radically with the arrival of the predator. A hunted animal cannot relax, it feeds less and spends more of its energy looking over its shoulder so that even if it never ends up actually being eaten it becomes more susceptible to disease or may have lower breeding success. So it seems that the greys' days are numbered. While it continues to spread in areas where pine martens are not yet common there is no reason to believe that the advance of the carnivore is slowing down. In hindsight it now seems clear why the greys never got a foothold west of the Shannon. As the authors of this groundbreaking research point out, squirrels can swim and on narrow stretches of the river the tree canopy meets over the water. It was the standing army of pine marten on the far shore that kept them at bay all along.

The broader view of this dynamic is also revealing. Grey squirrels entered an environment that was degraded by the absence of predators. This made invasion a piece of cake. All the evidence shows that healthy ecosystems, with all their species intact, are more resilient to the inevitable changes that come about whether through the introduction of new species, disease, climate change or extreme weather events. The restoration of a key element in the food chain has sent the invader packing, and a semblance of the natural order has been restored. It is just another example of how important individual species are to a healthy environment.

There is always a practical impact when a species is removed from the ecosystem in which it is adapted. It is not always easy to identify what this is as many species do not have the obvious role in their environments that wolves and pine martens do. I don't know how the extinction of black-necked grebes is affecting the integrity of Irish wetlands. Like taking the screws out of a bookshelf, however, it does seem obvious that the removal of each one somehow weakens the overall strength of the system, and that eventually the whole thing will come crashing down.

The logical argument for saving species has always seemed weak to me in comparison to the ethical and cultural value of saving nature. In losing a species we lose forever a chunk of our heritage. It is not so much like wilfully knocking down a Norman castle but it is as if all Norman castles were to be razed to the ground. The crane was a common and familiar bird in Ireland up to the sixteenth century. Elaborate illustrations of these graceful and mystical birds adorn the pages of holy manuscripts, including the Book of Kells. They were associated with the afterlife and a carving at the base of the eighth- or ninth-century North Cross in Ahenny, County Tipperary, shows a crane leading a funeral procession.[63] In pre-medieval times it seems that eating crane was taboo although subsequent foreign invaders tucked in with relish. Perhaps the most interesting way in which cranes were a part of old Irish life was their presence as pets. Newly hatched cranes easily imprint on humans and their bones have been frequently unearthed among the remains of human settlements. Were cranes a status symbol or were their omnivorous diets just useful in disposing of unwanted scraps? Lorcán O'Toole's exploration of the crane's place in Irish ritual and culture (from where I found this information) suggests that the prefix '*Cór*' (in the Irish name) is actually a reference to a possible crane 'cult' and may not refer to the more commonly accepted meaning as a place with a round hill. He found that of the townlands with a '*Cór*' prefix only 40 per cent seem to be associated with a round hill.

Cranes are remarkable birds to see in the wild, a sight I was fortunate enough to witness by chance one April morning near a creek which joins the River Shannon in County Roscommon. It is an enormous bird, twice the size of a heron (a bird often referred to as a crane), but it was not its size that drew my attention that morning. Long before I saw it, its incredible cries, something between a lost child and a strangled cat, filled me with terror and curiosity in equal measure. Normally gathering in large flocks, this solitary bird may have been lost or looking for a mate. Overhunting and habitat loss are likely to have resulted in the crane's demise in Ireland and in losing the crane from our landscape we have lost one of our most intimate cultural links with nature.

Restoring the lost connections in our natural environment by bringing species back makes sense. Not only does it strengthen ecosystems, and help to restore a semblance of natural order, it also revives long-lost cultural links. Unfortunately, not everyone sees it that way. Take the wild boar, for instance. Remains of wild boar in Ireland date from the earliest times; bones over 9,000 years old have been recovered from excavations at Mount Sandel in County Antrim. The Irish word for wild boar is *torc* and this association with place names is to be found throughout the country, from Torc Mountain in Kerry to Kanturk in County Cork.

How wild boar got to Ireland at the end of the ice age is a mystery. It seems plausible that people might have brought captive boar on boats from Britain to populate the woods here, especially since there would have been little by way of large animals to hunt (research has shown that red deer were imported in this way). Boar are forest animals and it is also plausible that there was somehow a forested connection between Ireland and mainland Europe – although there has never been any evidence to prove this. Such a link would have allowed not only boar but other forest animals to colonise Ireland such as wildcat, lynx, badger, fox and pine marten. It is implausible to me that all these animals were introduced to Ireland by people. There is no scientific doubt that wolves were in Ireland at the time of human colonisation although there is actually no physical evidence to prove this either. It would, nevertheless, be a preposterous claim that wolves were introduced by man. Yet for wolves to be present in

A young badger in an Irish woodland. Between 6,000 and 7,000 are legally snared and shot each year to eradicate TB in cattle; however, TB levels remain stubbornly high. Hundreds are killed on our roads while unknown numbers are illegally persecuted.

ABOVE: A skylark, among the most familiar sounds of summer. Between 1998 and 2010 the Irish population fell by a third.

BELOW: Bumblebees are vital pollinators of wild plants as well as fruit crops. Half of Ireland's bee species have undergone substantial declines since 1980 and nearly one third are threatened with extinction.

ABOVE: Erris Head, County Mayo, where chronic overgrazing by sheep is leading to massive soil erosion. Policies based on food production only are leading to loss of livelihoods, landscape damage and habitat loss.

BELOW Lough Boora Parklands, County Offaly. Once a barren wasteland, in less than thirty years nature has bounced back. It is now full of wildlife and is a popular recreational amenity.

The sublime beauty of Killarney National Park in County Kerry masks urgent threats from fires, invasive species and overgrazing by animals including deer, goats and domestic sheep.

ABOVE: Pressure to comply with regulations is forcing farmers to remove scrub and drain land, like here in west Cork, to the detriment of habitats, water quality and landscape value.

BELOW: Free from pesticides, ploughing and fertiliser, roadside verges are now important habitats for wild plants and insects.

ABOVE: Yellowhammers have disappeared from much of the north and west of Ireland due to changes in agricultural practices since the early 1980s. (IMAGE BY MISS BATTERSBY, COURTESY NATIONAL LIBRARY OF IRELAND)

BELOW: Contents of a prawn trawl from the Irish Sea. The level of discarding from bottom trawling can be up to 90 per cent; many of the juvenile fish and other organisms are dead when thrown back to the water. (COURTESY JOHNNY WOODLOCK, IRISH SEAL SANCTUARY)

ABOVE: Contrary to popular belief, trees grow well in the west of Ireland if given a chance, like here in Connemara. Too much grazing, as well as traditional views which hold that the landscapes should be treeless, prevent their regeneration.

BELOW: Under current rules the land on the right is worthless, despite providing wildlife habitat, helping to prevent flooding, cleaning water and adding scenic and amenity value to the countryside.

'Wild Nephin' in County Mayo, the epicentre of Ireland's only 'wilderness' experiment. Could we see the reintroduction of lost animals such as wolves, bears, lynx, capercaillies and wild cats? Where traditional livelihoods are declining there are opportunities to reinvigorate rural economies through 'rewilding'.

the absence of deer leaves a conundrum – what did they eat? Paddy Sleeman of University College Cork suggests that without wild boar in Irish forests there simply would not have been the quantity of prey needed to sustain a wolf population (which he estimated as being in the region of 2,000 animals) along with the three other predators believed to be around at the time: lynx, stoat and wildcat. 'Wild boar are essential if this ecosystem worked in the way that ecosystems work today.'[64]

Because domesticated pigs are direct descendants of wild boar it is impossible to distinguish their bones from archaeological sites. Domestic pigs arrived in Ireland during the late Neolithic Age and the jumble of pig bones means it is not even clear when wild boar disappeared from this country.[65] The lack of any usable genetic material (which might come from a surviving boar population) means that the type of tests carried out to determine the provenance of living red deer cannot be repeated for boar. We can always hope for some new evidence to be unearthed but until then it will be impossible to say with certainty whether wild boar were or were not introduced by humans. But this argument rages for practically all of our animals and, for me at least, there can be no doubt that the boar belongs as much in the Irish countryside as badgers and foxes.

My colleague, the mammologist Dr Daniel Buckley, has argued strongly for the reintroduction of this charismatic large mammal. He has pointed out that not only does the boar belong in the Irish 'ecological family' but that their presence would be beneficial for native habitats. Boar are quite disruptive animals. They dig and they root and they turn the ground over in their search for food. At first this might seem to be a destructive trait that we'd be better off without. But this disturbance of the soil is actually very healthy in a forest ecosystem as it promotes the germination of acorns and other woodland seeds.

In Sweden, wild boar are native animals where, like Ireland, they were hunted to extinction, but were accidentally reintroduced in the 1970s. A study in the 1990s, after they had become reasonably established, showed that the diversity of native plants in areas where boar were most densely populated was significantly *higher* than areas with no boar.[66] Native woodlands in Ireland today lack this type of disturbance and as a result many are very limited in terms of the plants that grow there. In many cases there are carpets of only one species, such as bluebells or wood rush. Because bluebells are so pretty most people see nothing wrong with conserving these springtime carpets of azure and topaz. Yet their monocultures hint that there is something missing from the ecosystem. Many species of orchid, for instance, depend upon disturbance for their life cycle and those species that specialise in Irish woodlands are either endangered or found in only a very limited number of locations, such as the narrow-leaved helleborine or the bird's-nest orchid. Could this be a result of boars being absent? Of course, wild boar also root and dig in farmers' fields and this would be a nuisance for some. Foresters don't like them either because they damage plantations. For any species to be reintroduced we have to acknowledge the economic and social context as well as the scientific one.

There is also an issue as to how their numbers would be controlled. Without wolves, we could have a situation like we have now with deer, whereby populations have no natural limit to their numbers. Across Europe wild boar are hunted and eaten, and they taste good too, but even with hunting there are problems with too many boar in countries like Spain and France (although the increasing wolf numbers in those places may sort that out). At the end of the day, Dr Buckley, when he was chairman of the IWT, was clear that any reintroduction would need to be properly thought through and done in accordance with the appropriate guidelines. But before this debate could be had, wild boar found their own way back.

The first anyone heard about it was in 2009 when startled locals in the Slieve Bloom Mountains, which straddle Counties Laois and Offaly, reported seeing giant pigs staring back at them through their car's headlamps on dark country roads. It's not known for sure how they found their way into the wild; it's possible that some escaped from known boar farms. It is also possible that hunters brought them in illegally for something new to hunt. Either way it appears they are now breeding and according to the National Biodiversity Data Centre they have turned up in at least ten counties, including Waterford, Fermanagh, Laois, Offaly and Tipperary.[67] Genetic tests on some of these animals have shown that they are hybrids between domestic pigs and wild boar and that raises conservation questions: can they really be considered wild boar?[68] The official response has left us in no doubt that they were not welcome. In one article in *The Irish Times* they were described as 'really good breeders', although in nature bad breeders don't tend to last. One government scientist was quoted as saying, 'We are aware of the extensive damage they have caused [in the UK] and the huge risks posed. The diseases that they carry would be enough to curl the hairs on the back of my neck.'[69]

But every animal carries disease. Shortly after the presence of boar in Irish forests was recognised the authorities moved swiftly to brand them an 'alien invasive species' and legislated accordingly. It is now an offence under the Birds and Habitats Regulations 2011 to deliberately bring a wild boar into Ireland or release it into the wild. In a report commissioned by the state and produced as part of this process the argument hinged upon the lack of evidence of boar prior to the arrival of humans. This was seen as sufficient to label them a pest and mobilise government forces in their eradication. The report is full of references to the economic damage that boars could cause and the diseases they carry but there is little reference to their long pedigree in Ireland or the benefits they could bring to restoring our native forests.[70] The report concluded that: 'The Irish economy is heavily dependent on the agricultural sector and wild boar populations are known to significantly impact upon agricultural enterprises.'

Again, just in case there was any doubt about the virility of the swine we are told that boar are 'fecund and reproduce vigorously'. Most of the references used for the study were from research carried out in the Americas where wild boar were never native and really are an invasive species. Was this a political rather than a scientific decision? Research into the origins of red deer has found that they were brought to Ireland by Neolithic hunters about 5,000 years ago.

But in the case of red deer it was argued that '[the Kerry red deer] represent a unique population within the Irish context' and deserve 'special conservation status'.[71]

Deer also are fecund, vigorous breeders and carry diseases.

Species matter and missing species matter just as much. Modern-day extinction is rare enough, at least of large, familiar species. Indeed, of my 115 plus species that have been lost in Ireland it is remarkable how many of them have returned, either through deliberate reintroduction programmes (officially sanctioned or otherwise) or by natural recolonisation. These days, when a species is threatened, usually something is done to bring it back from the brink before it is too late. Legal protection helps a lot, as does general awareness-raising among ordinary people who have shown a willingness to act to prevent total disaster. We have seen that in Ireland with quite intensive measures introduced to stop the total disappearance of the corncrake. Corncrakes are still seriously threatened, having disappeared from all but a handful of offshore islands and some other remote locations, but the risk of them vanishing completely is easing.

Therein lies another issue and one that is just as worrying as extinction but which receives must less attention, i.e. the overall reduction in the amount of wildlife in our countryside. The corncrake is a good example: once abundant to the point of nuisance, it was known to nest not only in agricultural land across Ireland but also in coastal sand dunes and even the car park in University College Dublin. Its population nosedived during the 1960s and continued to do so well into the turn of this century. The reasons for this are still unclear; changes in agriculture were probably a large part of the problem but surely not enough to see corncrakes retreat on the scale they did. Despite concerted efforts to keep them, the corncrake disappeared even from its stronghold along the River

Shannon in 2012, perhaps a victim of a series of atrociously wet summers. Since then their numbers are stable and even increasing in Donegal and Galway where there were in the region of 130 pairs in 2013. This seems like a positive conservation story, and it is in a way, but we are still left with the fact that the total number of corncrakes is at a tiny fraction of what it once was. It is part of a trend whereby species decline and decline until they teeter on the precipice of extinction, only to be saved at the last minute. One by one, these animals and plants are pushed into a corner and we seem satisfied that they don't go over the edge, but the broader message is being lost – what is happening in our country that is making it uninhabitable for so much wildlife?

As a child in the 1970s and 1980s my summer holidays were always to a mobile home near the sandy shores of County Wexford. No motorways then, the route was along what today are secondary or even tertiary roads. Although not even worthy of remark at the time I look back and think about the state of my father's Toyota Corolla at the end of each journey. The front grille and windshield were routinely caked in the remains of splattered invertebrates. Flies mostly, but larger wings of butterflies, dragonflies or moths were not unusual in the hotchpotch of insect body parts that were pressed against metal and glass. I also remember the communal shower block and the cloud of moths that were nightly attracted to the light bulbs. It wasn't unusual to hear *Psycho*-like screams from within on foot of some airborne beast that discommoded the nightly rituals, while the next day dead remains of exhausted moths littered windowsills and porcelain washbasins.

Yet mysteriously, these things never happen any more. As an adult I don't remember ever having to wipe large numbers of dead bug remains from my car. Despite my ongoing fondness for camping I rarely have to call my kids to see strange and wonderful moths flying into the toilet block light bulbs. Where have all the insects gone? Traditional methods of surveying wildlife have focused on determining whether or not a species is present in a particular location and, using this data, distribution maps are drawn. Changes in the distribution of particular species is often used as a proxy for how the overall population is doing but this rarely takes account of the abundance of a species within its traditional habitat. The scientist in me says it's impossible to be sure whether insects in Ireland, never well recorded until recently, have increased or decreased in abundance because the long-term studies are not there to prove it. Yet my windscreen tells a different story.

In the UK, meanwhile, there is data on moths going back over 300 years. Since 1964 the Rothamsted Insect Survey has been monitoring 430 sites across Britain for moth activity using specially designed light-traps and an army of volunteers. Because the traps have been run every night for such a long time, not only is there an incredible volume of information on the changes in types of moth (of which over 2,500 species have been recorded) but, because the total numbers of individuals have also been counted, we can also see how the overall abundance of moths has changed. It is the longest-running study of insects at this scale from anywhere in the world. It has found that across Britain, between 1968 and 2002, the abundance of moths decreased by 31 per cent. There was a marked difference between the north of Britain (from a line joining Lancaster to Middlesbrough), where there was no apparent decline, and the south, where the decline was an 'alarming' 44 per cent.[72] In Ireland monitoring of butterflies has been ongoing only since 2008, so not far back enough to reflect what was happening during my childhood. But even in this short space of time a number of species previously thought of as common have gone into what the researchers describe as a 'steep decline', including the orange tip, speckled wood and wood white. A further five species are in 'moderate decline'.[73]

In 2014 and 2016 the WWF and Zoological Society of London published their *Living Planet* reports. They are among the few organisations that have sounded the alarm not only about saving species from extinction but also about the implications of an overall diminution of the amount of wildlife in nature. The findings were stark, including that between 1970 and 2012 there was a fall of 58 per cent in the populations of the species they looked at. In other words there were less than half the number of animals on Earth in 2012 than there were in 1970. The greatest fall was in freshwater species where the loss in numbers is a massive 81 per cent. Even in protected areas, National Parks and such, the decline in wildlife was measured at 18 per cent. The WWF/ZSL team looked at over 10,000 populations of 3,038 vertebrates (mammals, birds, reptiles, amphibians and fish) across the continents and at sea so we can rely on its authority. Its findings should be shocking and sobering:

> A range of indicators reflecting humanity's heavy demand upon the planet shows that we are using nature's gifts as if we had more than just one Earth at our disposal. By taking more from our ecosystems and natural processes than can be replenished, we are jeopardizing our future. Nature conservation and sustainable development go hand-in-hand. They are not only about preserving biodiversity and wild places, but just as much about safeguarding the future of humanity – our well-being, economy, food security and social stability – indeed, our very survival.

The loss of biodiversity, changes to our climate and converting atmospheric nitrogen into fertiliser are placing the welfare of people on this planet at peril. Ireland is shown as having the fourteenth-highest ecological footprint in the world per head of population. WWF talks about 'planetary boundaries', i.e. the ecological limits in which life exists, among them climate change, ocean acidification, chemical pollution, ozone depletion, the use of water, etc. In three of these, biodiversity, climate change and alterations to the nitrogen cycle, WWF believes we have already overstepped the Earth's capacity to sustain human activities.[74]

In Ireland the ongoing loss of wildlife is barely noticed. By the standards of environmental change these things are happening in the blink of an eye but over a human lifetime they can seem barely discernable. But when we go to look the evidence is all around us. BirdWatch Ireland lists twenty-seven bird species which have undergone significant declines in the past twenty-five years.[75] These include some familiar birds such as lapwing, woodcock, red grouse and herring gulls. In 1900 Ussher and Warren wrote of the nightjar – a nocturnal bird of heathery habitats: 'Considering this is a nocturnal bird which haunts unreclaimed uplands, and is seldom noticed, we have a great mass of evidence about it from all the provinces. It is probably most common in Munster, extending to Eastern Clare and to Kenmare and Caragh Lake in Kerry. It breeds on the mountains and waste lands of Leinster ... It is found in Down and Antrim and many parts of Donegal ... It is also found through most of Connaught.'[76]

None at all was recorded in 2011.[77] Ussher and Warren reported that the ring ouzel, a relative of the blackbird and similar in appearance, was known from all counties bar four in 1900 (Meath, Westmeath, Longford and Armagh – because they lack mountains). By 1966 Ruttledge worried about its steady decrease but still it bred locally in Kerry, northern Tipperary, Wicklow, Laois, Sligo, Fermanagh, Down, Derry, Tyrone, Donegal and, 'perhaps', Limerick and Antrim.[78] The latest information on the ring ouzel is not good. In 2014 BirdWatch Ireland reported grimly that 'observations of breeding birds continue to decline, with no reports from outside counties Donegal and Kerry'.[79]

Barn owls are found almost everywhere in the world and until the 1950s were at home in every Irish county and a number of offshore islands. They are one of our most beautiful birds with their creamy plumage and heart-shaped face. They are best known to my generation as the logo of *The Late Late Show* but their truly frightening nocturnal cry has been known throughout the ages and is probably the source of the banshee legend. Concern for barn owls was being voiced in the 1960s with Ruttledge reporting in 1966 that it was 'absent from many of its former haunts'. Since then their population has declined by about 40 per cent. They are now clustered in the south-west of the country with only scant numbers recorded from the north and east. The continuing perilous state of barn owl numbers is indicative of the overall health of the countryside as their demise has been linked to a number of problems. Chief among them are the use of rodenticides for controlling pests, the intensification of agriculture (leaving little to prey on) and, most recently, roadkill, particularly from the expanded motorway network. BirdWatch Ireland has reported that toxicology analysis on the livers of sixty-nine Irish barn owl carcasses revealed detectable residues of rodenticides in over 85 per cent of birds.[80] Barn owls are not in imminent threat of extinction but the pressures that have depressed their numbers since the 1950s are, if anything, growing.

The curlew, whose cry is such a distinctive part of the countryside, has declined by a whopping 78 per cent in a mere twelve years up to 2010.[81] Curlews once bred across the country in great numbers but now BirdWatch Ireland is forced to fence off known breeding sites with electric wires to keep out foxes. Grey partridges in Offaly, where they were reintroduced in 2002, are only increasing in number because rangers control predators, particularly foxes and crows. It is reminiscent of Joni Mitchell's line from 'Big Yellow Taxi': 'we took all the trees and put them in a tree museum'.

Even birds that are generally considered common are showing worrying trends. Across Europe the total population of birds has fallen by 421 million between 1980 and 2009 with significant declines in birds we assumed we didn't have to worry about, like jackdaws, house sparrows and grey herons.[82] This study concluded that 'European birds are declining at an alarming rate' and pointed to some worrying implications. While losing rare birds is bad enough, the remaining common species are left to do all the heavy lifting in terms of keeping an ecosystem functioning. They disperse seeds, keep pests under control and help in the breakdown of waste. If the common birds start becoming rare could we start seeing these ecosystems break down all together?

Some data has been gathered in Ireland on the total populations of birds in Ireland and between 1998 and 2013 sixteen species were declining out of the fifty-three that were looked at. Very cold winters were responsible for some of this decline but arguably severe changes in the landscape, especially in agriculture, occurred before 1998 and so this impact will not be shown here. All the same this period has seen falls in the total numbers of robins, starlings, grey wagtails, song thrush, skylarks, ravens, grey herons and kestrels.[83]

Why has the countryside become so inhospitable for birds? Fully 128 species of bird were identified by BirdWatch Ireland as being under critical or moderate threat out of a total of the 202 that were assessed. That's a massive 63 per cent of all the regularly occurring birds in this country. Each bird is different in its requirements and when you drill into the data there seem to be different reasons behind the losses of each one. Many are migratory so that the pressures may be occurring thousands of kilometres away and have nothing to do with factors here in Ireland. Yet one reason pops up again and again: changes in how the land is used and how the countryside has been farmed. Wetlands drained, the use of fertilisers, silage cutting where once there was a hay meadow – all leading to the homogenisation of farmland from low-input farms with a diversity of life, to high-input, intensively managed land.

The trend in the loss of our birds is acute but is seen across nearly all groups of animals. The Red Data books that have been published for butterflies, non-marine molluscs, water beetles, mayflies, dragonflies and damselflies, mammals, bees, birds, reptiles and amphibians, and freshwater fish show that on average one third of all species in Ireland are threatened to a greater or lesser degree. No group has escaped. Our rivers once teemed with Atlantic salmon. From earliest times the salmon has been important to people in Ireland, particularly as a reliable source of food through the meagre winter months. It is no coincidence that the ancient settlements of Newgrange, Dowth and Knowth emerged along the banks of the Boyne, one of the richest of the salmon rivers. Even the River Liffey was regarded as having an 'intense fishery' during medieval times[84] while up to the 1800s salmon were still much valued and sought after: 'There is also a considerable salmon fishery on the river Liffey, belonging to Sir Wm. Worthington where eighteen men are employed from the first of January to Michaelmas, it is divided into the upper and lower fisheries, viz. in the river six men and at Poolbeg, twelve. They catch, in the above time, from 90 to 200 each week, which average in the market to 16 or 18/-. This fishery extends from the weir at Island Bridge to the Lighthouse at Poolbeg.'[85]

In 1820 there was this report from Islandbridge, now well within Dublin's urban fabric and up the road from Heuston train station:

> There is a salmon fishery at Island Bridge, which is rented for £200 per annum and which during the year 1816 produced 1,762 fish weighing from five to 30 lbs each. The salmon here taken are in greater esteem among the inhabitants of Dublin than those caught in other Irish rivers so universally prolific of the species, but this arises, probably, from their freshness alone, those brought from the Barrow, Suir and Shannon on the roofs of the mail coaches from Ross, Waterford and Cork, being considered by many to excel them in quality.[86]

The Liffey, if anything, had been much abused by this time, quay walls built where open floodplains once spread out and receiving all the unattenuated filth of the city. It is not even a long river by Irish standards so imagine the unquantifiable bounty of the Nore, Shannon, Erne, Boyne and Bann river systems.

Fish were so easy to catch that they were hand-speared with tridents from the banks of the Shannon. But salmon have had it rough since then. It was estimated that in the 1960s there were 8 million salmon in the North Atlantic but that has shrunk to only about 3 million today.[87] Most of this loss has been felt in the southern part of the salmon's range, i.e. Ireland, France, Britain and Spain. Because salmon travel so far out to sea, and penetrate so deeply into rivers and lakes, they are subjected to a whole range of pressures. When the ocean feeding grounds of the salmon were discovered some fifty years ago directed fisheries were quick to deplete the stocks. When this stopped, but fish numbers failed to rebound, it was discovered that myriad other factors were at play. For nearly 200 years we have been draining and polluting our rivers.

Dams and weirs have physically prevented salmon from travelling upriver, something that wiped out the fisheries along the Shannon and Erne – two of our most important rivers that are now virtually devoid of salmon. Illegal fishing and the practice of drift netting at the mouths of rivers were capturing too many salmon returning to spawn. This practice was banned in 2007 and still fish numbers have not rebounded. Despite some efforts to revive salmon populations by opening up rivers and improving water quality, it has emerged that their survival at sea is now unsustainably low. For Irish salmon this has plummeted from a range of 15–25 per cent to 8–12 per cent and even as low at 5 per cent. This means that up to 95 per cent of the young salmon leaving Irish rivers never make it back to spawn.[88] Climate change is shifting ocean currents and the nutrients that these carry so it may be that fish are turning up to find no food in their usual haunts. The latest data on salmon in Ireland show that of 143 rivers being surveyed by Inland Fisheries Ireland only 55 are reaching what is termed the 'conservation limit'. This is the limit at which fishing could be allowed without depleting the overall population (and is likely a lot lower than what a maximum population would look like). In the Liffey there were estimated to be only 640 fish in 2014, about a third of what were caught there in 1820.[89]

While there are many species that have gone extinct in Ireland, or are on the verge of extinction, it is sometimes hard to point to the concrete consequences of this. For many of us it is just another piece of bad news from the world but daily life is not affected. However, the salmon is not one of these species. For millennia, livelihoods have depended upon the salmon and it has a deep association with the mythology of our ancestors. The ban on drift-net fishing in 2007 hit coastal communities hard and left them with little to fill their nets. Most fishermen are leaving the industry all together or remain to pursue crabs and lobsters. The closure of the fishery was estimated to have put 900 fishermen out of business: 'It's the only way of life we've had down in Mayo that we've had for the last fifty years,' said one.[90]

Wild Atlantic salmon as a food has disappeared from our plates and all we find in supermarkets is the dyed flesh of mass-produced farmed fish. These are not only a poor substitute for the real thing but contribute to the demise of their wild cousins. Those few returning to rivers are forced to pass pens with parasite-infested fish, which increases their mortality even further. In early 2014, despite government/industry assurances that it couldn't happen, nearly a quarter of a million farmed fish escaped their pens in Bantry Bay in Cork after being battered by storms.[91] There is a serious risk that these fish will go on to 'pollute' wild salmon genetically by interbreeding with them. Here at least is one clear example of how the diminution of a once-bountiful species has left us drastically worse off.

The salmon is the great link between the land and the wide oceans that dwarf our island. The eel, another fish that shares with the salmon its epic migration (although the eel breeds at sea while the salmon breeds in freshwater) is now classed as 'critically endangered' in Ireland (and globally) and commercial fishing for eel is now also banned.[92] It shows that the loss of wildlife is not confined to the land. Indeed, the enormous abundance of the sea, once thought to be inexhaustible, has been whittled away to an alarming degree. A report by WWF in 2015, looking at life in the seas, found that between 1970 and 2012 the abundance of marine vertebrates (animals with backbones such as fish) fell by 49 per cent.[93] Another study found that seabird populations globally have declined by 70 per cent since the 1950s.[94] In their press release the lead author of this paper said: 'Seabirds are particularly good indicators of the health of marine ecosystems … When we see this magnitude of seabird decline, we can see there is something wrong with marine ecosystems. It gives us an idea of the overall impact we're having.'

Large-scale industrial fishing since the 1800s has been largely responsible for the dramatic loss of fish abundance in the sea (see chapter 2). By value, most of the fish caught in Irish waters are now hauled in far from shore by giant factory ships while the fleet of small boats which fish close to the shore no longer catch the range or abundance of marine life they once did. Yet it would be unfair to place the blame solely on industrial fishing as this epic depletion has a history long predating the steam engine. Other marine creatures which were never targeted by trawlers or seine nets have also been brought to the brink of annihilation. Some of this was down to old-fashioned hunting with spears. Basking sharks are the second largest fish in the sea and known for their docile nature. So little was known about their habitats that in the early 1900s it was said 'to feed on seaweed and other soft stuff, including eggs; in fact it is mainly a vegetarian shark'.[95] Until recently nobody knew where they went after they disappeared from our shores in summer. We know now that they feed on the plankton, tiny animals and plants that float in the water column, and they languidly sieve the water with their capacious mouths. Still, it was only recently that scientists discovered the extent of their migrations across the Atlantic, largely thanks to the work of the Irish Basking Shark Study group.[96] In the days before petroleum-based products, the livers of large animals were highly sought after as they could be rendered down to yield their oil. People along the west coast of Ireland hunted basking sharks from open boats, as in this account from Killybegs in Donegal in 1793:

> They are taken in the Hot Season in the Months of June and July, in this manner. As they sleep on the Surface of the Sea, they are Discovered by their Fin, which being extended above the Water, resembles the Sail of a Boat. They lye in this posture, til the Fishermen, making up to them, strike them with their Harpoon Irons ... til at last Dying, they Float on the Surface till the Fishermen come along their side, and cut out the Liver, which affords several Barrells of Oyl.[97]

Spearing 'sun-fish' by hand was laborious and dangerous. But the advent of the swivel gun-mounted harpoon, invented by an Irishman, Thomas Nesbitt, made it much more effective. We can have no idea just how abundant basking sharks were in Irish waters at the time although the aforementioned Nesbitt was reputed to have killed forty-two in one week in the 1780s.[98] Hunting for basking sharks in Irish waters did not stop until 200 years later, largely because by then they were too hard to find. Today, it is illegal to land them; however, incredibly, Ireland is the only country in the range of this great fish where it is not legally protected. Indeed, no fish or marine invertebrate is afforded legal protection under the Wildlife Act of 1976. Ireland is the only country bordering the North Atlantic where marine life is not officially acknowledged as wildlife.

Irish waters are home to approximately twenty-eight species of sharks and rays. This ancient order of fish, the skeletons of which are made not of bone but cartilage, are diverse in shape and size – from the aforementioned basking shark, second largest fish in the sea, to the spotted dog fish and rays that commonly end up wrapped in batter and served with chips at Dublin fast-food outlets. One of these, the angel shark, is an odd fish and does not look like a shark at all. A 'really ugly beast' with 'features only a mother could love' as our foremost fish expert Prof. Ken Whelan has described it.[99] It is flattish in appearance and although it was once to be found all around our coast now it is listed by the IUCN as 'critically endangered'. It is so rare that in 2013 it was believed there were only twelve left in Ireland.[100] They are all in Tralee Bay in Kerry which also seems to be an important spawning ground for other endangered species such as the white ray and common skate. Tralee Bay is a co-called Special Area of Conservation, designated under EU law for the protection of its unique nature.

Yet the use of tangle nets, a harmful form of fishing which consists of a clump of old netting that is sunk to the sea floor, has been devastating. It is used to target crawfish, a lobster-like crustacean which is itself in danger of extinction, but anything that wanders into it is likely to perish, including seals and dolphins. Its use is a major threat to the very existence of the angel shark as an Irish species, yet little is done to introduce conservation measures. In some countries, such as Iceland, the use of tangle nets is banned; why do we not see measures in Ireland to help fishermen move to more sustainable ways of hauling their catch?

Sharks and rays are fascinating creatures. Perpetually on the move, sculpted by evolution for unhindered movement in water and the pursuit of prey, many even give birth to live young. Compared to creatures found on dry land, however, we know very little of their comings and goings. People's view of sharks tends to be inseparable from the great white shark, a species not known from this part of the Atlantic, yet most species are much smaller and inhabit very deep waters far off shore. None of the sharks around Ireland could be considered dangerous but then our chances of encountering one these days are pretty slim. It was not always like that, however.

The porbeagle shark is very much built from the same mould as the great white, albeit on a much smaller scale. It is a taut, silvery barrel of muscle, its eyes, discs of the inkiest blackness and with rows of inward pointing, needle-sharp teeth. It's a sizable beast, up to 2m long and the record from these waters is 165kg, taken by rod and line in 1932 by Dr O'Donal Brown off Achill Island in County Mayo. Perfect quarry for the adventurous sea angler. Indeed, 'porgies' were perfect even for landlubbers as they were regularly found very close to the shore. Hotspots were off north Antrim and around the coast to the rocky shores of Clare. Liscannor in County Clare was particularly famed and sharks in excess of 45kg were regularly taken from the rocks during the 1960s. Anglers at this time tended to kill anything they caught, whether for the pot or not.

To register your haul in any competitive way the fish had to be weighed and there was no way of doing this while keeping it alive. And what's the point in landing a sea monster if you can't tell the world? Today, this practice is viewed as shocking waste, and is especially detrimental to conservation as hunters target the largest specimen fish – which also tend to be the most fertile. Larger fish are also likely to have a greater impact upon the ecosystem in which they live as they keep their prey in check. Porbeagles vanished from the coast of Clare (and pretty much the rest of the country) during the 1990s. Long-line fishing, whereby lines of baited hooks were set for days, by French and Spanish boats were blamed but the anglers themselves played their part. Once found all around our coast, porbeagle sharks are now listed as 'critically endangered' by the IUCN.

The role that sea anglers were playing in the depletion of certain species was being highlighted by the late 1970s. The unfortunately named common skate can grow larger than a car bonnet and lurks in the sand and mud of shallow waters. Like the porbeagle, it too is a predator and historically much sought after by the angler because of the impressive size it can reach; the largest Irish specimen is over 100kg. But also like the sharks, these large predators tend to live long and breed slowly. It is a feature that makes them very vulnerable to over-exploitation. Today, the common skate is also listed by the IUCN as 'critically endangered' and is a rarity in Irish waters where once it really was common. By the end of 1976 the Irish Specimen Fish Committee temporarily stopped taking records for this fish because of the dramatic fall-off in the catch rate. Here are some extracts from an article which appeared in *Sea Angler* magazine in 1978 where Des Brennan describes the collapse of the skate population off the south-west coast (between Valentia in Kerry and Kinsale in Cork):[101]

> The committee was in a unique position to judge the effect that intensive angling pressure and the unnecessary killing of fish was having on the species in certain localities as it had full record of no less than 461 specimen common skate which it had authenticated and accepted during the period 1956 to 1975 inclusive. This number, however, is only the tip of the iceberg. It took no account of the very many specimen skate which were rejected as the claims did not meet the stringent rules of the committee. Specimen skate are the cream or 'top layer' of the skate population. Just how many 'ton-up' skate which did not reach the specimen size or indeed actual specimens for which anglers did not submit claims were killed during this period we shall never know.
>
> From just one specimen [>55kg] in 1959 [taken from boats out of Kinsale] the figures rocketed to 12 in 1960, 25 in 1961 and

22 in 1963 before falling to 4 specimens in 1964. It then peaks again to 12 specimens in 1965 before falling once again ... How many others were boated and killed before anyone was aware that skate were a vulnerable species? And all this, remember, was done by anglers. The common skate is not a commercial species and was only sought by rod and line fishermen. So the blame must rest firmly with us and no one else.

The remedy, Brennan insisted, was simple: 'put them back alive'. This may seem like a no-brainer. Today the Irish Specimen Fish Committee is more conservation minded and can take records from flesh samples and good-quality photographs. On its website it states: 'The Irish Specimen Fish Committee encourages all anglers to return specimen fish alive whenever feasible'.

The rules of the Irish Federation of Sea Anglers competition states: 'every effort must be made to ensure all fish, with the exception of those to be retained for culinary use, or considered to comply with record or specimen regulations are returned alive'.

Today's sea anglers are at the forefront of conserving these magnificent species, frequently tagging and weighing sharks and skates before releasing them back alive and relaying the data to scientists. But the lack of legal protection for marine life leaves a certain ambiguity. Terms like 'whenever feasible' and 'every effort' leave room for those who would not be so scrupulous.

In 2009 a fisherman in County Clare was photographed lording it over the corpse of a blunt-nosed six-gill shark weighing half a ton. The dead specimen was ingloriously hoisted onto a forklift truck on the quay so the victor could pose for a photo.[102] This species of shark is listed by the IUCN as 'near threatened'. It was an old, fertile female – a bastion of the world which she inhabited. There was an outcry among sea anglers and nature lovers but the fisherman in question did nothing illegal and suffered no sanction.

This argument could be approached from a totally different tack. Instead of fixating on species that have gone extinct from Ireland, this chapter could have been a celebration of the arrival of hundreds of new species that have made their home here in recent times. After all, about half of all the types of plants in Ireland have been brought here by people. We have rabbits, brown rats and dormice, muntjac deer and slow worms and all of these new arrivals surely increase the diversity of life and the level of wildlife in our countryside. Invasive species aside, this is probably true. We recoil at the thought of rats but they are food for many creatures, particularly our birds of prey. Many of the non-native plants are visited by bees and other pollinators. The landscape of west Cork is unimaginable without fuchsia (a native of South America) and an Irish spring just wouldn't be the same without daffodils (originating in South Asia). In addition, some species have greatly increased in number in recent decades. Foxes and woodpigeons seem to have the run of the country while deer are pouring down from the hills of Wicklow to feast on suburban gardens. This may also be true. Yet the losses have far outweighed the gains.

From a cultural perspective no amount of pheasants or mandarin ducks, pretty as they are, can make up for the disappearance of the crane from Ireland. Nothing comes close to replacing the job of the wolf in putting manners on deer and foxes. As for the species that are booming this may even be something we need to worry about. The dominance of a few species at the expense of the many hints at a greater loss of equilibrium in the environment. Fulmars are a delightful seabird that bred for the first time in Ireland only in 1911. They are now found in great number all around the coast wherever there are suitable cliff-nesting sites. Research has pointed to this expansion being due to the gargantuan quantities of fish waste that is thrown away every day from the back of trawlers.[103] In nature, everything is tightly interwoven.

The good news is that, with the exception of the giant auk which will only return if genetic cloning can find a way to 'unextinct' it, there are ways of getting our wildlife out of the intensive care unit. The successful reintroduction of white-tailed eagles and the rebounding of the red squirrel population give us hope. With the right policies at government level and informed debate with (and respect for) those whose livelihoods depend upon the land and the sea, there is no reason why we cannot be more hospitable to the creatures with which we share our country. In 2011 the government published its second National Biodiversity Action Plan in order to provide just these policies. The then minister in charge, Jimmy Deenihan, introduced the plan with an exhortation to rescue our 'rich natural heritage' against a 'background of growing urgency': 'Biodiversity provides us with food, clean water, building materials and other essentials that we simply cannot live without. It underpins vital economic sectors such as agriculture and tourism. For these reasons alone we should strive to protect what we have.'[104]

Whether the political or social will exists to implement these plans, however, still remains to be seen.

5

Culling: the urge to kill animals

> '"What is not game must be vermin" was the simple rule which guided many of our old-fashioned keepers in distinguishing between friend and foe.'
>
> Major Maurice Portal, 1930[1]

THERE ARE NO VOTES IN WILDLIFE. Clearly, the animals themselves don't vote but at election time politicians do not talk about nature and it is unlikely to be a common issue on the doorsteps. The only time I remember a wildlife issue making an appearance during an election was in 2011 and that was when candidates in certain parts of the country were climbing over themselves to assure voters that we could continue cutting turf on a small number of bogs that had been designated for protection. In 2013, with boil-water notices in County Roscommon and other parts of the country there was nevertheless great opposition to an inspection regime of private septic tanks. Fees were waived and sweeteners promised; those passing the legislation assured people it was only being done because bureaucrats in Brussels had nothing better to do than tell the people of Ireland how we should run our affairs. In the end, half of all the septic tanks tested turned out to be faulty and were polluting groundwater. No, in the main environmental issues are not high on the agenda of Irish politicians. Worse still, many of them have a habit of saying silly things about wildlife. Senator Ned O'Sullivan (Fianna Fáil) in 2014 used his speaking time in the Seanad chamber to clarify that he had nothing against pigeons but was 'very much against seagulls': 'I think something needs to be done to address the seagull problem in this city ... It seems that the seagulls have lost the run of themselves completely. In the apartment block I live in, it is impossible to get a night's sleep ... I saw that they're getting so cheeky now that they attack young children and dispossess them of their lollipops and stuff like that. It might be funny to many people but it is a serious issue in the city. They really are vermin.'[2]

In 2015 councillor Johnny Healy-Rae (Independent) waded into a debate about hen harriers and the lack of meaningful compensation being awarded to farmers with land in areas designated to protect them, a serious issue that warranted serious debate. Healy-Rae wondered (aloud): 'why can't we put the hen harriers into the national park, like the eagles?'[3] The same politician was keen that hedgerows be cut throughout the year, despite laws prohibiting this to allow birds to nest and rear young: 'To think that any bird would be foolish enough to build her nest where she'd be blown to pieces and the feathers blown off her and she to be left standing naked in the middle of the road never made sense to me.'[4] 'Birds have brains as well,' he informed the council chamber. In 2016 he won a seat in the Dáil.

In 2009 councillor Michael Newman (Fine Gael) became exercised in the chamber of Westmeath County Council where there was a debate about the designation of the Clonmacnoise monastic complex as a UNESCO world heritage site. He veered somewhat off topic after fulminating that he had had a 'bellyful' of 'interference' in the everyday lives of country people. He went on to take his anger out on pine martens, which he was quoted as saying to be 'the most nasty, vicious [animal] you have ever seen', adding that 'they were never in Ireland but have been introduced'.[5] In reporting of the incident the media wrongly accused him of referring to the pine marten as a type of bird. He set the record straight with a letter to *The Irish Times*: 'I suggested that there have recently been seen in the Midlands ravens, eagles and pine martens, and that it's conceivable that all these had been introduced from other parts of this country by well-meaning environmentalists who should know better. In over 40 years in this area I had not once seen such creatures until very recently.'[6]

Sections of the media thrive on comments like these and enjoy stirring things up with stories of the small animal (usually weighing no more than 2kg) stalking the countryside threatening livestock and even our children. 'They're killers – nothing is safe' squawked a headline in the *Irish Independent* in 2016.[7] The pine marten is a native animal that has been with us for millennia, although calls to have them culled pop up with increasing frequency as they recover from near annihilation.

Of course, most politicians don't come out with clangers like this but although many are sympathetic to wildlife there are few who can speak fluently on the issue. It was not always like this. One of our founding fathers, Pádraic Pearse, was well known for his love of the outdoors and promoted nature studies prominently in his school at St Enda's in Rathfarnham in Dublin. Reading past statements from government ministers in the 1970s up to the early 1990s showed that many of them took our heritage seriously, and spoke eloquently of the need to protect it. Taoiseach Charles Haughey made no secret of his passion for our wildlife and would pro-actively establish Irish waters as a whale and dolphin sanctuary in 1991. Much later, in 2009–2011, John Gormley, then leader of the Green Party, brought a refreshing command of his brief as Minister for the Environment. It was a blip in an otherwise barren political landscape for those hoping for progressive conservation policies.

Ministers who oversee the NPWS, the body responsible for wildlife conservation, usually discover this is part of their remit only when the European Court of Justice comes calling or they are asked to launch a report. In 2015 responsibility for the NPWS lay within the Department of Arts, Heritage and the Gaeltacht, itself a seat in cabinet that holds all the authority and gravitas of a seven-stone weakling in a rugby scrum. The best it can do for itself is stay out of the way and hope it doesn't get hurt. This lack of political heft matters. The NPWS gets shunted around between government departments like a stray dog that no one wants but can't be put down. Funding for the conservation body is never a priority. It never benefited from the Celtic Tiger boom but nevertheless its budget was slashed during the recession from a high of €34.87 million in 2005 to €13.62 million in 2011. After the 2016 general election the department morphed into the Department of Arts, Heritage, Regional, Rural and Gaeltacht Affairs, the task of nature conservation dissolving further into total obscurity.

Meanwhile, pressures mount from the other side. Time and again the government signs up to international agreements but doesn't dedicate the resources needed to implement them. The EU Habitats and Birds Directive, the Water Framework Directive and the Marine Strategy Framework Directive are all legal instruments that our politicians have gaily signed us up to before walking away in the hope that the civil servants will come up with some cost-free way of implementing the provisions. This approach has been going on since 1979 when the first of these, the Birds Directive, was introduced. It has landed the state in hot water with the European Court of Justice so often that in 2009 Ireland had, at fourteen, the highest number of infringements to its name (i.e. where the European Court of Justice had not only hauled Ireland into the dock but had found against it).[8] In fact, it is only when legal judgements have been brought against us that the government feels the need to act. At that point the overriding argument for implementing new laws to protect the environment hinges on the massive financial penalties that would otherwise ensue. Lack of leadership brings with it immense cost and not only of the type that can be counted in euros and cents. The debacle over the protection of the tiny fragment of remaining raised bog habitat was surely a low point. Here, the absence of sensible direction sent ructions through rural Ireland. NPWS staff were threatened and abused, Gardaí with better things to be doing were deployed to bogs where they watched large crowds assert their 'right' to cut turf. Decent citizens were being hauled through the courts amid widespread confusion. Michael Fitzmaurice TD (Independent), chairman of the Turf-cutters and Contractors Association (TCCA) went on RTÉ Radio 1 in May 2015 to say that it was turf-cutters that had 'minded the bogs and kept them in such pristine condition for hundreds of years'.[9] In the end, the area of intact raised bog, a unique habitat with which Ireland was over-endowed, went from 8 per cent of its original extent in 1990, just before the Habitats Directive was introduced, to 0.6 per cent in 2015.[10]

Irish people love wildlife, nature and the spectacular outdoor landscape with which we are blessed. This is evident from the avid participation in voluntary activities from the Tidy Towns contest, which sees over 800 towns and villages enter annually, to community-led initiatives to clean beaches or develop local walking trails. The problems arise where there is conflict between people and wildlife, something which is inevitable given how widely spread our population is. For instance, Ireland is the only country in Europe where hare coursing is permitted. This is the practice whereby men of a certain age gather to watch hares being chased around a field by dogs. The Irish hare is a unique species, closely related to the mountain hare found elsewhere in Europe (except that Irish hares do not turn white in winter due to our lack of snow). For coursing events hares are trapped but, being a 'protected species', a licence is required for this from the NPWS. The Irish Council Against Blood Sports (ICABS) has documented evidence of 'blooding', an illegal practice that allows a dog to maul a hare that has been tied up. These days the dogs wear muzzles but nevertheless the hares can sustain horrific injuries. Even when released back into the wild the hares may suffer long-term consequences from the trauma. This is why it is banned everywhere else, including in Northern Ireland. There are those who argue that hare coursing is a part of the tradition of rural Ireland and calls to have it outlawed are just a part of an urban conspiracy to undermine country living. The Minister of State within the Department of Agriculture Tom Hayes (Fine Gael) told *The Irish Times* in 2013: 'It's part of a culture in a rural part of the country that people have been part of for generations and nobody is more protective of the welfare of animals than the people that are involved in those sports.'[11]

Badger baiting, dog fighting and cock fighting could also be considered part of this 'tradition' but they are banned and we haven't seen the fabric of rural life disintegrate because of it. A serious attempt to ban hare coursing was made by Maureen O'Sullivan and Clare Daly – two independent TDs who distinguish themselves in the Dáil for their habit of speaking out on wildlife issues – during a debate on the Animal Welfare Bill in 2013. The motion failed: in my view, not because there is a groundswell of support for hare coursing among Irish people (in fact, it is falling in popularity) but because politicians were generally apathetic. Deputy O'Sullivan tried again in 2016, hoping a free vote would allow TDs to vote with their own conscience. The Taoiseach of the day Enda Kenny, after all, had previously told ICABS, 'I am opposed to the practice of live hare coursing.' But the party whips intervened and the final vote was decisively in favour of resisting change. So this medieval practice continues. Supporters point to the fact that hares are not endangered and this is true. But is that really a good reason to mistreat an animal that is a unique part of our natural heritage?

In fact, not being endangered can carry its own problems. Being on the verge of extinction brings with it certain advantages; at least people can see a reason to act and a sense of urgency can even bring special protections or conservation initiatives. Being common and widespread carries with it the burden of apathy. Even our official conservation assessments, enshrined in Red Data books produced by government scientists, refer to these species as being of 'least concern'.

Among our fish of 'least concern' is the pike. A fearsome-looking fish of freshwater lakes, it is the top predator in its aquatic environment, growing up to a metre in length and weighing up to 18kg. Until recently it was assumed that this fish was introduced by people and so not only is it of 'least concern' but it is also ascribed the label 'non-benign'. However, recent genetic research has revealed a more complex picture, showing that pike have, in fact, been in Ireland a very long time, perhaps having colonised our lakes under natural conditions.[12] Pike eat trout and because trout attract anglers, more pike mean fewer trout for those with rods and hooks. This has been reason enough for Inland Fisheries Ireland (IFI), the state body charged with managing our inland fishery resource, to remove pike across ten lakes, mostly in the west, including Loughs Corrib, Mask and Conn. Figures I received under the Freedom of Information Act showed that between 2010 and 2014 IFI spent €725,037 removing 35,738 pike using a combination of gill netting and electro-fishing. The practice has attracted vocal protest in recent years, not least because anglers themselves will pursue pike in their own right.

The need to deal with perceived wildlife threats is not confined to animals. Between 2001 and 2005 Kildare County Council spent €275,000 removing ragwort from the side of roads, according to a Freedom of Information request. While ragwort is defined in law as a 'noxious weed', and is poisonous to horses if ingested in hay, it is not so common to see livestock grazing on the side of roads. What's more, ragwort is a vital wildflower for our beleaguered bees and other insects; at least thirty species are entirely reliant on it for food.

One persistent worry is that animals are at risk of being 'out of control' – a phrase that carries emotional weight even though the control of most species is far beyond our powers. Following the extermination of the wolf in the 1700s fox numbers have gradually grown, and although there are no figures to estimate their population, there certainly seem to be lots of them about. They're not protected and landowners are free to shoot them as and when they see fit. There is no evidence that they do any harm to people or livestock and yet the urge to bring them 'under control' is irresistible for some. In April 2015 councillor Dan McCarthy (Independent) urged the government to award a bounty for the killing of foxes and mink which, he estimated, were causing €1.3 million worth of damage to sheep flocks in Kerry alone (to arrive at this figure he calculated that foxes and mink were responsible for killing 26,000 lambs at €50 a head – an enormous number with no supporting evidence).[13] He believed that public money should be used in this endeavour, another common thread of many of these debates.

Another common assumption is that animals are to blame for the decline in people's livelihoods. Rarely do we see the close examination of government policies that have led to chronic problems for both wildlife and people. One example is from the fishing industry, which is mired in a long-term decline that has witnessed the disappearance of fish stocks and coastal communities. This squeeze has led to conflict between fishermen and seals ever since these marine mammals were given protection under the Wildlife Act in the 1970s.

There are two types of seal in Irish waters, the common seal and the larger grey seal. Both have increased in number in recent decades for a variety of reasons and are considered to be of 'least concern' from a conservation perspective. There is no longer legal hunting of seals, they have taken advantage of less disturbance on offshore islands where the human population has declined or disappeared entirely, and they may be taking advantage of easy meals in the huge quantities of fish and other sea life being discarded by the industrial fishing fleet. For reasons unknown, seals have no natural predators in Irish waters even though killer whales, or orca, prey on them elsewhere. They are clever animals and can frequently learn to minimise their own effort in catching fish if someone else will do it for them. Fishermen setting gill nets or tangle nets, which are fixed in position over time rather than towed along, provide easy feeding stations for seals. The seals have learnt how to follow the fishermen and will even take the fish as it is being hauled into the boat to save them the bother of diving for it. Faced with an abundance of food the seals will often take a bite out of the tail where ordinarily they would eat the whole thing. In this way a single seal can ruin an entire catch, to the great cost and frustration of the fishermen.

Experiments have been done with lights and alarms in an effort to scare the seals away but these have only been used by the seals to their advantage and had all the impact of a dinner bell, saving the animals from even having to look for the nets. Recent evidence from Scotland is also showing that plans to rebuild cod stocks, which were decimated by overfishing, are being undone by seals.[14] It's an

Culling: the urge to kill animals

Grey seals on the Blasket Islands, County Kerry. They are popular with tourists but less so with local fishermen because they steal fish from catches.

example of what happens when an ecosystem is badly damaged. Under healthy conditions there would be lots of fish and certainly too many for seals to make a dent in their numbers. Perhaps also killer whales kept seal numbers in check (although sightings of killer whales in Irish waters are common, I can find no records of any preying on seals). On the other hand, research carried out by Queen's University Belfast and University College Cork found that, in the south-west of Ireland, seals were having no impact on 90 per cent of fish species which are commercially exploited.[15] Nevertheless, for decades there have been repeated calls to have seals culled.

An Bord Iascaigh Mhara (BIM), the state board ostensibly established to ensure the sustainable exploitation of our seafood resource, a goal in which it has a miserable record, has historically allied itself with fishermen in this regard. Writing in the IWT's *The Badger* newsletter in 1996 BIM outlined the case against seals, stressing that the cost to fishermen in 1991 was estimated by them at IE£7.5 million and affecting the income of 1,800 fishermen: 'Left alone, the continued growth in the grey seal populations could result in the further depletion of inshore stocks and fisheries, putting more inshore traditional fishermen out of business.'[16]

When adjusted for inflation the value of Irish fish landings peaked in 2001 at €450 million and have been on the slide ever since.[17] The real cause of fishermen going out of business has been the mismanagement of the seas and the over-exploitation of sea life, especially those species habitually exploited by small boats operating close to the shore (see chapter 2). But perhaps it was easier to blame the seals. There has never been an officially sanctioned cull of seals, something that may be due to the strict protection afforded them under EU law. It may also be due to clearer heads in government departments who know that culling programmes like this rarely, if ever, work and that allocating more taxpayers' money in this way would be unacceptable. Unfortunately, this has not stopped illegal culling.

Ireland hit international headlines in November 2004 when French film-maker Jacquie Cozens arrived on the Blasket Islands to make a wildlife documentary. On approaching the small island of Beginish she and her partner, Neal Clayton, noticed something strange: the seal pups beached on the shore, 'white coats' as they are known, were not moving. On approach she found seventeen dead seals, mostly pups which 'had a lot of blood coming from their noses and mouths and around their eyes. Some had small holes in their skulls and one or two had large chunks of skull missing.' It was not immediately obvious what could be done, as Jacquie, a resident of Dingle at the time, described:

> At this stage we had no idea what to do. We have seen the occasional seal floating in the water or on the beach with a clear bullet hole through the head, but despite them being a protected species, it seems to be an acceptable practice here. Everyone turns a blind eye. Whether that is because they believe

that killing seals is justified or whether it is just because no one likes to make a fuss, I don't know ... Being relative newcomers to the area, Neal and I canvassed a few opinions from friends and the feeling was unanimous, summed up by the statement 'Oh that's awful, but it happens; it's best to leave these things alone, you will be very unpopular if you make a fuss.'

But Jacquie and Neal did make a fuss. They found that something similar had happened ten years before and were determined that it would not be allowed to happen again. On returning to the island with the authorities they counted a total of sixty dead seals, which had mostly been shot but some had been bludgeoned. It was shocking, but was it really something new?

Nobody goes around to the Blasket Islands at this time of year [which is the breeding season for grey seals]; the tourist season is over and most of the time the weather is too poor for anyone to get out there. We just happened to go and we just happened to see it, but we could easily have missed it or decided not to go ashore. Perhaps it happened last year too, but nobody reported it ... I've since been told that the biggest mistake the perpetrators made was mistiming – they usually wait until a big storm is on the way, which will wash away all the evidence.[18]

There was wide reporting of this slaughter in the media and a general sense of outrage that this goes on. To my knowledge no one was ever charged or prosecuted and nothing on this scale has been witnessed since. Sadly, that does not mean the practice has stopped. In June 2012 the disembodied heads of two seal pups were nailed to the entrance of the Dingle Wildlife and Seal Sanctuary, an animal welfare centre and tourist attraction. The words 'RIP' and 'Cull' were daubed in red paint. Again, there was widespread reporting and general outrage but no charges brought. I have come across bodies of dead seals while walking along our coast, one even had its head removed.

It is hard, if not impossible, to say how or why it happened and organisations like the Irish Seal Sanctuary (ISS) rely on volunteer vets to carry out an autopsy where a corpse like this comes to their attention. In 2012 there was a spate of mysterious seal deaths off the coast of Waterford and at least two were confirmed to have been shot. The same conclusion could not be made of other bodies that were washed up with holes in their skulls, simply that their wounds were 'consistent with a gun shot'. From April 2012 the ISS has been gathering data on dead seals as part of its work with fishing organisations and state agencies. Their 2015 report, which relied on 'citizen scientists' making reports of a dead seal rather than any dedicated search, recorded a total of 171 carcasses in this time. Most records were in the south-east with Counties Wexford and Waterford providing the majority share. Seventeen bodies were reported to ISS as having been shot, twenty-four were reported to have been found with no head, while twelve were reported to have had 'damaged heads'. ISS concluded that: 'many of the bodies reported to us certainly were suspicious in that it is unlikely that they died of natural causes' while 'the actual mortality rates of seals could well be many times more than we are recording'.[19]

Meanwhile, calls for an organised, and presumably taxpayer-funded, cull continue. It is astonishing how little relative effort goes into rebuilding fish stocks, although to their credit BIM has been working on technical solutions, particularly deterrents to keep seals away from boats, and of late these trials have met with some success.

While seals are safe for now from official culling, elsewhere wildlife of 'least concern' has had less fortune. And nowhere is this more evident than in the case of officialdom versus the badger. Most people would be surprised to hear that a large-scale programme of removing badgers is under way in Ireland. Certainly that was my experience when in the IWT, at a time when we were trying to raise the profile of the cull and our opposition to it. The cull is significant; it snares and shoots around 7,000 animals every year and has done so since the early 1980s. By 2013, over 100,000 badgers had been killed in this way. Badgers are snared and shot by licensed contractors on a year-round basis – i.e. there is no closed season like there is for duck shooting or deer hunting. The badger-culling programme and the reasons behind it are complicated. The simple version that we hear from the Department of Agriculture and the big farming organisations is that badger culling works in reducing the level of bovine tuberculosis (bTB) in cattle. The headline from a Department of Agriculture handout that I was given by officials in that department at a meeting in 2014 said it all:

- Recent evidence that the badger-culling strategy has contributed to a significant reduction in the incidence of TB in Ireland.
- Cattle herd incidence has fallen from 7.5 per cent in 2000 to 3.9 per cent in 2013.
- Number of TB reactors [i.e. cattle testing positive for the disease] has declined from 40,000 to 15,600 during the same period.
- Lowest recorded since the commencement of the TB eradication programme in the 1950s.

In 2015 the Minister for Agriculture, Food and the Marine, Simon Coveney, told the Dáil: 'It would be dishonest of me to say that I do not think that the badger-targeted cull programme is being done in as humane a way as we can do it. If there are other suggestions as to how we can do it better, we would happily take them on board but to suggest it is not working would be wrong. This has been a hugely successful programme.'[20] Very impressive and seemingly incontrovertible. But, like all bald statements from government sources, it's always worthwhile doing a little digging to find out what is behind these confident assertions.

The Department of Agriculture's TB eradication programme is a bit like a fat ball that people hang in their garden to feed the birds. It's a mixture of science, politics and money all held together with a mush of unpleasantness which most people would prefer not to think about. Most people have at least heard of tuberculosis, or TB for short. Like cholera, whooping cough and smallpox it conjures images of the past, when people got old quickly and sea bathing was a common prescription for medical ailments. Thankfully today, in our part of the world at least, TB is a rare affliction in people. The disease itself is caused by a group of bacteria, known only by its scientific name *Mycobacterium*. Different strains, or species, of *Mycobacterium* are recognised in different host animals. *M. tuberculosis* causes TB in humans and it is believed that this 'jumped species' around the time when cattle were domesticated. Known as *M. bovis*, this cattle strain went on to reinfect humans and other mammals with which the cattle came into contact. TB was passed from cattle to humans through unpasteurised milk, and in 1950 there were 900 deaths from the disease in Ireland. The introduction of pasteurisation effectively eliminated TB in humans but did not address the disease in cattle.[21]

TB is difficult to detect. Animals can carry the disease and show no outward symptoms. Lesions usually form on the lungs and only in very advanced cases are lesions externally visible, usually on the udders. The milk yield from dairy cows is reduced and the animal itself can suffer laboured breathing, coughing up blood and a premature death. Today, cattle are randomly tested for the disease and farms found with an infected herd go into lockdown. Sick animals are immediately slaughtered and are prohibited from entering the food chain. The remainder of the herd is quarantined until they can be confirmed to be disease free. No animals can be sent off the farm and no new animals can be brought in. Farming operations are effectively suspended and although farmers receive compensation from the state, it is a traumatic event to have your livelihood go into a deep freeze. There may also be the added stress of not knowing how your cattle were infected in the first place and living with the risk of reinfection. TB is a problem in all parts of Ireland but in recent times has badly affected County Wicklow – this from an Irish Creamery Milk Suppliers Association spokesman speaking of the situation in that county in 2015: 'I know farmers that have been locked up for years on end. They end up living off their compensation, which means that they have nothing left to buy replacement stock.'²²

The body set up by the Department of Agriculture to deal with the problem is known as the Eradication of Animal Disease Board, and it is clear from the title that the goal of government has never been anything short of eliminating TB from the national herd. This is a key difference from controlling or managing disease. The department's website lists twenty-five different afflictions of domestic animals but in only two cases, TB and brucellosis, is the goal of eradication made so explicit. At the time the TB eradication scheme commenced in 1954, this clearly did not seem like an unreasonable target. Other countries in Europe and North America also suffered from TB in their livestock and had managed to eliminate it entirely. Back then up to 80 per cent of Irish herds were infected but simply by introducing a routine testing regime that rate had plummeted to only 2.8 per cent twelve years later.²³ Eradication seemed imminent.

'Unfortunately the impetus up to that time has not been maintained in later years' is how the Economic and Social Research Institute (ESRI) delicately put it many years later. Post-1968 infection rates started to climb again and by the end of the 1980s were, at 2.39 per cent, just fractionally lower than they had been when victory was so tantalisingly close. This ESRI report is important in that it was an independent evaluation of the situation at that time. In the decades since the cull commenced the independence of state publications has been absent as the information entering the public domain comes from those departments, particularly the Department of Agriculture and its agents, which are deeply involved in it. This ESRI report formed the basis for the policy which has changed little since it commenced. It is remarkable because so little has changed since it was written in 1989. Note how the 3.9 per cent infection rate that the Department of Agriculture was boasting about in 2013 was over a third *higher* than in 1968

(2.39 per cent), before the widespread killing of badgers got under way. Back then the ESRI were less confident about the role of badgers in spreading the disease:

- it should be stressed ... that the badger is not the sole or, indeed, primary source of bovine TB in many areas of the country;
- with the high residual level of TB in the cattle population, the most serious risk of cattle infection in most areas is from direct or indirect contact with infected cattle;
- eradication of the badger population would not eradicate bovine TB in the country.

It also concluded, however, that 'removal of the badgers combined with intensive testing has, in most of such cases, been attended by a marked decrease in bovine TB'.[24] It is on this basis that the badger-culling programme was initiated in 1989 and has been ongoing now for nearly three decades with no end in sight.

Badgers are remarkable animals. They are social, skilled engineers and a key inhabitant of woodland ecosystems. Their great downfall is that they are rarely seen, except perhaps as a heap of bruised remains on the side of the road. If they visited gardens in the way that birds do they would no doubt have more human friends. On only two occasions have I been fortunate enough to see one alive. The first time was on a rainy day walking in woodlands when, practically enveloped in rain gear, I nearly stepped on what must have been a cub. I froze in delight, hoping to prolong the moment and assuming it would bolt into its sett; I could see the entrance not far off. But it didn't even look at me. I spent the next fifteen minutes standing by a tree watching it sniff around, snacking on slugs and snails that had emerged from the damp soil. Its wet fur was matted but its stripy black-and-white snout was unmistakable.

There was no hint of aggression or fear. Shortly thereafter I had the pleasure of watching a family of badgers that was living in a park not far from Dublin's M50 motorway. They were being fed each evening and the lady feeding them was kind enough to let people share in this wonderful experience. These badgers were city slickers, used to the noise of the motorway and round-the-clock artificial lighting. They tolerated us at a discreet distance but we didn't have to conceal ourselves, just keep our voices down. We don't have very many large animals in Ireland – spotting any is a treat – but seeing a live badger is something you will always remember.

Badgers used to be eaten in times gone by and were considered of particular merit to women who had recently given birth. They were the source of some odd beliefs including that there were two types of badger: carnivorous, corpse-eating 'dog-badgers', and vegetarian 'pig-badgers'. Only the pig badgers were safe to eat. They were believed not to have a backbone and could tuck their snouts into a special pouch under their tail in order to roll down hills.[25] Folklorist Niall Mac Coitir, who has documented the relationships between the Irish and our plants and animals through myths and legends, managed to unearth this particularly colourful story from County Clare:

> A curious folk tale from County Clare concerns a child born out of wedlock to cowherds, and then raised by a badger. According to the tale, the parents left the child in a badger burrow, hoping the badgers would eat him. Instead one of the badgers adopted the child as her own, and brought him to her den where she suckled him. The child stayed there until he was a few years old and found by a local man. The child acquired the nickname *Gárlach Coileánach* ('the foundling whelp') and grew up to be very glib-tongued and crabby.

Far from raising our young today *broc* (the Irish for badger) is more likely to be seen as a threat, a nuisance, or not thought of at all. Although illegal for over 200 years, persecution of badgers continues, including the medieval practice of badger baiting. A survey of setts across Ireland found 'disturbance', i.e. digging with shovels or blocking entrances with rocks, at 15 per cent of them.[26] One man was caught on camera in County Wicklow in 2015, in army-style fatigues and balaclava, tearing an animal out of a sett. The badgers in question had been part of a study by Department of Agriculture vets and the NPWS. A severed GPS collar was later found nearby. The *Irish Independent* reported that:

> The baiters trap the animals by releasing dogs into the badgers' setts. Then, when the terrified creatures are cornered, they dig down and drag them out. Superintending Veterinary Inspector Peter Maher from the Department of Agriculture said: 'One badger was chipped in East Leinster but found dead in Myshall, in Co. Carlow. A post-mortem showed scars that suggested he had been used for baiting ... The badgers are often transported to different localities to facilitate the blood sport. Most of these badgers will end their lives in an enclosed ring being mauled by dogs.[27]

Badgers are also routinely killed on our roads. Over 1,600 roadkill incidents have been logged on the biology.ie website since it started recording in 2007. The government is obliged to collect data on roadkill but little is done to prevent it happening. In the round, then, badgers are under pressure from a number of sides and you can't but feel sorry for them. With their communal lifestyle, industrious habit of digging complex subterranean structures and fastidious level of hygiene, it is hard not to relate to them. Although powerfully built, and no doubt able to assert themselves in a bind, there is no evidence that badgers are in any way dangerous or aggressive. They just seem to want to go about their business and get on with things with a minimum of fuss, like the rest of us, really. Certainly they can't be blamed for the TB problem. Badgers were not found to have TB until one tested positive in west Cork in 1975 so it is people, via our cattle, that spread it to them.

So why is that badgers have been given such a bad reputation among farmers? Why are they considered responsible for the failure to eradicate TB in cattle? Is shooting 7,000 badgers every year really helping in this goal? And what, if any, effect is it having on the badgers themselves?

Suspicion started to fall on wildlife when the randomised testing regime, which had been so successful initially, started to plateau and the goal of eradication lay stubbornly at arm's length. There were reports of previously healthy cattle herds becoming infected when movement of a diseased animal or contamination from the neighbours could be ruled out. Testing by regional veterinary laboratories showed that visible TB lesions were present in 20 per cent of snared or roadkill badgers and by the mid-1980s the disease was considered to be endemic, i.e. widespread, in the badger population. As the people of 1950s Ireland knew well, TB is a problem not only in cattle. It can be carried by any mammal and has been recorded in foxes, deer and others. It can be carried by the farm dog or cat. However, for the disease to travel from one species to another there must be a means of transmission. For TB to travel from one host to another the two must come into close contact. This could be through breathing each other's air (such as might happen in a barn or an underground sett), but it could also happen through ingestion (humans caught it through drinking contaminated milk) and the bacteria that cause the disease survive well in urine and faeces.

Since badgers in Ireland are mostly found on farmland it was decided that of all the likely wildlife hosts this was the one mostly likely to come into regular contact with cattle. In the UK it was concluded that deer would also share this trait but in Ireland in the 1980s deer were not as common as they are today and were generally confined to uplands so the likelihood of them coming into contact with cattle was slim. Therefore, in so far as a wildlife reservoir of TB could be identified which, in turn, could spill over into the national cattle herd, all fingers were pointing at the badger. The problem is that in thirty years of research into this subject, both here and in the

UK, there has never been any conclusive evidence to show how the disease is actually transferred from one species to the other. That the disease can, in theory, be transmitted between the two species, was demonstrated by experiments which placed an infected badger in a covered yard with disease-free calves. All calves contracted TB, but only after six months. A similar experiment over shorter periods of one to four weeks resulted in no cross infection. So while TB can be transmitted between badgers and cattle this doesn't mean that it is, and the conditions in this experiment are highly unlikely to be replicated in nature.

That cross infection could occur under natural conditions was demonstrated by a further study at Woodchester Park, Gloucestershire, UK. Gloucestershire is a TB hotspot in England to this day but it also has extraordinarily high densities of badgers, up to 20 per km^2 (by comparison, densities in Ireland are estimated at around two per km^2). The Woodchester Park study was closely monitoring disease-free cattle in a landscape known to contain diseased badgers. Special precautions were being taken to control factors that could spread disease and between 1981 and 1987 there were no breakdowns in the herd. That year a serious outbreak saw four farms coming down with TB and one farm having to slaughter forty-six cattle. A dead badger, with visible lesions, was discovered within a farm building. It has therefore been concluded that cross infection during the natural course of events can occur. The ESRI report (from which I am gleaning all of this information) was clear to point out, however, that this study also proved that diseased badgers can live in high density and close proximity to cattle for a long time, in this case six years, with no cross infection occurring.

In theory, badgers could pass on TB to cattle in a number of ways. One is through direct contact, i.e. badgers and cattle coming into close contact in the pasture. For the disease to be transmitted outside in the fresh air the two animals would need to come into contact for a prolonged period so that the cow would literally inhale the vapour from the badger's breath. While this all sounds very romantic, and cattle are indeed curious creatures, it's totally implausible that this could account for the rates of transmission needed to cause the infection rates that actually exist. In any case, badgers have been shown in repeated trials to avoid paddocks with cattle.[28] In a blog discussing an ongoing study in County Wicklow, one of the authors from Trinity College Dublin concluded that: 'it seems unlikely that direct contact between grazing cattle and healthy badgers under a paddock grazing system is a major route of bovine TB transmission in Ireland. Future strategies for controlling cross-infection between badgers and cattle may need to focus on the behaviour of badgers with advanced generalised TB, indirect contact between badgers and cattle or contact between badgers and cattle in farmyards or farm buildings.'[29]

Which points to an alternative source of transmission that focuses on the unusual behaviour of badgers with severely advanced TB. A number of cases have been recorded whereby these very ill animals have moved out of their normally tight-knit social groups and taken up residence in farm buildings. Indoors, where the air is still and warm, provides greater opportunity for other animals which may also be in the vicinity to breathe in the disease-causing agents. This is what caused the outbreak at Woodchester Park. Research carried out by Timothy ('Badger') Roper in the UK to investigate this further found that some badgers were indeed prone to entering farm buildings where they were feeding on cattle feed, grain and silage. Badgers have even been found to enter food troughs and Roper's book features a photograph of badger faeces in the trough mixed with the cattle food. Other research from Warwick University found that the areas around badger setts were heavily infected with the *M. bovis* bacteria, as were badger latrines (badgers like to defecate in the same places and these latrines usually double up as signposts to warn other badgers that they are entering another's territory).

But if these are likely routes of transmission they also seem to be the easiest to deal with. Better management of farm foodstuffs, or installation of barriers to prevent badgers entering buildings and troughs in the first place, would be cheap and easy to implement. Badger setts are very obvious features and a farmer would find it easy to identify a sett on their land. Even the latrines are routinely in the same places. Keeping cattle away from the sett and the latrines would not be an excessive challenge and while a fence would do it, the old-fashioned way of planting stock-proof thorn hedges would mean a lifetime of protection with minimal maintenance costs.

However, these cheap and simple 'biosecurity' measures are not promoted. In fact, the section on biosecurity on the Department of Agriculture's website mentions nothing about precautions to be taken on farms to stop the spread of disease. There is nothing about how the main route of transmission of TB is from cattle to cattle. Nothing about maintaining old-fashioned, stock-proof hedgerows between farms that prevent cattle from neighbouring farms from coming into contact with one another. Contrast this with the detailed but accessible booklet published by the Department of Agriculture and Rural Development in Northern Ireland which discusses these very things. The fact that these are even guidelines is surprising, given the amount of taxpayers' money that goes into TB control (calculated by the *Irish Farmers Journal* at €62 million per annum): surely these kinds of measure should be part of the farm-inspection programme? Instead, all the weight of blame is placed on the badger.

With direct badger-to-cattle contact all but ruled out at this stage as a means of transmission, the other commonly implicated vector is through urine and faeces. The bacteria causing TB are excreted in large numbers in badger waste, especially urine, and were cattle to eat grass contaminated in this way then there is a clear route of transmission. However, most diseased cattle have been found not to have caught the disease in this way as lesions are found on the lungs rather than in the gut. Those who have faith in this route of infection should also consider that the bacteria survive well in cow slurry, which is spread in copious amounts on farmland. Furthermore, diseased badgers tend to display lesions in their throat and head region, identifying 'the respiratory route as an important route of exit' according to a vet working on the cull programme from University College Dublin.[30] So despite all the 'clear' and 'definitive' evidence that we hear about, some very basic questions have still to be answered.

It points to an issue that is at the heart of this debate to the present day and which has yet to be resolved: that is, to what extent can badgers be held responsible for the prevalence of TB in cattle? The studies that the ESRI described were enough at least to point to the fact that, when testing of cattle had done all it could to reduce the disease, all that remained was down to the wildlife reservoir, i.e. badgers. Carrying out field trials where badgers are removed would reveal the true extent to which badgers are responsible.

In its presentation of the evidence the ESRI looked at a number of areas where levels of TB in cattle were high, mostly in County Offaly but also in Counties Galway, Longford and west Cork. It cited data from six regions, one of which comprised two areas in Galway where there was no culling, and it said 'may be taken as controls'. In each of the five areas where badger culling took place it was clearly seen that after badgers were removed the levels of TB were reduced. In the so-called 'control' this effect was not seen. I put the words 'experiments' and 'control' in inverted commas here because these were not experiments in the scientific sense. As anyone with an idea as to how the scientific process works will know, correlation does not mean causation. In other words, the fact that TB incidence fell following the badger culls does not necessarily mean that one was as a direct result of the other. Remember, the primary cause of TB was, and still is, cattle-to-cattle contact. For these to be experiments with meaningful results we would need to know that such contact had been controlled. Field experiments like these are notoriously hard to establish due to the range of variables, which include the weather and the way that people behave when they know they are under scrutiny. The data presented also showed that in many cases the herd prevalence went both down and up when badgers were removed. This point was not lost at the time. In a letter to *The Irish Times* in December 1988, Dr J. M. Barry of the Irish Wildlife Federation (predecessor to the IWT) wrote: 'there is no evidence whatsoever to connect the badger with herd breakdowns in the Galway/Offaly/Longford area. The so-called trials and research which were carried out would be laughed to scorn in any scientific assembly.'[31]

To be fair, more thoughtful efforts did follow which attempted to be more scientific in their approaches. Between 1989 and 1994 badgers were removed from a 528km² area of County Offaly, and results were compared to a surrounding area where there was some badger removal. Known as the East Offaly Badger Research Project, it found that by 1995 the incidence of TB in cattle in the core removal area was 76 per cent less than in the nearby area where badger removal had been more limited, and 86 per cent less than in the Republic as a whole. However, this study was criticised for its relatively narrow scope (being just one area) and despite its seemingly conclusive results, it was effectively flawed from a scientific perspective. In an effort to overcome these difficulties another trial was devised, this time covering four geographically distinct areas, each with a removal zone where badgers would be proactively removed (regardless of outbreaks in cattle) and an adjacent control area (where badgers would be removed reactively, i.e. where there was a local TB outbreak and badgers were suspected of being responsible). The so-called 'Four Areas Project' ran from 1997–2002 in Counties Cork, Kilkenny, Monaghan and Donegal. The clear results from this project showed that the risk of a TB outbreak in any cattle herd was significantly lower in the areas where there had been proactive culling when compared to the reference areas where only reactive culling had taken place. However, this is not the same as saying that culling was responsible for the declines, and the study again lacked scientific rigour.

Meanwhile in the UK a truly massive experiment was under way. The Randomised Badger Culling Trial (RBCT) was designed to be scientifically rigorous: ten areas in total, each with areas of proactive, reactive and no culling control areas (it is of note that in none of the Irish trials had there been comparisons with these true control areas where no culling whatever would be carried out). It lasted from 1997 until 2007 and cost a fortune, £50 million. The results were surprising, to say the least. The first surprise came when, in 2003, the levels of TB in the zones where reactive culling was taking place *increased* by 27 per cent. This element of the trial promptly came to an end and reactive culling was ruled out as part of any future policy solution. The proactive element continued, however, and by 2006 it was reported that in these areas TB prevalence had been reduced by 19 per cent. Badger culling did seem to be having an impact but, after ten years, this kind of reduction is small.

But there was a further twist. While TB levels were reduced inside the cull zone they had correspondingly shot up, by 29 per cent, in a 2km band outside the cull zone. Any benefits of culling were therefore cancelled out by negative effects beyond the cull zone. But why would this be? It had been proven previously that badgers like to stay in very stable social groups, not wandering far from their main setts. Main setts can be occupied for a very long time (hundreds of years perhaps) by the same badger family. Of course, some badgers will always move outside of this zone to find mates or establish new territories, but by and large the degree of movement is very limited. However, when badger densities are thinned out, or when established territories are vacated the disruption causes a much greater degree of movement across larger areas. This in turn is believed to result in greater prevalence of TB in badgers and a greater degree of contact between badgers and cattle. This has come to be known as the 'perturbation effect'. It is anecdotally borne out from numerous farmers I have spoken to here in Ireland who will testify to the fact that they have had the same badgers on their land for decades and have never had a problem with TB. Meanwhile, back in the UK it was found that even where benefits accrued from the proactive culling, after four years TB levels were back to where they started. At this point it was concluded that the costs of culling would outweigh any benefits over the long term. The headline conclusion from this mammoth undertaking was that 'badger culling can make no meaningful contribution to cattle TB control in Britain'.[32]

Until recently British authorities had abandoned the notion of badger culling, largely on foot of the findings of the RBCT project. The results were, however, entirely lost on the Irish authorities. It has always been held that the so-called 'perturbation effect' does not occur in Ireland and indeed they are correct in emphasising the differences between badger ecology and behaviour between our two islands. Badgers in Britain live generally in woodlands, at higher densities and seem to feast entirely on earthworms. Their Irish cousins are much more at home in hedgerows, have smaller family units and dine on a much wider range of animal and plant material. The context of the trials between the two islands were also divergent; in the UK the culling was disrupted by protests, saboteurs and even farmers who did not want to participate; here, farmers were much keener on complying with the programme while society at large hardly raised an eyebrow. Nevertheless, studies in Ireland have shown that when badger numbers are reduced locally their territories correspondingly increase in size. A radio tracking study in Wicklow that is currently under way recorded how quickly other badgers moved into a territory that was suddenly vacated when a mature male was killed. Therefore the perturbation effect does occur in Ireland, yet official policy continues to deny its existence. Unlike the RBCT there have been no Irish studies into the impacts that the 'perturbation effect' may be having on TB levels in cattle. In a Master's thesis by Fintan Kelly, who studied the cull for the IWT during 2013, he raises the possibility that the 'perturbation effect' was not seen in either the East Offaly Trials or Four Areas Project because they had built-in safeguards. The areas chosen were specifically drawn so that they were bounded on all sides by large rivers or roads or the sea, significant barriers to passage for a badger. Perturbation can only occur where it is possible for other badgers to move into the vacated territories but, in the Irish experiment, the barriers seriously reduced or prevented immigration – scenarios that are not reflective of conditions in the wider countryside.[33]

All of the foregoing should at least demonstrate the difficulty in drawing any clear conclusions when it comes to wildlife, disease and animal husbandry. The ability of science itself is sorely tested. Politicians and farmers (not to mention the taxpayers who ultimately fund these programmes) want clear answers and expect science to be able to deliver these. Unfortunately, such answers are sorely lacking. This should be borne in mind the next time you hear anyone say that there is firm scientific evidence which concludes that culling is effective in reducing TB. For example: 'It has been proven without any shred of a doubt that infected badgers have been one of the main root causes as to why bovine TB has remained an ongoing problem on many Galway and Irish farms.'[34]

What has been proved is that there is a link between badgers and TB in cattle; however, there is no proof that badgers are the root cause of this ongoing problem. It is nearly thirty years since the badger-culling policy was put in place and it goes on, unmodified, to this day.

I find it odd that a policy that has not reached its stated aim after such a long time continues to be seen as such a success. If we go back to the years when this issue was being debated it is possible to sense a reluctance among those involved to start killing badgers. We tend to feel these days that attitudes to the environment have improved over time, and that these improvements continue as the years roll by, but here we actually see a deterioration of attitudes. This is evidenced in the opening address delivered by Seamus Kirk TD, then Minister of State at the Department of Agriculture and Food, to a conference on the badger organised by the Royal Irish Academy in 1991. He opened by talking about the importance of protecting wildlife and the commitments Ireland had made nationally and internationally in conservation. Only after this did he move to the central issue of TB in cattle, saying: 'It is increasingly accepted that the badger is associated with TB infection in cattle; the question requiring clarification is the degree to which the badger is a spreading agent. We should, however, emphasise that in no sense should the badger be regarded as the sole cause of the spread of infection. The factual position is that, over the forty-odd years of the operation of the Bovine Tuberculosis Scheme, a number of sources of infection have been identified and monitored.'

He pointed to a forthcoming National Badger Survey which would, he hoped, 'provide an accurate assessment of existing population' and that 'the research will discover a means of effective vaccination to ensure the health status of the badger species.'

He finished by emphasising the place that the badger has in our countryside, literature and heritage. 'In conclusion, I would like to emphasise that my department ... will continue, in conjunction with the Wildlife Service, to seek ways and means of ensuring a healthy badger population.'[35]

How the years have flown! Whole careers have been built around the badger-culling programme, between vets, Department of Agriculture staff and contractors who do the dirty work. No longer is there any feeling that badgers are anything *but* the sole cause of TB. In correspondence received from the NPWS in 2015 under Access to Information on the Environment, and requested by the IWT, the Irish Environmental Network and An Taisce, it was revealed how licences for culling are rubber-stamped *en masse* on an annual basis. It showed how conditions attached to the licences are routinely ignored, while there is no ongoing assessment of the impact it is having on the badger population. The NPWS relies entirely on information given to it by the Department of Agriculture and there are no checks and balances to ensure that best practice is followed or that culling only takes place when strictly necessary. The original National Badger Survey wildly overestimated the badger population and, because the badger is not listed under EU legislation (i.e. the EU is not standing over us with the big stick) there has been no attempt to rectify this by carrying out an up-to-date survey. A conservation assessment of the badger as being of 'least concern' was not based on accurate census data.

Clearly, in 1991, badger culling was seen as, at worst, an interim measure until such time as the vaccination could be rolled out. So, in 2017, where is this vaccination? In November 2014, I was part of a contingent from what we called 'Team Broc' (a loose coalition of environmental groups and concerned individuals) which met with Department of Agriculture officials. Vaccination trials which have been ongoing in Kilkenny since 2009, we were told, were showing promising results but hard data was still some way off, maybe in 2017. Additional trials were also under way in a number of counties but, make no mistake, culling would still be needed. It is very hard to escape the feeling that there is no great enthusiasm for the vaccination solution. And it is not like we are waiting for a scientific breakthrough: the vaccination formula is the same BGC vaccine that is used on people. Yes, there are problems to be solved, like how to best deliver it but nothing that requires twenty-five years of head-scratching. In 2012, a total of 6,939 badgers were snared and shot. And the corresponding reduction in TB numbers? Fifty-five fewer cattle were diagnosed with the disease compared with 2011. In 2014 the cost of the TB control programme, excluding staff, came to over €33.5 million. Of this, a tiny fraction – just over €1 million – went into the vaccination research programme. That's a little over 3 per cent of the total. This tells me that vaccination is not a priority, if it ever was, and that, as was always suspected by those opposed to the cull, it is merely the fig leaf of respectability that allows the killing to continue.

Department of Agriculture officials will assure the public that this is all a good way to spend public money and that reviews and research are ongoing. And yes, the volume of reports would fill a small library. The problem is that they are all written by the Department of Agriculture or their agents. Irish people don't need to be reminded of the difficulties that arise when the organs of state are reviewing their own progress in implementing targets that they themselves set. Meanwhile, memories are short. In 2015 Minister Coveney went before the Oireachtas committee on agriculture, food and the marine to praise the success of the TB programme and appeared unaware of its original goal:

> We are making phenomenal progress in dealing with TB. It is a great story. We have saved a large amount of money, but, more importantly, we are moving towards trying to eradicate it. My personal ambition is to eradicate, in Ireland by 2030, which is possible as a result of conversations we have had with veterinary officers and so on. It is the first time ever that we are actively talking about eradicating TB entirely from Ireland. That is a statement that will probably be quoted back to me in

the future but that is what we would like to do. We will keep the pressure on to try to eradicate TB.[36]

In a rare independent review of the disease control policy in 2014, vets from the European Commission's Food and Veterinary Office concluded that 'it is still uncertain whether the currently applied measures are sufficient' and that 'the possibility of being granted the officially bTB free status ... for the Irish cattle herd is still far away'.[37] That would require a level of disease in herds to be 0.1 per cent – in 2015 the level was 3.37 per cent, a level which masks much higher levels in other areas, nearly 13 per cent in parts of Wicklow, for instance.

So what effect is all this having on our badgers? At the end of the day it is very hard to say for sure as no studies have been carried out to assess the impact the cull is having on them. The badger is protected under the Wildlife Act so the Minister overseeing the NPWS has a responsibility to ensure that appropriate conservation measures are enforced. Culling is carried out under licence from the NPWS but this is a formality, with no real scrutiny of the practices on the ground. Under Access to Information on the Environment rules, the IWT along with the Irish Environmental Network and An Taisce saw correspondence which showed that the NPWS had received no information on updates on the vaccination trials and had no role in supervising it. We also received emails from NPWS staff highlighting their concerns that licence conditions were not being followed and that culling was significantly affecting the viability of the badger population. Once the Department of Agriculture receives a licence, culling can take place within a radius of 2km around the farm where the outbreak occurred. There is no time limit on the licence and so culling can go on within this radius forever. Snaring can, and does, go on for years. In this way whole areas of the countryside can be kept permanently badger-free. There are serious concerns that this is actually happening – even Department of Agriculture staff have come to admit that 'culling isn't sustainable'.[38]

Despite this, there is no closed season for culling. Any conservationist will tell you that a closed season is essential in order to allow animals to breed and rear their young. This is why there is a closed season for shooting deer or ducks. There is a prohibition on cutting hedgerows during the spring and summer months to allow birds to nest and lay eggs. Allowing the destruction of a species during its breeding season is the kind of strategy that would be employed were the goal to wipe that animal out altogether. Snaring and shooting a lactating female means that young badgers are left underground to starve to death, a barbaric and inhumane practice.

The total badger population was recently estimated as being in the region of 19,200 social groups.[39] With an estimated average of nearly four badgers per social group in Ireland this would give a population in the region of 75,000.[40] However, these figures are based on models and ground surveys looking at the frequency of badger setts in the landscape. It doesn't take into account the fact that many setts might be empty because of culling. Certainly, where culling is taking place, badgers are being lost. 'Significant reductions in badger density occurred in areas where management had taken place' was the conclusion of one study in 2013.[41]

Are badgers about to go extinct? No. For a start, culling only takes place on farmland and lots of Ireland is not farmland. Badgers in Ireland are predominantly farmland animals but they live happily in woodlands, cities and even in marginal areas around bogs. To assuage fears that culling would end up wiping out badgers the public was always told that the programme would only ever be confined to 30 per cent of farmland. The government also told this to the Bern Convention, an international treaty to which Ireland is a signatory, and under which the badger is protected. However, this figure is practically meaningless as the numbers are easily massaged. In any case, in 2012 the government itself exceeded its own calculation of what 30 per cent of farmland means and faced no particular sanction. The Bern Convention itself was exposed as about as effective a tool in conserving wildlife as a sign in the ground saying 'please don't pick the flowers'. I was told privately once by someone from within the Department of Agriculture that there is nothing in the Bern Convention that could stop the government doing anything it wants with badgers. In 2013 the total number of badgers culled exceeded the 100,000 mark and approximately 7,000 are snared and shot annually. Clearly the current programme is not sustainable from the badgers' point of view and as things stand we face losing this charismatic animal permanently from large tracts of our countryside.

Assuming that most people would prefer this didn't happen, what can be done to have both a healthy badger population and a healthy cattle herd? Before badgers were ever implicated in the disease the main way of controlling the spread of TB in cattle was early detection through testing. Cattle showing early signs of the disease were quickly removed from the farm and slaughtered. This is what reduced the prevalence in herds from plague proportions in the 1950s to below 3 per cent by the late 1960s. The test that was used to achieve this massive reduction is called a skin test but, like all such tests, it is not 100 per cent accurate. In fact, about 20 per cent of cattle will not test positive even if they are carrying the disease. This is a large figure and, because only animals that have tested positive will go to slaughter, it allows infected cattle to remain in the herd, going on to infect others. This inherent failing of the skin test has allowed a reservoir of TB to remain in the national herd.

Since 2000 another test, referred to as the IGN-gamma test, has also been introduced in tandem with the skin test. Together these two tests increase the likelihood of detecting TB to 90 per cent. This new regime means that infected cattle have been less likely to slip through the net and it is no coincidence that TB prevalence in cattle fell dramatically after its introduction the early 2000s. This drop is frequently attributed, falsely, to killing badgers. Because there is still a one-in-ten chance of a diseased animal not being detected, improvements are still needed. In 2015 about a quarter of cattle were being diagnosed with TB at the slaughterhouse. This highlights the massive failings of the current testing regime. More research is needed into a more accurate test and a tighter testing programme on the farm. The benefits of this alone are clear to see.

One of the problems with the culling programme is that so many of the badgers that are killed are perfectly healthy animals. Each badger is examined by a vet post-mortem and this finds that between 80 and 90 per cent are free of the lesions indicating TB. Killing whole family groups was found to create the 'perturbation effect' and spread disease to surrounding areas. All agree it would be much better to remove only diseased badgers and leave the healthy ones where they are. In Northern Ireland (where there has never been any culling) a new approach based on this principle has been trialled since 2012. This is called the 'test and vaccinate or remove' (TVR) method. Live badgers are trapped, tested for TB in the field and, if found to be healthy, they are vaccinated and released. If they test positive for TB they are killed. After initial surveying the TVR project went live in two areas of County Down in 2014. Results from the trials are not expected until 2019 but it is a very positive step forward.[42] If successful, it will be a major advance in this long-running battle.

At the end of the day, badger culling in Ireland is not a policy based on science. Rather it is based entirely on politics and the need to be seen to be addressing the problem, not to mention to satisfy vested interests. When asked about the seeming ineffectiveness of the badger-culling programme in dealing with TB, Eddie Downey, then president of the Irish Farmers' Association (IFA) said: 'There's no point in having sick animals going around. We're farmers, we love animals. We're not here talking about taking out animals, taking out badgers. We're talking about controlling the numbers and making sure there are no diseased animals there. And then all the animals are healthy.'[43]

In the UK, where conservation voices carry much more weight than they do here, the cull is much more overtly political – at least in Ireland there is a veneer of pseudo-science to justify the policy. The badger cull in the UK was nasty, and aggressive confrontation between farmers and anti-cull activists was a prevalent feature. I am happy that we have not seen that kind of confrontation here. Agriculture in the UK is a far less important element of the national economy than it is in Ireland so relatively speaking Irish farmers have a lot more to lose than their British counterparts. Despite a recent spike in scientific publications from Irish authors relating to badgers and culling, none of these actually question the validity of the cull itself and generally originate from people intimately involved in what the government refers to as the 'wildlife programme'. Actually independent studies are more abundant across the Irish Sea. A paper in the prestigious journal *Nature* reviewed the options facing policymakers at the height of the UK's cull controversy in 2014. The authors' conclusions? 'Very few of the control options tested have the potential to reverse the observed annual increase [in cattle TB rates], with only intensive strategies such as whole-herd culling or additional national testing proving highly effective, whereas controls focused on a single transmission route are unlikely to be highly effective.'[44]

For 'single transmission route' read badger. It's what the Irish authorities specialise in. Culling entire herds of cattle would indeed be a drastic step, but if it is something that could work should it not at least be considered? Improving biosecurity measures on farms, more accurate testing and novel approaches like Northern Ireland's 'test and vaccinate or remove' get short shrift south of the border. Badger vaccination, that modicum of hope first touted in 1991, remains, for now, out of reach. When a breakthrough is announced will it be too late for badgers in many parts of the country? In Mr Downey's interview with RTÉ he added: 'It's actually a combination of different things. The cattle carry some of the disease, the badgers carry some of the disease, but the deer population is also a reservoir of the disease. If you want to control a disease you've got to deal with all sources of infection.'

This hints at the possible next chapter in this ongoing saga, as articulated in the *Irish Independent*: 'Urgent Government action to cull "out-of-control" wild deer has been demanded by the ICMSA [Irish Creamery Milk Suppliers Association], despite the Minister for Agriculture, Simon Coveney, claiming that badgers remain the key source of infection in TB blackspots. While ICMSA president, John Comer insisted that the wild deer herd had reached 150,000, and were "out of control" in Wicklow, Kerry and Clare, Minister Coveney said that TB levels in badgers were still higher in Wicklow.'[45]

In early 2016 a open pit filled with deer carcasses was reported to authorities in the Wicklow mountains. It followed repeated claims that deer are out of control and are causing high levels of TB in cattle there.[46] The Wild Deer Association of Ireland was quick to condemn the discovery: 'This incident highlights the level of vilification we have seen towards wild deer in the Wicklow area, where misleading statements have been made about deer and tuberculosis, resulting in deer, a protected species under our Wildlife Acts, being reduced to the level of vermin by some landowners.'

Deer are becoming a nuisance for many farmers because they eat the grass meant for cattle and can leap over fences with ease. Cattle TB rates in Wicklow are running at over three times the national average. The deer population in Wicklow is strong and probably increasing although there are no population counts to back this up. Deer, like any other mammal, can carry TB but there is no evidence that they are passing it on to cattle. They are classified as 'least concern' by the NPWS. Sound familiar?

6

The battle to save the bogs

> 'In the north part of Germany, where there are many turf bogs, the people have a regular plan for reproducing the turf. They cut it in square holes of a certain size, in which the bog water collects. In this manner, marsh plants spring up, and by their decay and deposits new layers of turf are gradually formed, which, after 30 or 40 years, can be cut again. Thus they possess an inexhaustible source of profitable fuel. In Ireland they know nothing of this.'
>
> From *Travels in Ireland* by J. G. Kohl, 1844

IN THE END IT WASN'T THAT hard to find what I was looking for. Blank, roadless spaces of green on Ordnance Survey maps are usually a giveaway. The lack of contour lines hint at a flat expanse while railway lines, scattered like fallen matches, with neither stations nor destinations, confirm the presence of what normally remains unlabelled. Although I had been to places like it before I wanted to return to get a fuller sense of what remained. It will come as no surprise to anyone that industrial turf-cutting has been going on for decades. Yet for most people, especially those like me who live in the cities, the look and feel of a cutover bog is probably an abstraction. I remember my first impression of the scale of turf exploitation upon the launch of Google Earth in 2005. The great novelty, now banal, of fast and free aerial images provided a new perspective on the colours and textures of Ireland. The green fields blended into one, the towns and villages liberally sprinkled among them and the dark sinews of rivers as they meander across the land. The Shannon, the biggest of them all, appreciably broad as it cleaves the heart of the country. It is

The ecological devastation from industrial-scale peat extraction is a feature of many midland counties.

here that the cutover bogs adhere to the shores of that mighty river like scabs on a wounded landscape. The continual scratching and scraping away of the bog surface is evident in the parallel striations left by the heavy machinery. In high-resolution clarity, the utter transformation of whole swathes of the Shannon basin is as plain as it is devastating.

On the ground, as I slow the car along the narrow approach road, swirls of dust smudge the clear sky. Eventually, a clearing in the willow and birch trees provides an aperture onto the full scale of operations and a great expanse of brown meets the eye. Springy underfoot, the dried surface layer of peat is also crumbly and brittle. Great trenches are carved all around, hewn into the uniform earth, and each filled with a black, stagnant liquor. These are the drains, so essential in transforming the peat from a living, waterlogged sponge to a dry, lifeless and exploitable material. Of vegetation there is none, the song of no bird is to be heard, neither fish nor frog swim in the lightless depths of its drains. Indeed, it has been deserted of all life. A more grim and soulless expanse is hard to imagine anywhere on Earth. If it is comparable to anything it must be the vast open-pit mines of Australia or the filthy fields of tar sands in Canada. In fact, that is exactly what it is: open-cast mining with all the attendant pollution and degradation that goes with it. The scene before me is not unique but can be found throughout the midland counties from Longford in the north to Tipperary in the south, from Galway in the west to Kildare in the east. Anyone with a smartphone can survey the extent of the damage without reading a word of a report, just look for splodges the colour of fake tan. Surely we are looking upon the greatest ecological catastrophe to have befallen this small island since the arrival of the last ice age?

Peat is a type of soil and covers approximately 20 per cent of Ireland's land surface. It is poor in nutrients and consequently poor for farming, which explains why the land remains open, without the usual fences or hedges to enclose fields. It forms in waterlogged conditions which in Ireland can, broadly speaking, happen under two scenarios.[1] The first is where rainfall is excessive, over 1,250mm spread over more than 235 days of the year, and this degree of wetness can be found all along the western seaboard as well as upland areas across the island.[2] This is known as 'blanket' bog because it blankets the landscape, paying no heed to contours or obstructions. It is frequently found in a mosaic with lakes, rivers, fens, rocky outcrops and drier heath.

The other type of bog has more complex origins that stretch back to the melting of the glaciers. As the ice melted, depressions within the flat interior of the land filled to become lakes, with some coalescing and overspilling to form the newborn channel of the River Shannon. Over time, these lakes filled with vegetation, new plants piled upon the semi-decomposed remains of their antecedents – the sheets of open water ever diminishing until eventually disappearing entirely. These mires then took on a life of their own as a family of mosses, known as sphagnum, changed the soil and water chemistry from alkaline to acid (and so preventing decomposition), not only filling in what was once a lake but rising above the ground and spreading out across surrounding land to swallow woodlands which had fringed their margins. This incessant expansion was slow but steady: a rate of about 1mm per year. Until quite recently, and excepting changes to the climate itself, it had no opposing force in nature. Adjacent bogs merged into one while domes of saturated peat breached the landscape's more solid undulations. These are the raised bogs, sometimes called red bogs as the peat has a bronze shine unlike the matt darkness of its blanket cousin. These continual accumulations explain why Nobel laureate Seamus Heaney referred to bogs as 'the memory of the landscape', not only subsuming but perfectly preserving trees, antlers of giant deer, and the detritus of mankind from the first boats, butter and even human bodies. Less visible are the filigree layers of pollen that have been deposited upon the bog surface in an unending rain of tiny particles. These remarkably resilient and distinctive beads have allowed scientists to painstakingly reconstruct Ireland's ancient landscape – from first colonisation by juniper and birch to the establishment of high oak forests and the arrival of agriculture (and so grass) around 6,000 years ago.

The idea of bogs as worthy of conservation is new. As recently as 1981 legislation was being passed to provide government subsidies to exploit undeveloped bogs.[3] The concept of nature conservation had yet to gain purchase in the public mind while notions of 'carbon sinks' or 'ecosystem services' were still a long way off. Kohl, with whom I opened this chapter, sums up well what could be described as the attitude to bogs from man's first arrival in Ireland to at least the early 1990s: 'The bogs are ... at once a source of wealth and of poverty; for whilst they supply fuel, they at the same time cover much fertile soil, which they withhold from cultivation; they spoil the waters of the rivers, fill the entire atmosphere with a turfy smell and infect the air with foul exhalations; are an impediment to traffic, and have long supplied a protection and a refuge to the thieves and robbers of Ireland.'[4]

The value of turf for fuel has been appreciated for some time. The earliest account among Celtic peoples comes from Germany during the first century AD and charred remains of turf in Ireland have been found under 8m of bog at Drumkelin in County Donegal, suggesting it is many thousands of years old.[5] References of peasants being so wretched that 'they burn their very earth for warmth' suggests that the preferred source of fuel – wood – was not available. As the forests of Ireland were cleared the harvesting of turf in medieval times became widespread and those bogs near centres of population such as Dublin were cleared out by the fourteenth century.[6] The population increased during the eighteenth and early nineteenth centuries and so too did the demand for fuel. The completion of the Grand Canal in 1786 provided a direct route to Dublin for midlands turf and, in the early nineteenth century, 30,000 tons of dried sods were being shipped annually to the capital from the immense Bog of Allen in Kildare.[7] In Connemara, turf was transported across Galway Bay to bogless north Clare and the Aran Islands by hooker (the traditional sailing boat of the area), a trade that continued right up to the 1940s.[8] In the early part of the last century, droves of Dubliners headed to the Wicklow Mountains to cut their winter fuel and the elevated turf banks can still be seen today around the Sally Gap.

In parallel with demand for fuel was the demand for land and great energies were expended in an effort to turn the unproductive bogs into profitable grazing land. Cheap and abundant labour allowed landlords the opportunity for digging extensive networks of drains and deep ploughing to mix essential sources of lime, such as was available from gravel deposits or coastal sands. Sir Robert Kane, writing in 1844, saw no obstacle in converting all of Ireland's peatlands to productive land with 'scarcely an acre to which the name of incapable of cultivation can be applied'.[9]

In coastal areas seaweed was piled on the wet peat once any surface vegetation had been scraped away. The vegetation itself was frequently burned to utilise best the available nutrients in the ash. Ploughed land was planted with oats or potatoes while the drier margins could, with perseverance, be transformed into hay meadows. Indeed, today the zone where bog meets mineral ground is abrupt and there is scarcely a handful of places where the natural interface can be seen. In John Feehan and Grace O'Donovan's book *The Bogs of Ireland,* an exploration of the human and natural history of Ireland's bogs, the dramatic effects of the reclamation effort were highlighted around Monivea Castle in County Galway where drainage works were undertaken by the French family:

> The castle was surrounded by two deep and virtually impassable bogs, which had grown so much since the castle was built that they blocked the view from the ground floor. Robert French diverted the river which flowed between the castle and the bog, and he cut wide, deep drains through the bog – wide enough for boats to ferry manure and limestone in to the castle farmyard, and out to where they were needed on the bog. Several years on, the bog had subsided by 15–20' [4.5-6m], and the view from the ground floor was restored.[10]

The lowering of the bogs is an impressive testament to their incredible retentive powers. Fresh turf is approximately 90 per cent water, making it more liquid than solid. The sphagnum mosses that carpet the surface of an active bog, and lay down the peat which enable its growth, have an absorption capacity up to twenty-six times their dry weight. The hydraulic deflation of the raised bogs was an essential precursor to their subsequent exploitation. Feehan and O'Donovan even describe how, during the construction of the Grand Canal, the bog at Edenderry, County Offaly, dropped by a massive 13m so that the canal, which originally was depicted on drawings as sunken into the land with bog on either side, eventually became elevated above it.[11] The effect on the landscape was like taking the stopper out of an airbed. Today, there are no Irish bogs left in existence where we can imagine how these great domes once appeared.

The massive expansion of the Irish population, which doubled in size from the 1780s to before the Great Famine in the 1840s, an increase to nearly 7 million people, would see tremendous activity on the bog. The abundance of innumerable toiling hands, coupled with the success of the potato on reclaimed peat, both facilitated the growth in population and the demand for greater use of uncultivated land. Official enthusiasm for the exploitation of the bogs had long been in the hands of individual landowners but crystallised in the early 1800s as Napoleon's armies advanced across Europe. It was the urgency of imminent invasion that spurred the parliament in London into action. The pivotal role of naval warfare during this period led to a frantic scramble for new sources of hemp and flax for the manufacture of sails and led to the passing of an Act in 1809 which would establish the Bog Commissioners. Their task: 'to inquire and examine into the nature and extent of the several Bogs of Ireland, and the practicability of draining and cultivating them, and the best means of effecting the same'.

With the urgency of war the ten appointed engineers and their teams were charged with submitting their reports by 1811.[12] The results of these meticulous surveys, decades before the establishment of the Ordnance Survey, are an incredible series of maps along with the minutiae of costs and construction methods. Here is an extract from Richard Griffiths, who was charged with surveying the great Bog of Allen. More a sprawling cluster of amalgamated bogs, this enormous area covers much of County Kildare and parts of the neighbouring County Laois:

> The first and most essential step towards the complete reclamation of Bog is drainage, and the second is the best mode of causing a decomposition of the surface of the Bog

when drained, whether by long exposure to the atmosphere, burning, or manure, so as to alter the quality of the boggy soil by any of these means, and thereby render it capable of affording nutriment to a variety of useful plants.

... the practicability of draining 77,505 English Acres of Bog, which is the total amount contained in the eastern and western divisions of this District, has been I trust satisfactorily proved.[13]

Despite the scale and depth of the Bog Commissioners reports', however, the changing fortunes of war and, ultimately, the defeat of Napoleon at Waterloo in 1815 meant that the grand plans for the transformation of the Irish bogs would not be realised, for now at least.

The Great Famine began a depopulation of rural Ireland and the abundant labour required for the maintenance of bogs began to trickle away. Neglected drains filled in, rushes and sphagnum mosses took over and, in many cases, the bogs resumed where they had left off. It would not be until the Second World War that significant attention would once again be brought to bear on the bogs, as coal became scarce and self-sufficiency became a necessity. This task was not insignificant as turf was now needed to power the entire country.

Mile after mile of new bog face was opened up, and already by 1941 around 1,000 bogs were being worked, in every county in the Republic. Even bogless Wexford, which had used up all the accessible turf from its bogs on the Blackstairs in the 19th century, did its best, cutting away the blanket peat from Moneer Bog, 2,000 ft above sea level on Mount Leinster ... In the summer of 1943, 160 men, boys and girls were busily at work on top of the mountain. Along the County Clare seashore at Seafield and Kilkee, and in several other places even sand-covered intertidal peat deposits were exploited during the War – a particularly laborious task since the rising tide filled the excavated trenches anew with water every day.[14]

This wartime scheme employed vast numbers of people during a period of general depression while also providing an indispensible source of warmth and energy.

For millennia it was only by the power of hands and backs and feet that turf could be hewn from the earth and stacked to dry. Despite the best efforts of some, technological progress never quite managed to supplant the manual task. As the machines grew, so did the capacity for all-out development of the bogs, something that had been envisaged in the first Bog Commissioners' report of 1811. Particularly for supplying fuel for the bigger towns and cities, mechanisation seemed the only way of meeting demand. In 1936 the state acquired 40,000 acres of bog at Clonast in County Offaly and set about digging 800km of drains *by hand*. Production began in 1939 with German machines ill-suited to the exceptionally wet Irish conditions.[15] The enterprise mostly failed in its ambitious targets but the path to large-scale industrial extraction was now set.

The post-war period saw the establishment of Bord na Móna as the single state authority that to this day is charged with the commercial management of the nation's peat resource. Output would increase from tens of thousands of tons of peat per annum to 2 million tons in the 1950s. The production model would change from cutting sods to milling the peat and the marketing of new products, such as briquettes and moss peat for horticulture (and later gardens).

The sense of national pride that Ireland, a newly modern and forward-looking nation, had finally 'conquered' the bogs, should not be underestimated. In an 1970 article for *Ireland of the Welcomes,* a magazine which has been going out to the world since 1952, a piece entitled 'The Bog Transformed' summed up the sense of achievement: 'The patriots' dream has come true. Research and mechanisation have turned turf production into a large-scale industry. Bord na Móna has had a more dynamic effect on Ireland than practically anything that has happened in its history.'

There was no downside: 'Has the landscape been impaired? By no means. Turf production leaves no slag heaps or disfiguring debris. But the landscape has been enlivened by men and machines. Man's work and machinery are easily dwarfed by high skies and spaciousness.'[16]

By 1973 all the bogs which were considered at the time to be commercially exploitable were being mined. The oil crisis of 1976 expanded this further so that in total 89,000 hectares were within the ownership of the company. It would be the second largest peat-producing operation in the world.[17] This was predominantly of the midland raised-bog variety as the western blanket bogs were generally ill-suited to machinery while in any case the depth of peat there is far less. These massive developments were, like many government projects at the time, outside the planning laws, such as they were. To date, none of the industrially worked-over bogs have ever received formal planning permission. Environmental Impact Assessments, first brought into law in the EU in 1985, have never been applied to these substantial operations. In 1994 alone 6.5 million tonnes of peat were extracted by Bord na Móna and private companies combined.

All the while, small-scale cutting of turf, mostly for private use or local enterprises continued apace. The back-breaking work of manually slicing sods of turf with a *sleán* (a spade with an extra cutting side) went into rapid decline from the 1960s although the process of 'footing the turf', that is, arranging the sods in little pyramids to dry them out, is still done by hand. So-called 'sausage-cutting' machines can be operated at small sites, or diggers scoop out the wet peat to be piled into hoppers. The effects of this kind of cutting is less dramatic than the large-scale industrial kind as the peat largely retains a covering of vegetation – purple heather on the dry top of the peat bank, cottongrass or tormentil on the wetter floor. It is not unusual to find lots of wildlife in these places, from frogs and lizards to the many birds that nest in the brambles, willow and birch that spring up on the dried-out fringes.

The right to cut turf has been acknowledged in 'customary rights' at least since the beginning of the eighteenth century. These rights were codified by the Land Commission after the establishment of the Free State in 1922 and individually owned plots were marked in numerous thin strips on minutely detailed maps. Rights for cutting, shooting or grazing frequently overlapped so the system of ownership was complex, to say the least. To this day turf-cutting this way continues and has been at the heart of the more recent controversies surrounding the implementation of conservation laws. Despite the tone of many of the statements made during this row, there are, and never have been, plans to ban outright the cutting of turf for private use.

In the west of Ireland or upland areas elsewhere, the impact of centuries of human activity on the bogs is today hardly noticeable at a distance, except perhaps where they have been planted with stands of spruce or pine trees. The same uniform, dreary brown monotony of Connemara, the Wicklow Mountains or Mayo is, superficially, much as it was 200 years ago. The drastic alteration of the landscapes of the midlands counties meanwhile, and the scabs where there were once raised bogs is today surprisingly easy to ignore on the ground. The worked-over bog at Kinnegad is one of the few places where a major road, the M6 motorway, passes a site where the visual impact of this industry can be fully appreciated, even at 120km/h. The vast majority, however, are down quiet country lanes or screened from passing motorists by the lines of willows and birch trees that have conveniently established themselves. By the 1970s it was well within the technological powers of the state and private turf producers to work over every remaining square metre of raised bog left in the country. The National Soil Survey of 1973–1974 calculated that while the majority of the blanket bogs remained intact, only 19.6 per cent of the raised bog surface could be described as such.

The 1970s were pivotal years in a global ecological awakening and this didn't bypass Ireland. A 1971 report into potential areas for conservation in Galway highlighted how 'raised bogs, with their different species composition are also becoming increasingly rare'.[18]

While in County Offaly: 'Another habitat which has been drastically reduced is that of the peatland. Offaly is covered by acres of raised red bogs or "hoch moor" as they are technically known. Much of the bogland has recently been excavated by Bord na Móna.'[19]

However, no action was taken in Ireland at this time to protect the remaining areas of bog. It would be a Dutchman, Matthijs Schouten, whose work to protect these unique habitats would ultimately prove decisive. In his native Holland all the bogs had been exploited by the 1950s and by the 1970s large-scale bog restoration works were under way at great public expense. He arrived in Ireland as a research student and was shocked by the scale of commercial exploitation here. He saw that without urgent intervention the same was about to happen in Ireland as had happened in Holland. The remaining, relatively intact Irish bogs, provided an invaluable reference that would allow the Dutch to recreate what they had lost. The work of Schouten was, at that time, beginning to reveal just how the centuries of drainage, ploughing and cutting had altered the hydrology as well as the communities of plants and animals that depend upon the bogs: '... contrary to what was generally assumed until then, relatively few bogs in Ireland had remained intact. Peat extraction, afforestation and overgrazing had destroyed a large amount of the once very extensive boglands, and many of the remaining undisturbed sites appeared to be also under threat, while at that time only a small number of bogs were actually protected.'[20]

Clara Bog in County Offaly, for instance, although far from pristine, still had great expanses of undisturbed and actively growing peat. Two hundred years ago a road was sliced through its centre, a remarkable feat which, it is estimated, saw the height of the bog fall by 10m. Nevertheless, it still had great domes of water-filled bog on either side. In 1982 it was listed by the newly formed National Peatland Conservation Committee (NPCC), a non-profit organisation, as important for the extent of its relatively intact surface. Nevertheless, between 1983 and 1984 Bord na Móna dug approximately 400km of drains through the bog in preparation for all-out industrial extraction. The surface of the bog subsequently slumped even more as the water poured out.[21]

In 1983 Schouten helped to found the Dutch Foundation for the Conservation of Irish Bogs which set about raising funds to buy three Irish bogs of conservation value. They held a benefit concert, an art auction and appealed to the Dutch branch of the World Wide Fund for Nature for assistance. A year earlier the Irish Peatland Conservation Council (IPCC and successor to the NPCC) was founded with the aim of raising awareness among the Irish public of the value of conserving what was left of this unique heritage. This must surely have been a monumental task. How could we possibly switch from seeing the bogs as the source of 'wealth and poverty', as Kohl has described it, to a part of our heritage to be cherished like Newgrange or the artworks of Paul Henry? As the next thirty years would show, it would not be easy. But change did happen.

The Dutch were ultimately successful in their lobbying and by 1987 had raised sufficient funds for the acquisition of three Irish sites – these were at Scragh Bog in Westmeath, Clochar na gCon in Galway and Cummeragh River Bog in Kerry. At a conference that year in Baarn, The Netherlands, Prince Bernhard handed ownership to the Irish state, then represented by Noel Treacy who, as head of the OPW, had responsibility for nature conservation. Whether it was the intention of the Dutch or not, the transaction served to shame the Irish authorities into taking conservation seriously. The talk at the time was forward looking; in a letter to his Dutch counterpart in 1987 Treacy wrote: 'Sound hydrological management is an essential element of these reserves. It is therefore crucial that the water management of and the problems in and concerning particularly raised bogs be analysed.'[22]

At about this time David Bellamy, botanist at the University of Durham in the UK, was carrying out some of the first scientific research into peatlands around the world. While he was not on his own, he is significant because of his undoubted communication skills, which he deployed on our televisions with great aplomb. There he was, knee deep in a swamp or getting face to face with a lump of soil from which a tiny plant was germinating. Watching him doing his thing it sometimes felt that he might fall through the screen into the living room itself! Parodied by Lenny Henry for the way in which he seemed to mash one word into the next, he was not beyond appearing on *Blue Peter*, a children's TV show, in an orange jumpsuit, singing a song about a brontosaurus. His sometimes childish overenthusiasm belied his academic background and penchant for serious environmental activism, which included a PhD, numerous books and scientific publications, and even getting arrested in Australia in 1983 for demonstrating against a dam construction project. It is unfortunate that, despite his credentials, he is best known of late for his belief that climate change is not man-made. However, at a time when global warming had yet to register as an issue he was forecasting the vital role peatlands were playing in regulating carbon in the atmosphere: 'It is obvious that the mires of the world play a significant role in the global cycles of water and carbon dioxide ... It is easy to speculate on the effects of say, "the combustion of the total peat resource": 500×10^9 tonnes of carbon dioxide would be released into the atmosphere and could increase the "greenhouse" effect altering the overall pattern of macro-climate.'[23]

As such, Bellamy is a key figure in demonstrating why bogs are worthy of conservation. He also pointed out the important role active peatlands play in regulating the flow of water off land. Today this would be referred to as an 'ecosystem service' but in fact it was known for centuries. During his stint as a Bog Commissioner in 1811, Richard Griffith put his finger on it: 'In the event of a general drainage of the neighbouring Bogs, these floods would have a tendency to rise much more quickly than they do at present, as the water would not then flow slowly from the surface of the Bogs, but would be very rapidly discharged by the several Drains into the River.' Without a corresponding expansion of the river channels, he wrote: '[the damage] the vallies would sustain by the almost instantaneous rise of the floods, might in some instance, equal the benefit which might otherwise be derived from the drainage of the Bogs.'[24]

Or, as Bellamy put it in the 1970s: 'Peat acts as a reservoir, increasing the surface retention (storage capacity) of the landscape.'[25]

But he did more than this in bringing home the wonder of plants and the urgent need to conserve our bogs before they were gone entirely. In fact, he had honeymooned in Ireland, a fortnight of which was spent exploring bogs:

> It was January 1959, and the frost jewelled both the callow and the bogs from the Lakes of Killarney where we stayed, clear up to Tullamore and beyond. The surface of each bog was frozen stiff like the crust of new baked bread and it seemed sacrilege to break such perfection. We waited till the warmth of a duck-egg-blue day had melted the rime. Then across the steaming bog we saw a sight of immense beauty and rarity. Skein upon skein of Greenland white-fronted geese rose up from the bog where they had spent the night in safety and made their way down to the callow to graze.
>
> The bogs of Ireland are now the only winter home for more than sixty per cent of all that is left of the world population of this rare goose. If the bogs are all cut away, where will they go? To extinction?[26]

The bogs are indeed home to a great diversity of plants and animals. Since these words, the geese population of the bogs has collapsed. Many flocks that are left are small, fewer than fifty birds in many cases while a number of sites have been abandoned altogether.[27] Bellamy introduced his book dedicated to the Irish bogs with a plea for intervention:

> The Boora complex of peatlands which stretched from the Shannon estuary almost to Dublin is no more. The Great Bog of Allen has itself been swallowed up by the gourmands of Philipstown, Tullamore and the Irish economy in their need for electricity. The signs of Bord na Móna, its great machines, electricity sub-stations and grid lines dominate the dead brown

scene once dominated by the living cupolas of the great red bogs of the central Irish plain ... if something isn't done, and quickly, a unique part of the world's living heritage could be lost forever.[28]

Between them, the Dutch scientists and David Bellamy inspired a sea change in how we see our bogs. Yet this would be just the beginning of a conservation battle that continues to this day. When the state's Forest and Wildlife Service looked again at the quantity of intact raised bog in 1985 they were shocked to discover that of the 65,000 hectares recorded in 1974, only 20,000 hectares remained. The following year, prompted by pressure from local people, the bog at Clara was bought back from Bord na Móna and the year after that it became a National Nature Reserve. While Clara was now safe from outright industrial exploitation, bizarrely, private turf-cutting was allowed to continue. A more thorough examination of bogs, which was initiated in order to present to government a list of sites for protection, was published in 1990. It revised slightly the remaining area of intact bog to 23,000 hectares but its conclusions were stark:

- No completely intact raised bog remained in the country
- Only 7.4 per cent of the original area of the habitat remained and which were suitable for conservation.
- 93 per cent of the loss at conservation-worthy sites was due to peat extraction and drainage; 67–70 per cent of this was due to hand-cutting and the remainder due to Bord na Móna activities. Arterial drainage and afforestation were affecting some areas while all were affected by fire.
- At the rate of decline there would be no raised bogs worth saving by 1994.

Ireland's raised bogs are on the verge of extinction. Unless urgent steps are taken to conserve the few remaining intact examples we will witness, over the next 5 years or so, the final destruction of an irreplaceable resource which has taken over 8,000 years to develop.[29]

The highlight recommendation of this report was to preserve a *minimum* (the author's emphasis) of 10,000 hectares, or 3.2 per cent of the original extent. In a plea for adequate resourcing the author, presciently as it turns out, predicted that 'delay in protection will increase the eventual cost of acquisition of sites which, in the meantime, will have been damaged. This, in turn, will result in increased management costs.'[30]

The report listed twenty-six sites which were in need of protection to meet this target. Only four at that time were in the ownership of the state or conservation organisations and even these were not safe. Writing in the *The Badger* magazine in 1996, Dr Catherine O'Connell of the IPCC explained that this is because raised bogs are 'hydrological units'.[31] If you need to stop water leaking from a bucket, not being able to control the 1 per cent with the hole in it is going to be a problem. It was becoming increasingly clear that on those bogs left worth preserving, private turf-cutting was now the major threat to their continued existence.

Even in the mid-1990s the legal grounds for protecting anywhere for nature conservation was shaky. The original 'Areas of Scientific Interest' had been shown to be unconstitutional and so these were replaced with 'Natural Heritage Areas', but only *proposed*, as the legal framework needed to be clarified in the Wildlife Act. As O'Connell put it: 'sites earmarked for conservation ... exist in legislative limbo, as the Wildlife Act of 1976 is still under review – a process that had been ongoing for six years. In the meantime landowners are selling out to peat extraction and forestry companies who have a free reign [*sic*] to destroy our natural heritage. Many of them are doing so legally as they keep developments below thresholds required for Environmental Impact Assessment and in general do not require planning permission.'[32]

Meanwhile the IPCC estimated that the area of intact raised bog had slipped to 19,402 hectares. The amended Wildlife Act was not passed by the Dáil until 2000, ten years after its revision was first mooted. However, events elsewhere were to take the initiative away from government hands and ultimately lead to one of the greatest environmental confrontations Ireland has ever seen.

The EU published its directive on the 'conservation of natural habitats and of wild flora and fauna', better known as the Habitats Directive, in 1992. This was done with the approval of the government of the day but, as happens so often with EU directives, the politicians seemed to have given little thought to the ramifications of what they were agreeing to. Michael D. Higgins (Labour) being the responsible minister at the time, was charged with transposing the directive into Irish law and, part and parcel with this, designating conservation-worthy raised bogs (among other things) as Special Areas of Conservation (SAC). He signed the necessary regulations in February 1997. From day one there was a clear obligation on the state to ensure 'favourable status' of these precious habitats and, where damage had occurred, to initiate restoration measures. This meant stopping turf-cutting of all kinds on the thirty-two raised bog SACs that had been then identified. The general election in June of that year would see the Labour/Fine Gael coalition replaced with one dominated by Fianna Fáil. The new minister in charge of the heritage portfolio, Síle de Valera, initially announced that turf-cutting on blanket bogs would cease within five to ten years and that cutting of raised bogs would cease immediately:

> While it is possible to phase out the cutting of peat in blanket bogs over a period of years, the position on raised bogs is much more critical. While it is difficult to be precise, it is estimated that 40 acres of intact raised bog are lost, effectively forever, each year cutting continues. To meet our obligations under the directive to conserve this habitat, cutting has to stop now ...

> I know it is difficult for people to accept the kind of changes that are necessary to protect disappearing habitats, but the sad truth is that without change we will not be able to pass on the natural environment to our successors in anything like the state in which we received it.[33]

However, turf-cutting did not stop. Following 'consultations', an agreement was reached between the government and local landowners in 1999 that turf-cutting for 'domestic use' could continue for ten years but that so-called 'sausage machines' would be banned. It was very much a political calculation given that there was no scientific reasoning, little social rationale (given that it simply delayed matters), and that the costs of watching the bogs being cut away for ten years would entail significant extra restoration costs down the line. It was also a decision with no basis in law. Ostensibly the delay was to allow for an orderly transition with proper consultation and compensation. The government would later admit that in reality it 'allowed all parties to forget about the issues involved until the last minute'.[34]

De Valera herself subsequently made it be known that she had little enthusiasm for implementing the Habitats Directive when telling an American audience in 2000 that 'we have found that directives and regulations agreed in Brussels can often seriously impinge on our identity, culture and traditions'.[35]

Meanwhile the bogs continued to shrink. In 2001 the European Court of Justice found that Ireland had not designated enough peatland SACs to satisfy conservation needs and the government was sent back to the drawing board.[36]

Contrary to the popular belief that the European Commission (EC) enjoys wielding the stick, it is in practice very slow to bring legal proceedings against any state. However, when twenty years of an established law passes with no credible action from the country which signed up to it, it is given little option but to act. As the ten-year 'derogation' on turf-cutting elapsed, Minister for the Environment John Gormley (Green Party) was facing mounting pressure and threats of fines from the EC due to the lack of protection of the raised bog SACs. In 2010 he announced that the ban would be enforced and that compensation measures would be available for those directly affected. In 2011 the EC again initiated legal proceedings against the state for failure to protect raised bogs. This came on top of a ruling from 1999 that environmental impacts from peat extraction were still not being assessed as required under the Environmental Impact Assessment Directive, which, in 2011 had yet to be fully addressed. At this stage, the fourteen years since the regulations came into force had been well and truly squandered and there would be no room left for reasonable dialogue with turf-cutters or soothing of community concerns.

Furious in the face of ultimatums and threats of Garda enforcement, turf-cutters vowed to serve prison time to preserve their time-honoured traditions. Many seemed to be baffled that their small-scale operations could be posing a threat to the integrity of the habitat. Many, after all, were genuinely fighting for their love of the bog. The years from 2010–2013, nevertheless, saw ugly scenes. NPWS personnel were threatened with violence, sometimes with personal details such as home addresses or car licence plates posted on social media. The Minister for Arts and Heritage of the day, Jimmy Deenihan (Fine Gael), bemoaned: 'My sympathies are first and foremost with the turf-cutters, including members of my own extended family ... Part of me wishes that the portfolio had been kept to arts, sports and tourism, but that wasn't the case and I have to accept responsibility on behalf of the Irish State on this issue.'[37]

At a time when the economy was in deep recession, vast resources were spent on Garda overtime and helicopter surveillance. The public mood, meanwhile, was behind the turf-cutters. There was an impression that turf-cutting across the board was being outlawed while government forces could only mumble feebly about looming fines and pressure from Brussels. Not once did I hear any official defend the need for protecting a vanishing part of our heritage or own up to the huge errors in implementing the laws that they had agreed to. They simply seemed to lack the vocabulary. It was left to the voluntary sector to plead the case but at that stage it was an uphill battle.

The debacle prompted a flurry of new research on the bogs but the news just got more depressing. Between 1994 and 2005 the area of 'intact' bog within the forty-four raised bogs that had been 'protected' shrank by 36.8 per cent. From 2005–2012 it shrank by a further 1.5 per cent, admittedly a great improvement but at that stage there was increasingly little left to protect. While some improvement on individual bogs was seen, the report concluded that 'despite positive actions being undertaken, damaging activities continue impacting and threatening raised bogs SACs'.

And these were the areas with the highest level of protection; for bogs that had been designated as Natural Heritage Areas (NHAs) or not designated at all, the prospects were even bleaker. 'Intact', however, just means that the surface had not been cut into. 'Active' bog meanwhile is wet and still generating new peat. The area of 'active' raised bog was calculated at 1,955ha, or 0.63 per cent of the estimated original area.[38] In fact, the situation was even worse as the report highlighted that the figure did not take account of any damage that may have occurred since the actual fieldwork took place in 2011.

In 2012 Friends of the Irish Environment, a non-profit environmental organisation, conducted aerial surveys of twenty-two of the bogs which had, by then, been served with turf-cutting bans. It found turf-cutting activity going on at seventeen of them, eight of which had even been in receipt of funds for the restoration of previous damage. It seemed that after a slowdown during the 2000s turf-cutting was ramping up again, probably in reaction to official handling of the situation.[39]

At Clara Bog in Offaly private turf-cutting ceased completely only in 2011, thirty years after its importance was first highlighted. Today an attractive boardwalk guides visitors into a portion of the bog without trampling on the delicate mossy surface. Glossy information panels detail the unique nature of the habitat and the special adaptations of the flora and fauna to be found there.

Clara Bog in County Offaly where restoration is bringing the bog back to life. The new boardwalk is a hit with local people.

Nearby, in Clara village, a handsome visitor centre attracts tourists to the area and schoolchildren can hear the sounds of the curlew on an interactive panel. A giant, but realistically sculpted, dragonfly is suspended from the ceiling. On the day I was there during the summer of 2015 the enthusiastic guide from the NPWS told me how the boardwalk is a hit with the locals, many of whom use it daily. Drains dug in the early 1980s have been blocked with plastic dams and it is possible to see how the trapped water is being colonised with the all-important sphagnum mosses. Slowly, the bog is coming to life again. A new sense of ownership and appreciation for the bog has taken root and from this springs hope for the future. According to the government's National Peatlands Strategy report: 'Natural peatlands are considered amongst the most important ecosystems of the world, because of their key value for biodiversity, regulation of climate, water filtration and supply, and important for human welfare.'⁴⁰

In 2016 there were reports that turf-cutting was taking place on some SACs but the impression is that this was the dying spurt of a doomed campaign.

The government established an independent Peatlands Council as a forum for the interested parties and issued a National Peatlands Strategy, the first time a comprehensive approach has been taken to all our areas of bog and peat. It hopes to bring an end to the cycle of legal action taken against the state by the European Commission. The talk now is of tourism, carbon storage and wise use. Novel research is under way in the restoration of peatlands and there are many boardwalks, just like the one at Clara, that allow local people access to this wonderful part of our heritage. The battle to save the bogs is not over. Indeed, 2016 saw Minister Heather Humphreys 'de-designate' forty-six peatland NHAs. In other words, places that were of national value to our peatland heritage are now open for turf-cutting, conifer plantations, wind farms or one-off houses. Developments within the Peatland Council also clearly show that it is planned to interpret EU legislation in such a way that turf-cutting will continue even on some SACs. But after centuries of exploitation and a bruising battle for conservation, the future for at least some of our raised bogs is now bright.

Looking back it is fair to say that saving the bogs was never going to be easy. However, what started as a difficult task became nearly impossible. Despite all the expertise and information available there was a fundamental lack of leadership and willingness to engage from the offset. The whole nation paid dearly for the historic and chronic neglect of state heritage bodies, particularly the NPWS, which suffered massive, and disproportionate, budget cuts after the recession. The debacle has fed the belief that environmental protection is a threat to our traditions and the rural way of life. It has done immense damage to conservation interests and led to distrust of European institutions in particular. It highlighted the way in which national politicians sign up to EU law but then fail to implement it in a reasonable and rational manner. Instead, a last-minute approach which uses the EU as a crowbar to force through unpopular measures is preferred. It is a tactic not unique to the bogs issue, but can we hope that the right lessons have been learned in the right places? Incredibly, the answer may still be 'no'.

> 'Some people see only its bronze monotony and imagine it full of sinister morasses. But its colours and textures are like those of rich fabrics: velvety mosses of black and bottle-green, cushions of rose and gold moquette, swags of filigree grey-green lace.'[41]

Much greater in extent than the raised bogs, over twice the area in fact, at 774,000 hectares, the blanket bogs swathe great expanses of the uplands and low-lying parts of the west of Ireland. Cutting turf in the west is a part of a long and valued tradition. It not only provided a source of local and affordable fuel, it was a day out with family or friends. Hard work, fresh air, flasks of tea and fresh cut sandwiches. The sound of skylarks and the smell of the earth, immortalised in the paintings of Paul Henry, the poetry of Seamus Heaney and the postcards of John Hinde. But who better than Michael Viney, naturalist, beloved journalist and urban refugee to describe a day on the bog:

> In every house that saves its own turf there is a specially buoyant and purposeful air about the first day at the bog. Even at this point of Ireland's lemming-flow to the cities, most people have some link with the ritual, if only in childhood memory of buttermilk picnics, sweat and men laughing; it is an Irish card of identity ...
> What strange material it is! Cutting down from one spit to the next, each progressively wetter and softer, I was carving

gingerbread, marzipan, chocolate blancmange ... this was a totally tactile experience. Olfactory, too: the smell of sweet, dark decay.[42]

To cut the turf Viney was using a *sleán* – a sharp-edged, two-sided spade which slices into the wet peat, hewing the sod from the earth. These sods were then stacked in little monuments to dry (referred to as 'footing the turf'). Viney reckoned that a sod had to be handled eight times, i.e. eight different steps, between the cut and flame. It is this tradition that was evoked during the height of the turf-cutting battles of 2012 and 2013 and one which struck a chord for a lot of people, particularly older generations with first-hand memories of a trip to the bog.

If it didn't mean so much to my generation it is because the tradition is all but dead. Since 1990 the use of turf for domestic heating has decreased by 67 per cent and the use of the *sleán* has all but died out. Those with traditional rights, called 'turbary', generally contract out the hard work to small teams with heavy machinery. Contractors typically use one of two types of technique – the so-called sausage machine which drives over the surface of the bog extruding the peat behind it, or large excavators that scoop the peat into hoppers, selling it on locally. The government's peatland strategy highlighted the lack of regulation involved with this activity, not only in terms of the potential environmental costs, but also the tax that is not being paid, the turf plots that are being exploited without the consent of the owners, illegal extraction from protected areas, safety issues arising from the use of unsuitable machinery on unstable soil and the impact to local watercourses. Turf-cutting has ceased to be a small-scale, artisanal activity for domestic use, and is now a major, unregulated commercial activity.

Blanket bogs largely managed to escape the ravages of industrial-scale peat exploitation due to their inaccessibility and relatively shallow depth of peat. Still, only about 28 per cent of these bogs are now considered suitable for conservation as much of it has already been drained and reclaimed for agriculture, or planted with monocultures of non-native conifers. The bits that remain are subjected to overgrazing by sheep, erosion, repeated and uncontrolled fires and, of course, 'mechanical removal of peat'.[43] It should be obvious that stripping away surface vegetation counts as habitat destruction, indeed it seemed obvious even to the politicians when they were first charged with protecting the bogs. Fifty such SACs have been designated for this purpose, mostly across the western seaboard, yet the furore over the raised bogs forced a rethink. Eventually, the publication of the peatlands strategy aimed to clear up matters. It was decided that, with some exceptions, turf-cutting on 'shallow, marginal areas of dry/degraded blanket bog could be compatible with the objectives of the Habitats Directive ... It is anticipated that turf-cutting will be able to continue within blanket bog SACs. However, more sensitive areas will need to be avoided.'[44]

This jars with the finding of an assessment from the NPWS which found that the status of blanket bogs nationally was 'bad' and that, when listing the main threats, mechanical removal of peat was of 'high importance' and even hand cutting was of 'medium importance'.[45] Meanwhile, the Irish Peatland Conservation Council says that 'traditional cutting of the bogs for turbary is having a serious impact on blanket bogs'.[46]

Aiming to align these opposing forces, action A17 of the peatlands strategy states that: 'Ireland will devise and implement a system of management that will ensure that turf-cutting on blanket bog SACs continues in such a way that will not threaten the integrity of the sites. Management plans will be drawn up in consultation with local communities to ensure that these important peatlands are managed in compliance with EU law and to ensure sustainable use of the peat resource for the benefit of the community. Work will commence on preparing such plans in the coming months.'[47]

That was in 2013. That year I was in Glenveagh National Park and saw evidence of sausage machines on blanket bog inside the park boundary (see chapter 3). I returned the following year and found similar activity in a different part of the park. When I most recently checked the NPWS website I saw no mention of blanket bog management plans or the consultation process that is supposed to allow for conservation and turf-cutting to become miraculously compatible with one another. It seems that the long finger of prevarication is at work again.

Turf-cutting, however, is only one of the pressures on blanket bogs. For years these areas were grazed by cattle but since the 1980s these have been replaced with sheep. This is significant because sheep are much less fussy eaters than cattle – in essence, they will go anywhere and eat anything. They can be seen on their knees, their muzzles pressed into the soil, nibbling vegetation down to its roots. This is not a problem confined to blanket bogs but is widespread across commonage areas on mountains and coastal districts. In early 2015, on a visit to Erris Head in County Mayo I could see where coastal grassland was as smooth as a snooker table. Indeed here, where the soils are shallow, the ground was literally being washed into the sea, leaving only bare rock in its wake. On deeper peat soils overgrazing is evident where there is bare soil. The sheep droppings promote the growth of slimy algae, making the surface slippery to walk on. Cattle, in contrast, are pickier in what they eat and, having bigger mouths with fleshy lips, do not graze so close to the ground. A study being carried out by researchers in the Institute of Technology in Tralee, County Kerry, is re-pioneering the grazing of cattle on uplands. By radio-tracking the cattle, and matching their movements with detailed habitat maps, they have found that cattle avoid blanket bog altogether – instead keeping to drier areas of heath. Blanket bogs are unable to withstand much grazing at all, which is why sheep can cause so much damage even if there are quite few of them. Switching back to cattle in uplands, therefore, holds promise in providing sustainable farming livelihoods and conservation. Although cattle are associated with higher emissions of greenhouse gases, this would be more than offset by protection of the bog.

The other great threat to upland peat soils is repeated and uncontrolled fires. From late March, through April and early May, even as late as June sometimes, is fire season. Under the Wildlife Act it is illegal to burn vegetation between the months of March and August to allow for the nesting of birds and other animals. However, this frequently doesn't suit landowners as in February the scrub is still too wet to burn off. In lowland areas farmers burn scrub as a way to clear gorse and make room for grazing animals. In upland areas, however, the aim is to burn away the heather and promote the growth of grass which is better for sheep grazing. It is hard to overstate the damage that this kind of fire causes.

In Britain fire is used regularly as a tool for managing moors for grouse shooting. In this regime small patches are burned in rotation approximately every ten years, outside the bird-nesting season. The low duration and intensity of the flames do not kill everything, indeed they encourage new growth of heather which the grouse thrive on. This intensive management is contentious all the same. Management of habitats for a single species usually means the ecosystem as a whole loses out. In Britain it is also associated with persecution of predators, including birds of prey and even hares. Studying the landscape-level effects of this 'prescribed burning' a team from the University of Leeds made a number of startling findings. These included that burning changed the chemistry of the peat soil so that fewer nutrients were available to plants; rivers downstream became more acidic with higher concentrations of potentially toxic elements such as aluminium, manganese and iron; the water table dropped significantly, drying out the upper surface of the peat and resulting in carbon emissions and alterations to vegetation.[48] It is possible to extrapolate that the effects of repeated, uncontrolled burning of the Irish uplands during the animal-breeding season is catastrophic and is contributing to massive biodiversity loss, erosion, carbon emissions and pollution of entire river catchments. The year 2015 was one of the worst for wildfires with large parts of Wicklow, Kerry, west Cork, Waterford and Mayo going up in flames. Fires engulfed the national parks in the Wicklow Mountains and Killarney. These infernos can be intense. One NPWS field staff member told me of an instance in Wicklow where the peat itself had caught flame, scorching through the earth.

The issues of turf-cutting, sheep grazing and burning of peatlands go to the heart of how we, as a society, value our landscapes. Despite the easy talk from government ministers about sustainability, there is one example where the simple calculus of economics shows where the real priority lies.

> 'The potential economic, environmental and social benefits and costs of peatland uses will be considered in policy and decision making, along with full consideration of the interests of the wider community.' General Principle P2 of the National Peatlands Strategy, which is to apply to 'all public bodies'.[49]

This is the tale of a bird of prey, a marginalised and declining rural community, and how simple economic realities are putting government promises to the test.

The hen harrier is a mid-sized bird of prey whose range spans most of the northern hemisphere. The female is dull and brown but the male shines like a knight in silvery white livery with black trim. It is a spectral presence on the treeless uplands: silent, swift and ever watchful. During courtship, food tokens are passed from male to female in mid-air, earning it the nickname 'sky dancer'. Open, heathery bogs and moors are where you will find the hen harrier sweeping over hillsides in search of prey. Its fortunes in Ireland have waxed and waned over the years. Before the turn of the twentieth century, it was considered to be 'resident, but decreasing' although breeding in up to sixteen counties. Kerry was a 'stronghold', where it bred on the mountains on both sides of Dingle Bay as well as

An icon of the hills, hen harrier is in decline even in areas set aside for its protection. COURTESY MARIO MCRORY

near Killarney. In Limerick and Tipperary it was 'not uncommon'. Connemara and the Slieve Bloom mountains of Laois/Offaly were among its 'chief homes'. In the mid-1800s it had been considered 'common' on Donegal's Inishowen Peninsula and known to be breeding in Wicklow but, fifty years later, was rare in both places.[50] Like all birds of prey at the time, the hen harrier was considered vermin, being 'most destructive to grouse and partridges', as detailed in a chapter entitled 'Vermin and the Sportsman' in J. M. Seigne's *A Bird Watcher's Notebook* from 1930. At this time, according to Seigne: 'owing to merciless persecution, hen-harriers are very rare in Ireland. Their habit of methodically beating to and fro across a moor at no great height and dropping every now and then in search of rabbits, rats, moles and mice, which are their principal food, makes them an easy mark for the keeper's gun. Only in wild districts, like parts of Kerry, do these birds still survive.'[51]

In fact, it was believed that it had become extinct entirely as a breeding bird around this time, with the last record from Kerry in 1908 and the Knockmealdown Mountains in 1913.[52] This is not unusual insofar as a number of birds of prey were lost as this time, including the eagle and buzzard. The tumult of the independence struggle led to a diminution of the landed class which, in turn, meant fewer estates with attendant gamekeepers. Kennedy records that in 1954, after a lapse of forty-one years, hen harriers once again began to breed and subsequent years saw a steady recolonisation of many of its former haunts, a testament to its resilience and ability to colonise new areas.

The earlier period of independence, followed by the meagre Second World War years left an Irish government focused on self-sufficiency. Just as Bord na Móna was established to put the midlands bogs to use in providing a reliable source of home-produced energy, so, too, attention was drawn to any land not in 'productive' use. The new state inherited a virtually treeless land. At the turn of the century forests covered only 1.5 per cent of the land area and early attempts to increase this fell foul of financing or wartime demand for timber. Only with the Forestry Act of 1946 did the state embark on a truly ambitious programme to increase substantially the level of tree cover in Ireland. The government's afforestation plan of 1948 envisaged the creation of 400,000 hectares of new woodlands and inevitably this could not come at a cost to food production. As a result the blanket bogs and heaths of the uplands became the prime target for tree planting.

While the government never quite met its own targets, by the late 1980s over 300,000 hectares were under trees.[53] It has always been a bone of contention among wildlife lovers that this wonderful opportunity to enhance habitats and landscapes was not fulfilled. The trees chosen were fast-growing softwood varieties, particularly the Sitka spruce and lodgepole pine, species native to the western

Clear-felled plantations near Connemara National Park, an area of high scenic and recreational value.

coast of North America. Planting non-native trees as closely packed even-aged stands does not create a forest. A real forest is a mesh of interactions between myriad species in a functioning ecosystem; these new features on the Irish landscapes were merely another type of crop. What made matters worse was the manner in which the trees were harvested – clear-felling everything in one go – which has led to pollution of rivers through acidification, sedimentation and other run-off. These practices continue to this day, although, in fairness, there is more sensitivity in selecting suitable sites and assessing the impacts of clear-felling. A handful of these areas, now managed by Coillte, are popular amenity areas with considerable wildlife value. While new planting has ceased on peatland soils, already over 200,000 hectares have been planted on blanket bogs, including in some of our most scenic areas such as Connemara, the slopes of Ben Bulben in Sligo and around the lakes of Killarney.

Afforestation is cited as one of the principal reasons for the loss of blanket bog habitat in Ireland, with up to 27 per cent of the original area now under plantations.[54]

The recolonisation of Ireland by hen harriers coincided with the new plantations. The young trees provided good cover for their nests, which are nearly always on the ground. Safe from predators, they also had access to wide areas of open ground for foraging, particularly for meadow pipits which are among their favoured prey. In Ruttledge's review of birds he states that by 1964 there were believed to be 'at least' thirty-five pairs that showed a 'marked preference' for afforested areas.[55] During the 1970s breeding pairs were back in Connemara, Wicklow and even Dublin but disturbance, including shooting, kept their numbers low.[56] In 1981 Gordon D'Arcy wrote that they had decreased again and 'only a few pairs now breed in the southern part of the country'.[57] By the turn of the century, there were estimated to be a total population of between 102 and 129 pairs, increasing to at least 132 pairs by the time of the second national census in 2005.[58] This study showed that by far the most popular locations for nesting were in the early stages of plantation forestry (approximately 60 per cent of all nests). However, the news was not all good. There were no hen harriers in locations known previously to be strongholds, such as north Donegal, Connemara or the Kerry peninsulas. The Wicklow uplands, despite being the largest area of suitable landscape, was devoid of the birds. Nevertheless, the trends nationally were on the rise.

So are conifer plantations good for hen harriers? The early stages of plantations, when the sapling trees are short, are similar to the habitats traditionally preferred by hen harriers – open and scrubby, with cover for nesting but good access to hunting grounds. These are very popular areas for nesting activity. However, as the trees grow access to the ground level diminishes, and with that their suitability for the birds. Once the canopy closes over, hen harriers (and most other wildlife for that matter) are excluded and there is no more nesting.

A repeat census in 2010 showed the population holding steady and again the majority of nesting was recorded from the early stages of plantations. However, the headline data masked a rather unexpected twist. Under the EU Birds directive, six Special Protection Areas (SPAs) had been designated for their conservation. In these areas there was found to be a dramatic 18 per cent *decline*. Why would it be that conservation efforts were actually resulting in the exact opposite effect? It is fair to say that the designation of the SPAs, particularly in the south-west of the country, was not popular locally. In 2003, when the proposal was first made, a dead hen harrier was posted to the office of *The Kerryman* along with a cutting from that newspaper referring to the conservation plans. A meeting of the IFA that year, which was addressed by its then president John Dillon, was reported by the *Limerick Leader* under the headline 'Shoot the Bastards'. Although Mr Dillon later denied using this phrase he was quoted in *The Irish Times* as telling the meeting 'when this fellow appeared in the sky, the first thing we got was the gun to sort him out'.[59] The local anger was reflected in other headlines at the time including from the *Irish Farmers Journal* in August 2003: 'Farmers – a "species under serious threat"'. The cause of farmers' ire was the implication of the SPA designation. Ultimately plantation forestry is not compatible with hen harrier conservation – farmers with land inside the SPA boundary would therefore not be eligible for the income, including grants and state payments, that comes from forestry schemes. Why was anger directed at the hen harrier (surely an innocent and oblivious party in the debate), rather than the (man-made) policies that were causing economic hardship for farmers?

Formal designation of the SPAs went ahead in 2007. That simple arithmetic was at the heart of this problem is evident to this day. Under the state's 2014–2020 Forestry Programme any landowner can earn up to €440 per hectare per year for planting fast-growing monocultures of Sitka spruce or lodgepole pine with a €3,500 grant up front. Added to this, forestry income is eligible for tax-free status under the budget of October 2015. Indeed a survey published in the *Irish Farmers Journal* told its readers that choosing forestry over traditional sheep farming can earn farmers 57 per cent more income.[60]

For farmers inside SPAs these grants are not available because of the likely impact on hen harrier populations. Under the Green Low-Carbon Agri-Environment Scheme (GLAS), a top-up subsidy paid to farmers for particular environmental measures, these farmers can apply for grants of up to €370 per hectare for up to five years. In other words, the state recognises the role of traditional farmers in maintaining important conservation areas by paying them substantially less with significantly lower certainty of future income. The headlines say it all:

> 500,000 acres devalued by hen harrier diktat – Farmers say €1bn wiped off land values by tough EU laws to protect endangered raptor[61]
>
> Compensation or remove the designation – clear message from farmers at IFA's hen harrier meeting[62]
>
> Hen harrier SPA crippling farmers across east Kerry[63]
>
> Hen harriers making farms worthless[64]
>
> Protection of hen harriers costs farmers €22 million per year[65]

Considering birds don't make any decisions about how land is valued it can be startling how people can ascribe to them such sweeping powers. No headlines I am aware of were written along the lines of 'government policy making farms worthless' or '500,000 acres devalued by Department of Agriculture diktat' – even though these would have been far more accurate.

Indeed, the failure of the state to appreciate fully the social and environmental costs of its policies was highlighted by the heritage body An Taisce. In February 2015 they challenged the Department of Agriculture, Food and the Marine to explain why €400 million of funds allocated under the Common Agriculture Policy were diverted away from supporting marginal farmers and wildlife. Originally, a total of €528 million had been set aside for farmers in protected areas. These farmers, like the ones in hen harrier SPAs, usually have to forgo potential income from certain activities, like forestry, and it is right and proper that they be compensated for this. However, An Taisce accused the Department of spending only €93 million for this purpose and diverting the remainder 'elsewhere', much to the dismay of the NPWS.[66] A subsequent Oireachtas report summed it up: 'There is no question [but] that EU Funding can be drawn down for conservation purposes: the Habitats Directive, Article 8.4 states that EU co-financing measures are available for Hen Harrier SPAs. In fact, Ireland has already benefitted from this provision. However, most of the funds were not used for their intended purpose.'

A resolution to this ongoing issue is making some progress with the announcement in 2016 that a dedicated 'locally led' scheme was to be established to manage farming better in hen harrier SPAs, and we will have to wait and see whether this can be successful. But as long as conservation is seen as depriving people of their livelihoods, the future for these magnificent birds of prey will be bleak.

But there is another issue. Why would hen harriers shun their traditional haunts of open moorland in favour of these new man-made habitats? Why are they absent from great swathes of the country where they were once well known?

A positive side effect of the controversies is that quite a lot is now known about hen harriers and their ecology. Like all animals and plants, they require a specific set of resources to allow them to feed, find a mate, breed and disperse. In this way they are connected to the landscape and all the other organisms that inhabit it. Duhallow is a hilly area that straddles the border of east Kerry and north Cork. Aerial photos of the region show the mix of greens and browns that reflect the range of activities on the ground from intensive dairy farming in the lowlands to forestry and open peatlands on the higher ground. It is the source of the River Blackwater, famous for its run of salmon as it runs through the bucolic countryside of east Cork and Waterford. In so many ways the area is a microcosm of rural Ireland and could be anywhere on this island.

Since the revival of the hen harrier in the 1960s and 1970s Duhallow has been a stronghold for the bird and has been extensively studied by ornithologists. It was among the six areas designated for the protection of hen harriers because of its importance nationally. Between 1998 and 2011 breeding was confirmed in thirty-five separate territories, that is, breeding pairs successfully rearing young. An extensive survey was carried out in 2012 that encompassed 430km^2 and employed 875 observer hours across 113 locations (referred to as 'vantage points').[67] The surveyors logged everything the hen harriers were doing including where they were hunting, what they were catching and the success of their nests. The data would not only provide information on what was happening but would tell us why it was happening, and the results were startling. Of the thirty-five past territories surveyed only seven had confirmed breeding attempts, representing a massive 80 per cent fall-off. Worse, of the seven breeding attempts only five pairs of birds went on to nest and rear any young and only four of these were successful in fledging the nestlings. In other words, only four nests out of the seven managed to rear any young to the point where the chicks were ably to fly off and find mates of their own. One pair of birds nested and laid a clutch of five eggs, but this was mysteriously abandoned. For anybody rooting for the hen harrier the results were miserable, but what could explain this virtual collapse in population? As the author of the study noted: 'such population crashes must in some way be related to the landscape and its ability to support breeding hen harriers.'

A key finding of the research found that in 2010 hunting birds managed to bring a new item of food back to the nest every hour and half, but in 2012 this had lengthened to every 2.7 hours. It was getting harder for them to find prey, typically meadow pipits or small mammals. Not having enough food would explain why nests were abandoned or why six pairs in the survey area which could have nested didn't. All animals have to make these choices as not only do they have to feed ravenous chicks but they also have to feed themselves during these long and active summer days. Even of the chicks that hatched, the author noted that their health was not what would be expected, noting lower weight, poor bone structure and delayed feather growth. This only leads to another question:

where had all their food gone? Where were the meadow pipits and small mammals that make up the majority of the hen harriers' typical prey? This question in turn points to the overall health of the landscape and the answer could be summed up in one sentence from the research: 'On two occasions, individual male harriers were followed visually from their nest to hunt over 10km away, bypassing mature forestry and burnt moorland.'

Mature conifer plantations and scorched earth simply have nothing to offer. The longer a parent needs to travel to find prey the more energy they burn up themselves, the fewer hunting trips they can make, and the longer the nest is left unattended. This combination of repeated burning of bog and the maturation of plantations first planted in the 1990s have had a sterilising effect on the uplands, not only in Duhallow but across Ireland. The plantations have also made it easier for predators, particularly crows, foxes and pine martens, to find ground nests. For crows and magpies in particular the tall trees are excellent lookout posts where they detect nests at their leisure by watching the movement of the parent birds arriving with food. The final contributing pressure is the likelihood of deliberate persecution. As the study's author points out, 'a sustained anti-hen harrier sentiment pervaded much of the period between 2003 and 2007 in the lead-up to the designation of Special Protection Areas. Duhallow was one of the regions where SPA designation was most vehemently contested.'

By its very nature it is difficult to prove persecution but there can be no doubt it is happening. A report from the NPWS details thirty-five confirmed cases of illegal shooting or poisoning of birds of prey from across Ireland in 2015, the highest number since this monitoring programme began.[68]

The most recent census of hen harriers, published in 2016, showed no signs of a reversal in fortune for these wonderful birds. Their total numbers continue to fall, while those in protected areas are faring worse than those elsewhere. In the Duhallow area, in the Mullaghanish to Musheramore Mountains, where there was an 80 per cent loss in population, it is now 'at risk of extirpation'. At another, Slieve Beagh in County Monaghan, turf-extraction and recent burning means the habitat is so damaged that researchers concluded it is 'currently not suitable for nesting hen harrier'.[69]

Ireland's uplands are in a bad way and this has impacted both people and wildlife. Traditional sheep farming in the uplands is entirely dependant upon taxpayer subsidies.[70] Plantations of conifers have provided income but at great environmental cost while obliterating traditional livelihoods. Farmers in these areas rightly consider themselves an endangered species. Meanwhile, nearly all the birds which are distinctive of our uplands, including breeding golden plover, dunlin, curlew, snipe, merlin, hen harrier, short-eared owl, golden eagle, twite, meadow pipit, nightjar, red grouse and ring ouzel are listed as of either high or medium conservation concern in Ireland.[71] Nightjar has not been recorded in years and may already be extinct while ring ouzel, golden eagle and twite are hanging on in perilously low numbers. The fortunes of both people and wildlife in these areas are intertwined. Representative bodies, like the IFA, have chosen the wrong target for their ire. The state has abjectly failed to come up with policies that can provide a future for people while acknowledging the importance of uplands for recreation, carbon storage, flood control and, yes, wildlife. The fluctuating fortunes of the hen harrier have mirrored the changes in these marginal rural areas and in its current phase of decline it is trying to tell us something. So far it seems, few are listening.

7

The myth of Ireland's 'green' farming

> 'It appears to farmers that the Department of Agriculture is more concerned about penalising actions not done than rewarding provision of public goods/ecosystem services. Budgets should focus on non-food products more.'
>
> A County Galway hill farmer (2014)[1]

A FIELD IN COUNTY MEATH in late spring is not a bad place to be, but this particular field made an exception. To the casual eye, perhaps driving by in a car, it was nothing unusual – green, with hedges and occasional trees – much like every other field in Ireland. But up close it looked ill. The ground was difficult to walk on, rutted and pitted by the churning of cattle hooves; my ankle twisted at every step and my wellies sank unpredictably in the heavy mud. It was a largish field by Irish standards but it felt larger because the view stretched far beyond its boundaries into the neighbouring fields and further.

Like most Irish fields, this one had a hedgerow border, features of such intrinsic value to our landscape that it feels like they are entirely natural rather than man-made. These hedges, though, were in bad shape, only about a metre and half high; they had been cut down and back so sharply they were little more than a bundle of sticks at the top. Large spaces between the skinny trunks of the hawthorns, and they were all hawthorns, meant you could see right through to the field next door without looking over the top. We value our hedgerows for the enormous importance they hold for our wildlife, maybe not the rare things that warrant action plans and EU Directives, but all the common stuff that gives us everyday contact with nature, like primroses and robins. But these hedges were scrawny with little life. An hour in the field looking for birds yielded only a woodpigeon and, later, a jackdaw. A tall ash tree in one corner was the last sentinel of its kind since these field boundaries were originally laid down, perhaps in the 1700s (although some hedgerows which mark townland boundaries may be many centuries older still). On the far side of the hedge a drain had been dug to hasten the water's flow off the land. It didn't seem to be helping much as mucky ruts of stagnant water were everywhere held back by the heavy, clingy soil. The ditch's stony bottom was coated in green algae, a sure sign of the nutrients, most likely from animal manure, that were washing off the land. The water itself was turbid with rain-washed sediment, there being nothing between the field and the ditch to hold it back. It was a grim and dispiriting sight despite the sunshine and the warming breeze.

This field was earmarked for houses and I couldn't help but think how in ten years' time there would be gardens maturing, and perhaps a park, which would act as refuges for small mammals like hedgehogs and maybe even a badger family. People would feed the birds and the ash tree would be cherished. Bats might roost in the cracks of its boughs, or even set up homes in the roof cavities of the nearby homes. It would be likely that the developer would include a small pond for collecting rainwater, providing a new habitat for frogs and dragonflies, while also keeping the water, which would ultimately enter the River Boyne not far downstream, clean and free of pollution.

Thankfully, most Irish farms don't look like this, but that is changing. Increasingly, the agricultural land of Ireland is coming to resemble a duoculture: one comprised simply of grass and cattle. With no practical need for the ancient hedges to be maintained, they are slowing dying away. A post-and-wire fence will do just as well. With farming subsidies, upon which most farmers rely for a living, geared towards producing the maximum quantity of food regardless of what the land is capable of, small-scale diversity turns to large-scale homogeneity. Irish food is sold as 'green', with our system of small fields and green grass having changed little through the centuries. In reality, however, farming is now divided between those who are connected to the global market for food, principally in the dairy sector, and those who cannot compete at this level, particularly those rearing beef and sheep. At either end the system is putting strain on our wildlife, landscape and wider environment.

In 2012 Ireland was in the bowels of the worst economic recession in a generation. The taxpayer had just given the banks €65 billion and unemployment was hitting 15 per cent. Time to go back to basics and promote the things we're really good at – like farming. But how to get the message out to the world that Ireland was open for business? 'Origin Green', conceived in the boardrooms of government offices somewhere between Kildare Street and the Grand Canal in Dublin, made its debut on YouTube and featured the actress Saoirse Ronan in a variety of bucolic settings. 'We did not inherit this world from our parents, we borrowed it from our children. One day we will return it to them ... when we do ... it should be every bit as bountiful as it was when we found it. That's what sustainability means.'

Sustainability, we're told, gives Ireland 'a momentous opportunity, maybe the defining opportunity of our time, because the rewards it can bring us all are [dramatic pause] breathtaking.'

But why us? Why Ireland?

> Think about it: our climate has always been this mild, our landscape this lush, our fields have always been this green ... and windswept ... and rain-washed. Our seas have always teemed with fish ... So, many of the things we need for sustainability are already in place.[2]

For anyone paying attention to environmental issues in Ireland over the past forty years, this message seemed absurd. Fact and fiction were blending into one. At the same time government ministers were publishing the Food Harvest 2020 plan – 'a vision for Irish agri-food and fisheries'. The ambitious vision predicted an increase in primary output by €1.5 billion, value-added outputs to increase by

€3 billion and to increase exports to €12 billion (a 42 per cent jump over 2007–09 period). That the colour 'green' would be central to the achievements of these targets is highlighted on page 1 of the Executive Summary – even coming ahead of 'growth'.

> Green. Capitalising on Ireland's association with the colour 'green' is pivotal to developing the marketing opportunity for Irish agri-food. This will build on our historic association with the colour and highlight the environmental credentials associated with our extensive, low-input, grass-based production systems ...
>
> ... consumers in key markets will learn to recognise implicitly that, by buying Irish, they are choosing to value and respect the natural environment.[3]

In the full 58-page report the word 'smart' appears 26 times, 'green' 37 times, and 'sustainable' 38 times. The word 'growth' appears 86 times.

I was involved in the preparation of a submission to the formation of the Food Harvest 2020 plan on behalf of the IWT. In it we emphasised our support for growth and jobs in the rural agri-economy. We expressed our backing for the 'Origin Green' concept and the goal of making Ireland a world leader in sustainable food production. However, we failed to see any actions in the plan which would back up the bold statements about environmental protection. Where was all the extra manure going to go? The EPA already told us that half of our waters were polluted and half of this was from agriculture. How was the plan going to reverse the precipitous decline in farmland birds like the corncrake, curlew or yellowhammer? The corn bunting had already become extinct because of changes in agricultural practices and according to BirdWatch Ireland 'as a group, farmland birds have experienced some of the largest population declines and range contractions of any bird species'.[4]

Where was the plan to reverse the dilapidation of the hedgerow network, or the ongoing degradation of uplands? What about the increasing greenhouse gas emissions from a national herd of cattle bigger than the human population? The EPA had warned that these amounted to over a third of our national emissions and we were not on track to meet legally binding targets agreed with the European Union. The Food Harvest 2020 plan contained no answers to these questions. Nevertheless the idea has taken on a life of its own. We even have a new term to ease its passage: 'sustainable intensification'.

In 2015 even the World Wide Fund for Nature (still known as WWF with its world-famous panda logo) got on side. In their spring bulletin, which goes to subscribers around the world, then Minister for Agriculture, Food and the Marine Simon Coveney said, in a piece entitled 'Ireland and the Future of Sustainability: A new model of forward-thinking agriculture': 'yes, it's an environmental crusade – but with a real commercial edge. Happy cows and green fields make a good image. But to back that up now we have the science to show the long-term benefits behind this traditional way of farming.'[5]

The science to which Minister Coveney referred is somewhat elusive. It is likely he is thinking about how the permanent pasture model of animal rearing produces relatively lower levels of greenhouse gases when compared with other models of corn-fed or shed-reared animals from other parts of the world. Fair enough, but it hardly amounts to an 'environmental crusade', particularly as it doesn't involve actually doing anything, and does little to address the serious environmental problems that we find ourselves dealing with.

Farming is deeply rooted in Irish history, folklore and society. Farming began to be developed here around 4,200 BC and gradually displaced hunting/gathering as a way of life. It expanded to such an extent that there is hardly a square kilometre of dry land that has not come under a spade or a hoof at some point. Tiny offshore islands out in the Atlantic, islets in larger lakes and even the tops of mountains were no barriers to the growth of farming. When the availability of dry land became an issue, wetlands, bogs, swamps and floodplains were drained and ploughed. The landscape that we know today, and indeed the plants and animals that live in it, are entirely a product of this farming activity. Today we mourn the loss of the corn bunting and scramble to save the corncrake and barn owl, but these birds would not have been in the densely forested land of the first hunter-gatherers. Rather, they originate in the open grasslands of Eastern Europe and moved west with the felling of the trees. The traditional low-intensity grazing of animals on permanent grassland has given us the botanical wonder that is the Burren, replete with dazzling orchids and hum of bees. The hedgerows that are still such essential refuges for nature in the Irish landscape were planted by people while the drainage ditches that accompany many of them can be home to frogs or fish. It is not that farming is necessarily good for nature as many of the changes that have occurred over time have had negative consequences for wildlife or water quality. But certainly it is not necessarily bad either. With 4.5 million hectares, or 65 per cent of the total land area of the country, used for agriculture, if we want to live in a healthy environment, farming must be central to our plans.

Farmers themselves occupy a special place in our society. For the centuries when food was actually a valued commodity, and not the cheap throwaway stuff that passes for food in most supermarkets these days, the farmers were the life-givers. Places, like France, that treasure their food also treasure their farmers. Food, after all, is not just sustenance but one of the joys of life.

In Ireland, like everywhere else in the world, agriculture has changed beyond all recognition since the end of the Second World War. A child from the 1950s would scarcely recognise the food we eat, the places it comes from, the way it is produced, or the manner in which we throw colossal volumes of the stuff away. Yet in our imagination the job of the farmer among our increasingly urbanised population has remained static. The perception of modern farming as having changed little, and maintaining traditional ways, has allowed the opportunity for marketing gurus to sell more produce to consumers who are increasingly sensitive to social and environmental pressures.

Ultimately, it is self-defeating. Eventually marketing juggernauts collide with the wall of reality. It's not that 'Origin Green' does nothing. Its Sustainability Report for 2015 is full of promising figures: over 474 food and drink manufacturers registered with the scheme representing 95 per cent of our exports; 128 sustainability plans submitted; 800 farms being carbon footprinted each week, and so on. But here's another set of figures: in 2014 An Bord Bia, the state agency responsible for the 'Origin Green' programme as well as marketing Irish food in general, received €27 million in taxpayers' money; the NPWS, responsible for the conservation of our natural heritage (much of which is on farmland), received €9.5 million.[6] Is this a case of promoting the message more than the facts?

Politicians and industry lobbyists have come to use the word 'sustainable' as though it were a punctuation mark. It is thrown about with such abandon that it has lost all significance and it is clear that the majority of those who use the word have little concept of its meaning. As *Irish Times* journalist Paddy Woodworth put it, in relation to the Food Harvest plans: 'If you substitute "super-profitable" for "sustainable" ... it makes a lot more sense.'[7]

This is unfortunate. Agriculture is something that can be practised indefinitely, providing food, jobs and economic opportunity without creating pollution, driving species to extinction or otherwise degrading the environment. This is what 'sustainable' used to mean, at least.

For most of farming history (i.e. until about fifty years ago) virtually all food production was local. Technology was powered only by the heft of people and their animals. Solutions to problems were found in nature because there was nowhere else to look. If the soil was exhausted it needed rest. If nutrients were low there was seaweed or animal manure but nothing more. If animals ate more than could grow from the land there was nothing for it but to reduce the number of animals; the alternative was starvation. From the mid-nineteenth century onwards, however, two things emerged that would start to tip the balance in the farmer's favour: technology and subsidies. Both have increased exponentially in this time. New technologies and/or the intervention of the state to override market forces have the power to do either good or ill. They would allow agriculture to increasingly step back from environmental boundaries and in turn provide greater security both for the farmers' livelihood and the supply of food for the consumer. At a global level they allowed the world population to increase from around 1 billion in 1850 to over 7 billion today. In Ireland the middle of the nineteenth century witnessed a population collapse from a high of 8 million people before the Great Famine to nearly 6.5 million today. It was during these desperate years of mass starvation that the concerted use of public money to expand agriculture began.

One of the biggest problems faced by small landholders was lack of suitable land arising from what many consider the bane of the Irish condition: rain. Famously shaped like a saucer with a flat middle and mountains around the edges, in Ireland rain mostly has nowhere to go. Along the Shannon and Erne Rivers, in particular, their courses are meandering and languorous. Large lakes slow the water further, spilling aqueous sheets across surrounding lands during times of flood. Heavy soils add to the problem, leaving fields waterlogged for long periods. In the western counties it was even worse as the vast quantities of precipitation promoted the growth of bogs, leaving the good land smothered and inaccessible. The solution to the 'drainage problem' has been engineering works to expedite the flow of water off the land and into the sea. Shallow, winding and slow-flowing river channels can be deepened, widened and straightened into fast-flowing ones. While individual landowners addressed the problem by digging their own field drains, draining whole river catchments was a task of greater magnitude. Although works had been carried out to improve navigation during the 1700s it was only during the 1840s that land reclamation really got going.

The Arterial Drainage Act was passed in 1842 while the Landed Property Improvement Act of 1847 made £3.5 million available in loans. The scale of these works, which were all carried out by hand using cheap and abundant labour, was colossal. According to John Feehan, in his book on the history of Irish farming, during the famine years the Board of Works employed over 700,000 workers in addition to 12,000 staff.[8] It is hard to image today the extent of floodplains and wetlands in Ireland before this work got under way. Not only was one fifth of the country covered with impassable bogs but lowland areas had vast marshlands and grasslands that were under water for large parts of the year. Feehan describes the impact of the drainage works on the upper part of the River Nore in the 1800s:

> At the time the upper Nore drainage project was undertaken [in 1885], the river was regarded as one of the most neglected in the country. From its point of entry into Laois at Quaker's Bridge it was a shallow, meandering river greatly obstructed by shoals and narrows, with numerous islands of silt and clay; in places the bed was no more than 1.5m wide, and no more than 60cm below the level of the floodplain. Consequently every rise in the river brought water on to the fields, where it remained all winter.[9]

The arterial drainage works would excavate the river to a depth of 2m below the land and widen it to between 3m and 6m. The works were effective in stopping flooding on the lands themselves but highlighted one of the major drawbacks to any kind of drainage project: 'one unexpected result was that the increased volume of water in winter caused flooding in the lower reaches of the river, while the decreased summer volume resulted in an insufficiency of water in these same reaches, making the river a "stagnant ditch".'[10]

Between 1840 and 1940, 182,000 hectares, an area equivalent to the whole of County Waterford was drained in this way.[11] However, the falling population and the reliance on manual labour meant many of these schemes were not maintained. A river has a life all of its own and will relentlessly shape its course to its own requirements. In the post-war period, however, technology improved and with the use of dragline excavators and floating dredges the new state was eager to encourage expansion of agricultural production. A new Drainage Act in 1945 sought to tackle whole river catchments. Between 1945 and 1980, thirty-seven river systems were drained, covering a land area of nearly 2,500km^2. And with the work went the subsidies. A total of £238 million (equivalent to over €7 billion in today's money) was spent through the OPW, accounting for 1.5 per cent of public capital spending.[12] The European Economic Community (EEC), which Ireland had joined in 1973, was also offering grants. In the west of Ireland, £19 million (€194 million in 2016) had been allocated to drain nearly 22,000 hectares and £1 million had been allocated for initial survey work along the River Shannon, the drainage of which was 'likely to benefit from EEC aid'.[13] At this time questions were beginning to be asked about the environmental implications of draining such large areas. A report by the Inland Fisheries Commission in 1975 rang alarm bells: 'The impact on the fisheries of recurring drainage maintenance work gives cause for anxiety, and it is essential that this work also have particular regard for the fishery requirements. Otherwise it will be a continuing disruptive force, renewing the damage to spawning beds and preventing the natural recuperation of river channels from the initial impact of dredging.'[14]

The OPW, after all, was a semi-autonomous unit within the Department of Finance. It did not need planning permission to carry out these works and was under no obligation to consider the views of others.

In 1968 the Department of Fisheries undertook a survey along the Trimblestown River, a tributary of the River Boyne that flows between the towns of Trim and Athboy (for this reason it is also marked on Ordnance Survey maps as the Athboy River). In 1948 this stretch was described in *The Angler's Guide* as 'worth fishing for trout' up to 2lb in weight.[15] The 1968 survey was designed to evaluate the impact of a forthcoming arterial drainage project and would include before-and-after investigations over a number of years. Photos from the report show a riverbank lined with trees and dense vegetation, while the water itself was found to be rich in mosses, aquatic plants and invertebrates – crucial fish food – including stoneflies and the now endangered white-clawed crayfish. Drainage works would remove the vegetation along the riverbank as well as excavating bedrock boulders and soil from the riverbed.

A stretch of the River Boyne in County Meath. Drainage has drastically altered many Irish rivers, leaving communities vulnerable to flooding.

This was done during the spring and summer of 1972: works both lowered the riverbed by 1.5m and widened it from an average of 6.1m to 6.8m. As you would expect, the short-term damage to the river was extensive, releasing enormous quantities of silt into the water and uprooting plants, but it had the desired effect of increasing flow rates through the affected area. In 1973 the surveyors returned and found that many of the invertebrates had recolonised the area but neither crayfish nor stoneflies were to be found.[16] It also found that the fish population had changed from one dominated by trout and salmon, to one in which much smaller fish such as stone loach and minnow were dominant.[17] In the short term at least the impacts of the works were recognised as being severe. In 1989, seventeen years after the initial works were carried out, fisheries inspectors returned to this stretch of the Trimblestown River to see what, if any, long-term effects had occurred. Perhaps surprisingly, it found that the salmon and trout had made somewhat of a full recovery in the intervening years.[18] Fencing of the riverbanks after the works had allowed vegetation, particularly trees and shrubs, to recover and this is recognised as an essential step as it provides cover for fish and a source of invertebrate food. Meanwhile, an entirely separate study along the Bunree River in County Mayo, which looked at the effects of drainage over a thirty-year period, reached a similar conclusion.[19] However, there is enough evidence to suggest that these relatively narrowly scoped studies were not revealing the full story.

Part of the problem is that so little hard data on before-and-after impacts has been gathered for these schemes. In particular, the Trimblestown and Bunree studies looked at short stretches of river while the effects of drainage works on whole catchments has never been studied with such rigour. The white-clawed crayfish, a protected species under the EU's Habitats Directive, which appeared in abundance in the Trimblestown prior to drainage there in the early 1970s, has not reappeared. A survey for this small, lobster-like crustacean in 2000 found none along the Trimblestown, or indeed many of the other tributaries of the Boyne where it had once been present.[20] A report on the effects of drainage, funded by the WWF in 1984, suggested that the grading of riverbanks would affect otters by making them 'less able to find appropriate sites in which to make their holts'.[21] Otter has not been recorded along the Trimblestown River since 1980. The wider affects of drainage schemes, although under-studied, are significant. It changes the characters of rivers from winding and diverse, with fast-flowing shallows and deeper pools, to straight and even with more or less the same habitat running along its entire length. The water flow itself changes from varying slowly with rainfall, buffered by the surrounding floodplain, to changing rapidly and with extremes of flow – torrents in flood and low flows in summer (extreme low flows result in higher water temperatures and lower dissolved oxygen, which many organisms cannot tolerate). This means the effects are felt far downstream and not only along the stretches where the works themselves took place.

The round-fruited rush is a small inconspicuous plant and was first recorded along the banks of the River Boyne in 1968 by the renowned Irish botanist Donal Synnott. He described the location as 'winter-wet alluvial pasture' and suggested it may have been growing here for a long time as there was no evidence of previous botanical surveys along this stretch.[22] It may have been the first and the last time it was seen in this area. Botanists returning in 2006 to the very spot searched in vain, suggesting the 'fact that they were not located at any of their sites could suggest that the management or ecology of the wet meadows found along the Boyne River has changed in recent times'.[23]

In other catchments a similar loss of species has been recorded. An extensive EU-funded drainage scheme along the River Blackwater, which drains into Lough Neagh, degraded what was described as 'a jewel of a river': 'Kingfishers, sand martins, dippers and grey wagtails were common. Otters flourished and the floodplain was important for breeding waders, especially snipe and lapwing. In winter the floodwaters attracted whooper swans and Greenland white-fronted geese.'[24] Despite mitigation measures to reduce the impact on wildlife, to this day 'the water remains persistently turbid. The re-profiled banks are generally steep, unappealing, and dominated by rank vegetation, such as nettles and reed canary-grass.'

Along the River Bush in County Antrim the drainage works excavated a population of the now critically endangered freshwater pearl mussel.[25] In County Galway drainage works were implicated in the disappearance of entire flocks of Greenland white-fronted geese from north of Spiddal and Gleninagh.[26]

David Cabot, one of our most respected ecologists of the past fifty years, has this to say about agriculture and drainage: 'The most damaging impacts on sites and areas of nature conservation importance have come from agriculture. Arterial and other drainage schemes have physically destroyed countless wetlands, particularly turloughs [a seasonal lake practically unique to Ireland] many of which were important wildfowl and botanical habitats.'[27]

While the slim available research suggested that fish populations recovered rapidly, the broader picture is less optimistic. Anglers at the time seemed to see it coming:

> Most Irish fisheries are very natural in appearance. River and lake shores tend to be wild and sometimes overgrown, as cultivation seldom extends right up to the waterside ... An exception to these remarks is the case of rivers that have been subject of arterial drainage schemes. These devastate the river bed and banks and do a great deal of damage to the fishing. River systems that have been subject to drainage will be named in the following notes; visiting anglers are advised to avoid them ...
>
> All of the Boyne was first class trout water as well as receiving excellent runs of large spring and summer salmon. At the time of writing [1980], it is the subject of a drainage scheme which has badly affected the trout fishing and the general appearance of the river. Salmon fishing is hardly worthwhile.[28]

Salmon stocks have collapsed in recent years with the numbers returning to Irish rivers falling by 31.3 per cent between 1985 and 1997, and a further 37.5 per cent fall up to 2009. The reasons for this are both complex and diverse with the more recent falls being attributed to effects far out to sea. While they are in fresh water they are affected by pollution and physical barriers such as dams and weirs but according to a report prepared for the North Atlantic Salmon Conservation Organisation in 2005 there were twenty-three river systems in Ireland where drainage/channel modification was threatening salmon habitat, including the River Boyne.[29] In their assessment of salmon stocks for 2015, Inland Fisheries Ireland (IFI), using automated fish counters in the water, calculated that the numbers returning to the Boyne were only one third of their 'conservation limit' – a number that implies safe limits to allow fishing and not necessarily a figure representing a healthy salmon stock.[30]

The eel population, like the salmon, has also collapsed in recent decades with the fall commencing in the 1970s.[31] The Trimblestown River itself was once home to a number of fishing weirs during the 1700s with 'large black trout, exceeding that of the Boyne, and also very good eels'. During surveys of the river in 2009 IFI found two eels along a 45m stretch, in 2012 only one.[32] IFI, to their credit, are frequently called in to carry out fishery enhancement works on these drained catchments along with local angling clubs. This may include creating artificial pools or gravel spawning areas. In 2014 the Trim Athboy District Angling Association started the first phase of rehabilitation works along the Trimblestown. According to their website they have increased salmon spawning areas by 70 per cent and introduced obstacles such as boulders to diversify the habitat. The river is now part of a European 'Natura 2000' site, encompassing much of the River Boyne and its tributaries, something that nominally affords it strict protection, although meaning little on the ground. Few of the riverbanks were fenced off and so vegetation never recovered. Today, because of drainage works, the Boyne and its tributaries more resemble canals than flowing rivers, cutting straight lines across fields and with little but nettles along their banks. Nor have the works prevented flooding of towns along its route, with Navan, Trim, Athboy, Kells and Drogheda all having been hit on a number of occasions in recent decades.[33]

Compounding the long-terms effects of river drainage on aquatic ecosystems is the fact that the OPW is required by the Arterial Drainage Acts to maintain the schemes so that they retain their benefit in keeping agricultural land dry. Indeed as drainage works promote erosion and siltation of rivers, so the need for maintenance increases. While the need to 're-drain' a particular stretch of river depends upon the particular circumstances, according to the OPW about a fifth of watercourses are worked on in any given year – approximately 2,000km of river channel. A total of 11,500km of river channel is subject to drainage programmes and nearly all of this was subject to maintenance works in the five years from 2005–2010. While the planning of these schemes is vastly improved, with consultation and impact reports to accompany works, these are flawed as the baseline against which impacts are assessed will only ever be the drained and degraded riverbed.

And what of the cost? The public purse picks up the entire tab for all works and fees during the planning stage. The OPW's annual report for 2014 includes arterial drainage within 'flood risk management' and came at a cost of €15 million. The report states that because of this 'more than 20,000 properties are sited in the area benefiting from the original [arterial drainage] schemes. The incidence of flooding to these properties is notably low and the Programme ensures that the risk is kept to this current relatively low level.' This is odd considering the aim of arterial drainage is not to protect property but to enhance land for agricultural purposes.

In the early days of arterial drainage there was no cost-benefit analysis. The increase in available agricultural lands seemed to be benefit enough, regardless of the cost. Only in the early 1980s did the eye of public scrutiny fall on the elaborate schemes being approved at the time. A report produced by the ESRI asked a bald but pertinent question: 'If drainage pays, why not let the landowner get on with the job, incurring the costs and reaping the benefits?'

In the cold analysis of accounts the ESRI found that drainage increased land values (by making more land available for farming) but this fell steeply with time as the main problem areas were addressed. For schemes that started after 1960, when land values were averaged between 1950 and 1980, the increase was little more than half the construction costs.[34] Along the Boyne the study found that drainage had increased the land value by 47 per cent but that the total increase only met 85 per cent of the costs – which were borne entirely by the taxpayer.

Of course, any analysis of this sort contains assumptions and a degree of imprecision but one thing that did not appear on the 'costs' column was the effect on the environment. It is never easy to put monetary figures on wildlife or vegetation but in this case it may have included the cost of downstream flooding, loss of angling potential or other forgone amenity values. The ESRI highlighted this deficiency in 1982 but while cost-benefit analysis is carried out for current drainage schemes by the OPW these do not factor in environmental damage. Nor do they seem to evaluate alternative ways to spend public money which may address flooding issues but in more sensitive ways. Flooding and land drainage, after all, are still major problems. A third of lowland soils are classified as 'wet' and farmers with fields full of rushes are not eligible for the single farm payment under current rules, something which promotes further drainage, flooding and habitat loss.

The effort to produce the maximum amount of food from every corner of every field is causing untold damage across the Irish landscape, no more so than in marginal areas which would be better used for low-intensity grazing or no grazing at all. According to the EPA we need to learn to live with flooding, which is likely to increase in frequency and intensity as a result of climate change.[35] Along the River Shannon, which has resisted catchment-wide drainage because of the scale of the task and not because of lack of demand, calls to have it drained date to the early 1880s and are repeated annually. Acknowledging that there are uses for land which provide greater value than food production, such as for wildlife, amenity, flood control, etc., is one of the greatest challenges we have in shaping future agriculture policies.

'Land is the major asset of the Irish economy' starts an essay on agriculture written for the British Association for the Advancement of Science in 1957.[36] And so it is today. Food production is one of a number of ways in which the land benefits the economy, and traditionally was probably the greatest of all, but fuel production from the bogs, plantation forestry and, increasingly, tourism and amenity use also play their role. In 1861 pastureland accounted for 3.3 million hectares in Ireland, approximately half the total land area. Hay and crops including corn, at 1.82 million hectares, accounted for over a quarter, while the remainder is described as 'grazed and barren mountains, bogs, marshes, water, towns and roads'. In 1955 the relative proportions had hardly changed. Grazing pasture then totalled a notch above 3.23 million hectares. In 1861 the country had 2.8 million cattle, 3.35 million sheep and nearly 1 million pigs. A hundred years later these figures were 4.4 million, 3.2 million and 800,000 respectively. The jump in cattle numbers largely arose from demand in Britain for beef during the post-war period but there was little change elsewhere in the dairy herd or other livestock. It is safe to say we have seen significant changes since then.

According to the Central Statistics Office, in June 2014 the national cattle herd stood at 6,926,100 animals, the sheep population at 5,096,800 while we now have 1.5 million pigs. The area of tilled land (mostly for wheat, oats or barley but also potatoes and other crops) amounted to 328,000 hectares in 2014, with 2.3 million hectares given over to grazing for animals. Hay, of which there was 764,000 hectares in 1955, barely registered in 2014 (218 hectares) – largely having been replaced with silage production. Silage is the fermentation of fresh green grass, wrapped up in plastic bales and fed to cattle in winter. Most farmers will get two crops of silage in a growing season whereas hay took all summer to grow before it was cocked into haystacks to dry, something that is only really seen these days in children's storybooks.

If we zoom out the broad picture is one of change from a reasonably diverse system of animals and cereals/vegetables to one that is now dominated by cattle and sheep. But if we zoom in we also see more subtle changes. For instance, the pig population has increased by approximately half over the past sixty years. Whereas in the 1950s these animals would have been widely spread (nearly every farm had a pig for eating the scraps) these days they are highly concentrated in a small number of intensive pig-rearing units. Sheep were primarily an animal of the lowlands during the 1800s whereas today they are synonymous with the uplands. Cattle, on the other hand, were widely used in upland areas but hardly any are to be seen on the hills today.

The breeds have changed also. There were considered to have been four native breeds of cattle in the early 1800s but the Kerry is the only one surviving.[37] Kerry cattle are small (about 1m at the shoulder) and hardy. These days half the herd are much larger Friesian or other breeds, which put on more weight faster. In recent years there has been a marked shift in the overall proportion of cattle which are reared for beef and dairy, as the dairy herd rose from just over 1 million in 2005 to 1.3 million in 2015. There has also been a marked increase in the quantities of cereals (oats, barley and wheat) sown in winter as opposed to the spring and these crops are mostly grown to provide feed for animals rather than people.

The pace of change in agriculture has mirrored that in global food markets. I once heard a conference speaker describe farmers as 'economically non-rationale actors', in other words farmers practise their craft out of love and devotion rather than economic self-interest. There is no doubt a lot of truth to this but at the same time farming is a remarkably adaptable industry, one that avails of new technologies that span the spectrum of applied sciences from genetics, soil chemistry, robotics, remote sensing and even drones. These changes have been necessary, essential even, for the survival of the sector as food is produced in line with the demands of retailers and ultimately consumers. To remain competitive, production must ever increase while profit margins are locked in the tightening vice grip of demand for ever cheaper food, a cost that has principally been borne by the environment and smaller farms.

The west of Ireland is not associated with intensive agriculture, quite the opposite, in fact. Rocky ground, waterlogged soil, weather to test the constitution. The small field sizes are not suitable for large machinery and sheds for animals are still rare. According to the Teagasc Annual Farm Survey for 2014 (Teagasc is the research arm of the Department of Agriculture), the west was the region with the lowest income (€15,000 per annum) and the second highest dependency on subsidies (which amount to 110 per cent of income on average). Perhaps not by coincidence it is also the part of rural Ireland most visited by tourists due to its stunning scenery. It also overlaps with the vast swathes of the land which are protected for nature. Among these, the great western lakes of Loughs Corrib, Mask, Carra and Conn mark the aqueous frontier between the fertile limestone midlands and the boggy moors to the west. Their crystal waters draw anglers from far and wide, eager to bag a salmon or trout. In the early 1900s it was said of Lough Corrib that 'the average annual catch of salmon by anglers has exceeded a thousand fish. Catches of twenty, and in some cases as many as twenty-five, by one rod in one day have been recorded; baskets of from eight to twelve in the height of the season are frequent.'[38]

Among their natural wonders were the Arctic charr, a fish of the salmon family and of such antiquity in Ireland that they were believed to have evolved into no fewer than six varieties depending on the part of the country it was found in. As its name suggests it is a fish of northern latitudes in both fresh and salt water. Its presence in high mountain lakes or deep inland suggests it was quick to colonise our waterways following the receding glaciers many millennia ago. A large population of the western variant, Cole's charr, still occurs in Lough Mask but during the 1970s those in Lough Corrib to the south and Lough Conn to the north disappeared.[39] The cause was effluent from farmland in the surrounding areas, cattle manure, which was entering ditches, then streams and ultimately the lakes.

Dung is a rich source of nutrients. On entering water bodies it acts as a fertiliser, boosting the growth of plants. Because it is mixed in the water column the plants that can best avail of this food source are the microscopic plankton which are present in abundance in fresh water. During their growth spurt oxygen dissolved in the water is used up, meaning there is less available for other organisms. All water bodies have a capacity to absorb pollutants but there comes a point where this capacity is exceeded and the ecology shifts to a new state. Rivers, being constantly on the move, particularly fast-moving ones, are constantly absorbing fresh oxygen and flushing contaminants out to sea. However, lakes do this at a much slower rate and so are especially vulnerable to this type of pollution, a phenomenon referred to as 'eutrophication'. Certain fish, like salmon and charr, require constantly high levels of dissolved oxygen in the water, as do their spawn. In extreme cases the quantity of plankton in the water can turn it from clear to sickly green. In these cases the pollution does not just affect fish but all life – certain organisms will thrive, perhaps multiplying in great abundance, but many cannot cope and die out. A complex ecosystem with many plants and animals is reduced to a simple one with a few hardy survivors.

Eutrophication reached its zenith in Ireland in the 1970s. For the centuries up to then, pollution of waterways from farms was practically non-existent. More nutrients were being removed from the land through crops and livestock than was added, meaning there was generally a deficit of key elements like nitrogen and phosphorous. Guano (excrement from seabirds, bats or sea lions) was imported from South America in the mid-1800s, bones were ground into meal or, if you were near the sea, seaweed was added to domestic supplies of dung. But for the most part, animal manure was a precious resource that was used judiciously and prudently. It was primarily used on root crops (turnips and potatoes) or cereals, rarely on pasture.[40] The development in the early part of the twentieth century by German chemists Fritz Haber and Carl Bosch of an efficient technique to turn atmospheric nitrogen (which forms approximately 78 per cent of the air we breathe) into ammonia would change everything. It led to the mass production of cheap fertilisers, which, along with the use of concentrate feeds for livestock, means that the quantities of nutrients entering the farm now greatly exceed those being taken off in the produce being grown or reared. The excess would largely end up in rivers, lakes and the sea.

Only tiny amounts of phosphorous leaching off the land are required to result in eutrophication in waterways – merely a concentration of 0.002 milligrams in a litre of water is sufficient to degrade the trout value of a lake. In Lough Sheelin in County Westmeath in the early 1980s the concentration peaked at three times this concentration.[41] In the early 1970s pollution in Lough Ennell, also in Westmeath, was so bad that visibility in the once-clear waters was reduced to one metre and what was described as the 'almost total destruction' of the unique aquatic flora.[42] Fish kills, defined as an incident where over 1,000 fish die, were averaging twenty-two per year from 1969–1974, but rose precipitously during the 1980s with a 500 per cent increase due to agricultural sources.[43] Such is the polluting power of manure that it was estimated that in the early 1980s the total volume from Irish livestock was equivalent to a human population of 100 million people.[44]

Calculations in the 1990s determined that for trout to thrive in Lough Ennell the loss of phosphorous from surrounding farmland should be no more than 150 grams per hectare per year. Given that an average dairy cow will poop approximately 40 grams of phosphorous *per day,* and with an average dairy stocking rate of two cows per hectare in 2014[45] this means that there will be around 29,200 grams produced in a year. So if even only 0.5 per cent of the manure produced in a year finds its way into a local watercourse, we can expect pollution problems. And this does not include any additional artificial fertilisers that may be added to the land. During the summer, when the cows are outdoors, this dung falls straight on the land, to be absorbed into the soil and, ultimately the grass. In winter, when the animals are indoors due to wet weather and poor grass growth, the slurry must be stored in underground tanks. It then must be spread on the land when the weather dries and the grass-growing season starts again. In theory, this is a valuable nutrient but in many cases the quantity of slurry is a problem, particularly as holding tanks fill up. In 2014 new regulations were introduced which prohibit the spreading of slurry from November through to January, the wettest months, when manure is mostly likely to be washed off the ground straight into the nearest stream. The regulations also stipulate minimum sizes for holding tanks and general farm housekeeping which would avoid slurry washing off yards and these rules are certainly a great improvement on the situation. But the figures show just how enormous will be the challenge we have in cleaning our water.

Agriculture is not the only source of pollution but it remains the biggest by far. We still have a major problem with untreated human effluent and a 2015 report from the EPA highlighted how forty-five urban areas across Ireland are responsible for discharging raw sewage straight into the environment. Half of these were in only three counties – Cork, Galway and Donegal – including such tourist hotspots as Clifden and Youghal.[46] But in many ways these issues are easier to deal with – the problem is widely acknowledged, technical solutions are readily available and all that is needed is the financial investment to get on with the work. Slurry, however, is much more difficult. Despite progress, the problem has yet to be fully acknowledged, while solving it will require a range of solutions, depending upon the circumstances. It is compounded by the government's targets to increase agricultural output enormously under the Food Harvest 2020 and Food Wise 2025 plans. Environmental impact reports for these plans, which were produced only under duress after pressure was applied from the European Commission, acknowledged that the quantities of slurry would rise but offered no solution to how this would be kept out of waterways.

In the early 1980s, 84 per cent of our rivers were assessed as unpolluted.[47] In the most recent report from the EPA on the quality of our water this figure had fallen to 53 per cent. In all, 57 per cent of lakes which are monitored and 55 per cent of estuaries (including Wexford Harbour and the upper reaches of the Suir) do not meet the requirements of 'good ecological status'.[48] According to the EPA over half of this pollution is attributable to agricultural sources. The report states that 'future pressures include the planned expansion in the agricultural sector' and highlights the ongoing problem of groundwater contaminated with faecal coliforms (bacteria from human and animal waste). It concludes by rather meekly stating that 'it is clear that additional measures may be required to ensure that Ireland's waters are both healthy and safe'. This is something of an understatement.

The insidious degradation of once-pristine waters by cattle run-off has been charted with scientific precision by Linda and Chris Huxley who moved to County Mayo in 2000.[49] They established themselves along the shores of Lough Carra, lured by the excellent trout and salmon fishing, and the chance to live at close quarters with what was then considered among the best examples of a calcium-rich lake anywhere in Europe. In the 1930s the botanist Robert Lloyd Praeger described the lake as 'delightful … Its waters, derived mainly from springs, are of a surprising pale green tint, and very clear. The unusual colour is enhanced by the fact that the stony shores and bottom are coated with a white soapy deposit of lime, which reflects the light, so that one can peer down into the clear depths, and see the water-plants rising like slender bushes or trees from the bottom.'[50]

Indeed, the water was so clear that it was possible to gaze into depths of over 6m. In the 1960s it was renowned by anglers for the immense clouds of mayfly that emerged from the lake waters in early summer. Its special status is acknowledged in its legal designation as a Wildfowl Refuge, a Natural Heritage Area, a Special Protection Area and a Special Area of Conservation; four separate but overlapping layers of protection that give the lake international importance. It has been the subject of a wide range of scientific surveys by state agencies, universities, non-governmental organisations and the Huxleys themselves. This means there is quite a good understanding of the changes that are occurring, crucial data which is hard to dismiss.

The first warning signs came from Central Fisheries Board researchers who recorded a decline in water clarity of 40 per cent between the 1970s and 1990s. Because nutrients entering a lake promote the growth of tiny plants suspended in the water, there is an established correlation between the clarity of the water and the degree of eutrophication. The Huxleys, meanwhile, noted the deteriorating situation by the increasingly muddy waters and occasional blooms of slimy green alga. The mayfly no longer hatched into the enormous clouds they once did and this is reflected in the fisheries value of the lake. The Huxleys suggest there is 'strong evidence' that the dramatic fall-off in catch rates of trout is a result of a 'substantial decline' in the population of that fish. The couple have charted this decline against the increases in the numbers of cattle (42 per cent), sheep (136 per cent), pigs (400 per cent) and chemical fertiliser use (90 per cent) between 1970 and 2003. They calculate that there has been a fourfold increase in the quantity of slurry generated as a result of this agricultural intensification, from 200 litres per hectare to over 800 litres per hectare. These changes have also seen the clearance of natural habitats in order to reseed grassland for grazing and the digging of drains in an effort to dry out wet areas.

It is not just the Huxleys who are concerned. A 2013 report from independent scientists, produced for the NPWS stated: 'it must be concluded that Lough Carra is under considerable ecological stress and the assumption that it is Ireland's best example of a marl [calcium-rich] lake may cease to be true in the near future.'[51]

Unless action is taken to incentivise farmers in the Lough Carra catchment to revert to less intensive forms of agriculture there is no chance of holding on to this very special place. Frustratingly, the multiple layers of legal protection seem totally inadequate in the face of this rising tide of effluent and habitat loss. Like many similar areas across the country, they are protected in name only and little is done on the ground to halt the deteriorating situation.

Nutrients are not the only pollutant to enter waterways from agriculture. The direct disturbance of land for ploughing, reseeding for silage production, 'cleaning out' of field drains or trampling by cattle along rivers' edges all result in soil and sediment being washed away. The clogging of gravel on a riverbed with silt can smother the spawning beds of salmon and reduce the amount of invertebrate life that would normally inhabit the small spaces between stones. In making the water turbid it can reduce the amount of light that can penetrate the water column and so reduce the potential for plants to grow on the riverbed. This, in turn, can lead to more erosion and the release of yet more silt.

Perhaps Ireland's rarest animal is the freshwater pearl mussel, a very long-lived shellfish which once lived in enormous colonies on the beds of fast-flowing rivers and streams. The hand-sized molluscs relentlessly filter the water passing over them, trapping their tiny plankton food as it is washed overhead. But this filtering action is their downfall if there is too much sediment suspended in the water. One population, in the River Nore in County Kilkenny, is considered a unique species globally but over-exploitation for its pearls and, more recently, water pollution have reduced its population from the millions to an estimated population of only 585 individuals. In fact, it may already be 'functionally extinct' as the surviving adults no longer breed and no juveniles have been recorded along its known remaining habitats. The NPWS says that 'its habitat is unlikely to be restored before the extinction of the wild population' and hope is now directed towards captive breeding where water quality can be controlled.[52] Other populations of freshwater pearl mussel are found throughout Ireland and great efforts are currently being expended to reverse their decline. One project has been trying to protect the species in County Donegal. Funded through EU money it has identified the main pressures on water quality in the catchment, which come not only from agriculture but also plantation forestry and one-off houses. When the researchers looked at how much sediment was settling on freshwater pearl mussel beds they arrived at an astonishing answer – a massive 3.5 tonnes per square kilometre of grazed land per year.[53] This is a problem not only for mussels. Soil is the very foundation of the land and the agricultural system itself – the farmers' greatest asset. To allow it to be simply swept away on this scale undermines the future viability of the farm. The move from spring-sown cereals to winter-sown crops, principally to provide feed for cattle, means greater areas of land are being exposed by ploughing just when the rains are heaviest and evaporation is at its lowest. Wetter summers, when fields are reseeded for pasture, adds to the problem. Vast quantities of soil, a non-renewable resource on any human timescale, is simply being carried away.

The myth of Ireland's 'green' farming

The Irish countryside is defined by its green grass, smallish fields and the hedgerows that hem them in. The trees and hedges that enclose our fields are so synonymous with our pastoral heritage that they seem hardly worth mentioning. They give the rural landscape a wooded feel in the absence of any real forests. It is only when travelling abroad, looking out of an aeroplane window perhaps, at the vast and open fields of southern England or northern France, or the desiccated plains of southern Spain, when we appreciate what we have at home. The lush summer foliage spilling onto verdant pasture is relief to the eye – reassurance even – that all is well in the world. In early summer the boughs of the hawthorn, still sometimes called the whitethorn or mayflower depending on where you are in Ireland, heave under the weight of blossoms. To stand close to a hedge on a warm spring day feels like standing on a balcony overlooking a vast orchestra busily rehearsing. There is an audible hum from the movement of aerial insects and the incessant movement of life all around. It is in the hedgerow that robins and blackbirds build their nests, that badgers dig their setts and trees from oak to apple send their gnarly roots into the rich soil. With no great areas of native woodland left in Ireland the hedgerow has become the new home to the vast majority of our wildlife.

There is no county that doesn't have an extensive network of hedges, although regions have their own distinctions. Fuchsia (an introduction from South America) brightens the lanes in Cork, soggy willows abound in Leitrim while the yellow flowers of the gorse thrive on higher or rockier ground. Some have smothered stone walls, many are accompanied by ditches and wet drains and a few mark ancient mass paths or cattle trails. While some date to the eighth century, the majority were planted during the 1700s.[54] All are entirely man-made. Indeed, it is the way they have been managed since they were laid down that has provided the resources for so many types of plants and animals. Enclosing fields was essential for separating cattle or sheep from the neighbours, so thorny bushes were planted to keep the animals on one side or the other. These were cut back every few years and this pruning ensured fresh, dense growth that maintained an impenetrable wall of vegetation.

In England, which has similar hedges, a man in Devon set about identifying all the species he could in an 85m length of field boundary on his farm that he described as 'nothing exceptional'. Over two years he clocked up an incredible 2,070 species, the vast majority of which were insects – sure to be the first study of its kind.[55] So ubiquitous are hedgerows that they receive relatively little attention from ecologists when compared to many of our habitats or rare species and there are no nature reserves or protected areas designated for their preservation. They do get special treatment under wildlife legislation which prohibits cutting them from the beginning of March through to the end of August. This is in place to give birds an opportunity to nest and rear their chicks but in providing this safeguard all the other animals and plants benefit. Of late, this closed period has come under fire, and during 2015 the IFA vigorously lobbied the government to shorten the closed season to between mid-March and the end of July. It was unclear why this was needed but nevertheless Heather Humphreys, Minister for the Arts, Heritage and the Gaeltacht (under whose remit the Wildlife Act fell) indicated her willingness to have the law changed. Over 23,000 people signed a petition organised by the IWT, BirdWatch Ireland, An Taisce and the Hedgelayers Association of Ireland but despite this the law was about to be passed when the general election of February 2016 intervened. The IFA's insistence that a change in the law was needed perhaps points to how the usefulness of hedgerows to farmers has diminished, and keeping them maintained has become a difficult task that outweighs the benefits. Why maintain a hedge when a wire fence can do the job just as well?

The evidence that hedgerows are not being maintained is everywhere to be seen. Frequently, it is not because hedgerows are cut too often but that they are not cut often enough. Without regular pruning of the thorny bushes the trees tend to lose their structure at the base. As the thorns die back the farm animals push their way through, making the gaps bigger and grazing areas that were once inaccessible. Trampling compacts the soil so it is harder for new growth to emerge, diversity dwindles and in time all that is left is a line of trees. This may not sound like a bad outcome but in fact it means that the habitat cannot support the great variety and abundance of species which it once did. It also means that once the trees die, as they eventually will, there will be nothing to replace them and no trace of the hedge will remain.

This effect is on view all over Ireland and a series of county-wide studies from the Heritage Council have shown that only around a quarter of hedgerows are in 'favourable' condition. One study from 2013, although regionally limited in its scope, further confirms these findings. It looked at an area of east Galway, thirty-two farms in all and 286km of field boundaries. It noted that without the field boundaries, there would be no wildlife habitats whatsoever on intensive farms. Of the hedges they looked at less than one third fell into a category they described as 'species rich' (containing four or more woody plants) while nearly half of them were 'not in favourable condition for wildlife'. The study concluded that the reasons for poor condition of hedgerows were 'gappiness, open and scrawny bases and the presence of more than 10 per cent non-native species'.[56]

The problem is not that the value of hedgerows is not recognised but that their benefits are widely spread. There is acknowledgement that they are important for defining the rural landscape, absorbing carbon, ameliorating floods, providing habitat for pollinators and other wildlife, and shelter for livestock from the elements. By separating neighbouring herds it is also likely that a dense hedge will help prevent the spread of animal diseases such as bovine tuberculosis. This was recognised in agri-environmental schemes which have been designed to give farmers top-up payments in return for pro-active measures to enhance habitats, such as the Rural Environmental Protection Scheme (REPS). However, there is no evidence that this did any good and a number of studies found that being in REPS had no influence on the value of the hedge for plants, beetles or birds.[57]

Digging up hedgerows entirely to make fields bigger has also been a concern for some time. In the first ever evaluation of the state of Ireland's environment in 1985 it was estimated that 14 per cent of hedgerows were lost between 1936 and 1985, with the author speculating that 'most of this loss has occurred since 1973'[58] (coincidentally, perhaps, the year Ireland joined the EEC). The advent of remote-sensing technology, such as satellite imagery has made estimates of land use considerably more accurate in recent decades. For instance, repeat surveys between 2006 and 2012 found that in this short time nearly 2 per cent of hedgerows were lost.[59]

The intensification of food production is a process whereby more is extracted from the same amount of land. It means bringing areas into production that were previously not worth the trouble, like the small patch of scrub on some rocky ground or a pond in a wet corner. It means digging and maintaining drains, adding extra fertiliser and using chemical sprays to limit the impact of pests. Concern for the environmental effects of intensification is nothing new – Rachel Carson's still-relevant book *Silent Spring*, on the effects of chemical sprays on humans and wildlife, was published in 1962. Thankfully, many of these lethal compounds have been outlawed. Once threatened with extinction from the effects of DDT, an insecticide which resulted in thin-shelled eggs that broke in the nest, peregrine falcons recovered after it was banned.

Huge quantities of herbicides and pesticides are still used in all aspects of agriculture, with the obvious exception of organic farms, but there is so far surprisingly little evidence that this is having unwanted side effects directly either on human health or the environment. Since 2013 there has been a National Plan for the Sustainable Use of Pesticides and regulations now require users to be trained, along with other sensible measures. There is something called 'integrated pest management' which recognises that the use of sprays needs to be minimised and done in a way that protects watercourses and other sensitive features. There are also controls on what pesticides can be used at a European level while manufacturers must conduct a raft of field and laboratory checks to determine their effects on plants and animals. Certainly, the direct threat posed by herbicides and pesticides is much less than it was and modern agriculture has come to depend heavily upon their use. However, with the pressure of increasing intensification, more and more sprays are required. Since 1965 the sales of pesticides in Ireland has increased six-fold and are rising.[60] There is now some worrying evidence that this is making its way into watercourses and drinking-water supplies with the EPA stating that the use of pesticides 'has emerged as a significant water quality issue in 2015'.[61] A leaflet from the organisation warns that 'a single drop of herbicide can breach the drinking water limit in a small stream for 30km'. In 2015, sixty-one supplies failed this limit, an alarming jump from twenty-eight only a year earlier.

Despite reassurances from the agri-industry as to their safety, these products do exactly what they are intended to do – kill pests and weeds, otherwise known as wild animals and plants. The lowly thistle, so recognisable with its spikes and purple flowers, and yet largely dismissed as a pernicious weed, is host to a bewildering array of invertebrates from gall mites to sap suckers and parasitoid wasps (a parasitoid lays its egg inside another living organism so that the hatchlings can eat fresh flesh). Up to thirty-seven different species of aphid, beetle, moth, fly, shield bug, ladybird and weevil spend at least part of their life cycle on or inside thistles, while hoverflies, bees and butterflies will visit the flowers to feed on nectar.[62] Over 100 types of insect regularly occur on the common stinging nettle, and indeed it is the principal or only food plant for a number of our butterflies, including the small tortoiseshell, red admiral, peacock and painted lady.

These tiny creatures, about which little is known, are the nuts and bolts of ecosystems. Collectively, they are the food for larger animals such as birds and some smaller mammals. Most are too obscure even to have common English names but some are better studied because they are important to aspects of our economy. Bees fall into this category, along with some other species which collectively are referred to as pollinators. They are not only much loved from Pixar films but bring home the bacon when it comes to helping our farmers. According to the National Pollinator Plan, the value of pollinators to the Irish economy is €53 million per annum, with crops such as oilseed rape and apples relying heavily on the work they do. But pollinators worldwide are under pressure and Ireland is no exception. Half of Ireland's bee species have undergone 'substantial declines' in their numbers since 1980 while nearly a third are threatened with extinction, according to the National Biodiversity Data Centre.[63] This is part of a wider pattern, indeed the loss of pollinators is one of the major drivers of concern for the global loss of biodiversity. The research suggests that this is happening as a result of multiple pressures working in concert with one another. The chief culprits? Parasites, pesticides and a lack of flowers, according to the School of Life Sciences at the University of Sussex, UK.[64] The pressure on farmers to maintain perpetually or increase yields is leading to ever more efficient monocultures, whether barley or grass for silage, combined with the loss of hedgerows and other 'unproductive' habitats, which, in turn, simply leaves little space for other plants.

One group of pesticides in particular, known as neonicotinoids, has come under intense scrutiny in recent years as it is shown to have harmful effects not only on bees but other insects and even birds.[65] When faced with a proposed moratorium on the use of neonicotinoids, government ministers and EU officials were lobbied intensively by both sides of the debate, including campaigners Avaaz as well as Bayer, Sygenta and the European Crop Protection Association, the umbrella group for pesticide manufacturers, before casting their votes.[66] Disappointingly, Ireland was one of only five countries to oppose the ban initially while in a subsequent vote it abstained. Evidence continues to mount on the negative effect of these deadly chemicals.

In 2015 Glyphosate, one of the most popular general-use herbicides, was labelled 'probably carcinogenic' by the research arm of the World Health Organisation. Meanwhile, as more and more pest species evolve resistance to chemical sprays, it would seem that their long-term use is unsustainable for both human health and the environment.

The impacts of agricultural intensification on bees is likely to be the tip of the iceberg. Insects and other invertebrates, which make up well over half of all known multicellular species, remain a mystery. Unseen to our eyes they are busy not only pollinating, but breaking down and decomposing our waste. Have you ever thought about where all the leaves go to in autumn? Or why finding the remains of a dead animal is so rare, even though thousands of animals die every day? Irish farmers find thistles a nuisance but when they were inadvertently introduced to Australia and New Zealand, they truly got out of hand. In these countries they are classed as alien invasive species. Without the dozens of sap suckers and weevils that limit their growth there is nothing to inhibit their spread. These other tiny, inconspicuous creepy-crawlies, in other words, do quite a good job at controlling pests.

There is one group of animals, though, that is well studied. Large enough to be noticed, colourful and vocal enough to be widely loved, ubiquitous enough to make use of every habitat that's going: birds. From the pied wagtails that bob around supermarket car parks to the redstarts that like their forests ancient and undisturbed, across land and sea the birds are messengers, sending us signals of the changes happening all around us. And we benefit greatly from the fact that birds have been remarked on for hundreds of years so that even in the absence of the hard scientific data we would like to have, we can get a glimpse of how our countryside has changed over these centuries.

In Ireland, 70 per cent of the land space is used for agriculture so changes in how the land is used are likely to show up in regional bird populations. In 2007, of eighteen Irish birds species then listed as being of 'high conservation concern', eight were associated with lowland agriculture – grey partridge, quail, corncrake, lapwing, curlew, barn owl, yellowhammer and corn bunting. The corn bunting has since disappeared entirely. The corncrake has vanished from its former stronghold along the River Shannon and is now hanging on at a handful of locations in the north-west. Breeding curlew numbers have crashed by 90 per cent since only 1990, lapwing plummeted from over 16,000 pairs in 1993 to only 2,000 in 2008, and redshank from 4,450 pairs to 500 in the same period.[67] Some birds, which are associated with marginal agriculture in uplands or coastal districts, such as the little twite or the blackbird-like ring ouzel, have gone from common and widespread to the verge of extinction in the space of sixty years.

Other species, such as woodpigeon, magpies, rooks and jackdaws seem to have benefited from the changes in agriculture so declines are not universal. But the overall trend is from a diverse range of healthy farmland bird populations to a smaller number of booming opportunists. The remainder are consigned to a type of intensive care, subject to specialist, and expensive, intervention measures designed to keep them from being wiped out entirely. Exactly why this is happening is itself obscured by a lack of data – annual counts of countryside birds in Ireland date back to only 1998, which postdates the greatest period of transition in farming which occurred after our entry to the EEC in 1973.[68] Maybe it is the increase in silage over hay, or the decline of spring-sown cereals in favour of winter-sown crops, maybe it is drainage works, pollution, herbicides, pesticides, artificial fertilisers or loss of habitats like hedgerows, or maybe it is a combination of all those things; what is incontrovertible and undeniable is that it is happening.

It is impossible to have any discussion about farming and agriculture without mentioning subsidies. The idea of transferring public money to farmers has been around since the Great Famine but the scale of these payments has ballooned since the introduction of the EU's Common Agricultural Policy (CAP). The CAP came into being in 1962, shortly after the EEC was founded, and in its latest round it guarantees approximately €1.2 billion to Irish farmers annually. This is supplemented by an additional half a billion euro per year under the Rural Development Programme. It doesn't include other subsidies, such as the €113 million used to fund Teagasc (the agricultural research arm of the state), the €27 million given to An Bord Bia for marketing Irish food, €15 million for compensation for farmers whose livestock is affected by TB or brucellosis,[69] or tax breaks such as those for the use of green diesel, tax reliefs for young farmers, stamp duty exemptions, etc. It is such a vast amount of public money that it is important to periodically remind ourselves why it is paid out.

The food-security argument dried up in the 1980s amid controversies over wine lakes and butter mountains. Even today, in a world of €3 chickens and where one third of produce is simply thrown away, there is a strong argument that we are *over*supplied with food. With food security no longer a high priority it was felt that farming and the rural economies that depend upon them needed protection from the forces of globalisation. It was recognised that market intervention would be required if we wanted to maintain high standards of animal welfare, food safety and environmental protection. According to the European Commission:

> The EU's common agricultural policy is designed to support farming that ensures **food safety** (in a context of **climate change**) and promote **sustainable and balanced development across all Europe's rural areas**, including those where production conditions are difficult.
>
> Such farming must thus fulfil **multiple functions**: meeting citizens' concerns about food (availability, price, variety, quality and safety), safeguarding the environment and allowing farmers to make a living.

At the same time, **rural communities** and landscapes must be preserved as a valuable part of Europe's heritage.[70] [their emphasis]

European consumers are spoilt with the range and low cost of food. Despite the odd scandal here and there our food is also safe to eat and European politicians have so far held out against the wholesale introduction of genetically modified organisms or the use of antibiotics as growth stimulators. At a social level the CAP works to a certain extent also; without it, there would be widespread depopulation of rural areas and abandonment of land. According to the 2014 farm survey, all of Ireland's sheep and beef farmers are entirely dependent upon subsidies to stay in business. Tillage farmers relied on subsidies for around 84 per cent of their income and only dairy farming provides a decent profit margin. In marginal areas, therefore, where sheep and beef farming is concentrated, the public is paying farmers to stay in business. According to Teagasc, nearly a quarter (23 per cent) of farmers earn less than €5,000 per year so in these cases farming is no more than income supplement (albeit a crucial one where incomes are low).[71] Subsidies may well therefore underpin the social fabric of counties where there is little alternative income. As the average age of farmers in these areas increases, however, how long this will last remains to be seen.

Which brings us to the final *raison d'être* of the CAP – safeguarding the environment and preserving treasured landscapes. The latest version of the CAP aims to do this in two ways – the first is by adding a 'greening' element to the 'basic payment' (traditionally known as the 'cheque in the post') for special environmentally friendly measures. This makes up 30 per cent of the €1.2 billion that goes to practically every farmer in the state. The move held great promise when it was first mooted but intense lobbying by the agri-food sector meant that the final measures are meaningless. According to Alan Matthews, Professor Emeritus of European Agricultural Policy at Trinity College Dublin, 'very little additional benefit for the environment will be obtained'[72] from the scheme and farmers will be paid for what they were doing anyway. This rightly leads to cynicism among both farmers and the public, what Professor Matthews refers to as 'just another set of bureaucratic hoops to jump through in order to receive the direct payment. There is no link made to encourage active management of the land to maximise the potential for biodiversity, nor to make best use of farmers' knowledge and experience as to what might be the most appropriate measures to take.'[73]

The other way in which the CAP is meant to protect the environment is through an additional payment scheme that is given to farmers for undertaking special measures, perhaps because they have valued habitats on their farm or to compensate them for certain restrictions in their activities, e.g. later mowing of meadows to protect nesting corncrakes, or avoiding ploughing to protect wildflower-rich grasslands. Ecologists tend to be very enthusiastic about these 'agri-environmental' schemes, as they are known, for they reward farmers who tend their land sensitively. In most instances this means farming using traditional ways that have been historically good for nature and wildlife. Farmers like them, too, because in low-income areas they can be a significant boost to their pay packet. These programmes have been around since the 1990s, first appearing as the Rural Environmental Protection Scheme (REPS), then as the Agri-Environment Options Scheme (AEOS) and currently the Green, Low-carbon, Agri-environment Scheme (GLAS). And what's there not to like? The current GLAS benefits around 50,000 farmers and has lots of eye-catching actions, from protecting watercourses, putting up bat boxes and maintaining cover for wild birds to creating habitats for bees. Landowners can choose from a menu of options from maintaining traditional hay meadows to laying hedgerows. It prioritises those farmers in areas already protected for nature or where land is in a vulnerable water catchment. Indeed, it is hard to find fault in any of the initiatives and after twenty years and at a cost of over €3 billion intuitively it must have had enormous benefit. Alas, there is scant evidence for this.

A comprehensive review of studies relating to REPS, which lasted from 1996 to 2010, and cost the aforementioned €3 billion, found 'insufficient evidence' on which to judge the impacts of the scheme.[74] This is mostly because there is no monitoring of the effects, no before-and-after studies which evaluate whether the measures are reversing declines in farmland birds or improving local water quality. Of the limited studies that were available, one found no difference in field-margin density (a proxy score for hedgerow quality) or bird diversity between REPS and non-REPS farms. Another found no effect on plant or beetle diversity. There is newer evidence that farms in agri-schemes have significantly lower rates of nitrate run-off and this is great.[75] However, in the latest assessment of our water quality by the EPA, 53 per cent of rivers were unpolluted – in the late 1990s it was 70 per cent.[76] So although things may have been worse without these schemes, REPS and its offspring have not improved water quality. Nor is there any evidence that it has halted or reversed declines in threatened species or habitats. Measures to protect corncrakes along the River Shannon, which incentivised farmers to mow hay later and cut from the inside of the field outwards (thereby allowing birds to escape the mower) were insufficient to save the species. Where once summer evenings were filled with the rasp of calling males, in 2015 there was silence.

Farmers will be as disappointed with this outcome as anybody. When it comes to the stated aim of protecting the environment, the billions of euro in subsidies have been a waste of money.

But it is worse than this. In fact, the basic subsidy scheme, which pays little heed to the particular circumstances or location of farms, is actively destroying the environment. Farmers are incentivised to homogenise their land in order to meet the Department of Agriculture requirements. The system pays out on a per hectare basis and 'non productive' land is not eligible. This includes rocky outcrops, wetlands and ponds, patches of scrub or 'unused areas' in arable fields. In upland areas heather is only eligible where it is at 'grazable height', and this creates an incentive to burn it off – a devastating tool which wipes out nesting animals and pollutes watercourses. The low profit margins on livestock mean in some areas too many animals are on the land. In hilly and coastal areas this leads to erosion, literally sweeping away the soil to leave only bare rocks and stones.

Farmers and environmentalists are too frequently seen as at odds with one another. Yet a closer inspection of the issues reveals this conflict to be unfounded. Many farmers are also environmentalists and show a concern for the natural environment that is not reflected in the policies of their representative organisations. There is acknowledgement that farmers must be part of the solution if we are to live in a country with healthy landscapes and wildlife. The greater division lies between farmers in marginal (i.e. non-profitable) areas which also happen to be of greatest scenic and wildlife value, and those in more profitable areas and who are better placed to benefit from global trade. The latter group is well represented by their lobbyists and have a ready audience among politicians and government ministers. Broad-brush policies are skewed to benefit this group while the less-well-resourced are left to fend for themselves. In a world where the marketing budget is three times that set aside for nature conservation, the message is the reality.

Life is all about choices. We don't have to accept that things will continue as they have done. Time after time, the choices made by Irish politicians have resulted in poor outcomes for our environment. And it is not only wildlife which has taken the brunt of these decisions, but people too.

8

A future for wildlife and people

> '... too beautiful to put into words ... It will certainly be the envy of Europe.'
>
> Michael Ring TD, Junior Minister for Tourism, in relation to the creation of 11,000 hectares of 'wilderness' in County Mayo[1]

PERCHED HIGH ABOVE THE Atlantic Ocean the Céide Fields of north County Mayo gaze across the restless seas. On approach there is little of remark, with only the visitor centre punctuating the tracts of featureless bog which fan out to the south. But beneath the bog lie the remains of an ancient and surprisingly advanced society. The site marks the location of a remarkable network of well-laid-out stone walls which are literally buried deep under the layers of turf. The walls mark the boundaries of field systems, laid down by some of Ireland's earliest farmers between 5,000 and 6,000 years ago. They are best appreciated from the balcony on the upper level of the visitor centre, allowing the viewer a bird's-eye view of the last remains of what was once a thriving and sophisticated dairy-farming community. Today, however, it takes a degree of imaginative gymnastics to visualise how any existence could be extracted from the barren land that stretches to the horizon.

First discovered in the 1930s by a local schoolteacher out cutting turf, it would be some decades later before the lines of carefully mounded stones now visible were excavated. Initially marked out by probing the soft earth with poles, later advances in radar technology have mapped the fields without the need for excavation or disturbing the bog. The known extent of these farming settlements now stretches far to the west and east with not only field boundaries but animal paddocks, tombs and also stone houses being uncovered. Pottery, fashioned stone tools and charcoal are some of the remnants of a society that numbered in the hundreds, one notable for its sophisticated organisation. The archaeological remains suggest that this pastoral existence continued for 500 years but then abruptly, perhaps over the space of only fifty years, it disappeared completely.[2] Why? The stumps of long-dead pine trees, buried deep in the peat, may give us some clues.

Neolithic farmers arriving in Mayo would have looked out on a very different landscape from what we see today. Instead of endless bogland there were dense forests dominated by pine trees, but also with some oak, alder, birch, hazel and willow.[3] The forests were soon cleared to make way for the people with their animals and this can be seen through the analysis of pollen layers which are laid down in soil like pages in a book. Here, the pollen layers dominated by trees suddenly shifted to reveal huge increases in the pollen of grassland plants like buttercups, docks and clovers. A dramatic change in climate took place 4,700 years ago which saw cooler temperatures and increased rainfall. This coincided with the spread of the bog, which relies on well over a metre of rainfall annually as well as permanently waterlogged conditions. The astonishing growth of the bog since then has left the fields over 4m below the current level of the land, rendering it unsuitable for dairy farming.[4] However, there's a twist. The evidence suggests that the demise of the society along the north Mayo coast occurred several centuries *before* the change in climate. According to an early guide to the Céide Fields:

> There is no general agreement among scientists as to what caused the bog to grow and envelop what was once fertile farm land. Some see this growth as the inevitable outcome of the incessant dampness of the west of Ireland climate. Others attribute the initial growth to man's activity in wasting the fertility of the soil through over-intensive use. A third explanation is that the pine forest with its closed evergreen canopy would have held up to 80 per cent of the total rainfall

> – when the forest was removed all this water came into direct contact with the soil and in draining through it, rapidly leached the minerals away. The mosses which are the basis of bog growth and which depend for their growth solely on the nutrients obtained directly from rainfall could have rapidly established themselves in this situation. It is possible then that man in his very first act of clearing the forest had triggered off an irreversible chain of events which led to his farm being engulfed by the bog which has continued to grow to this day.[5]

It is known from archaeological analysis that widespread deforestation had occurred not only in this immediate area but across the landscape more generally. Could it be that the farmers themselves had an unwitting role in their own demise? Trees intercept a lot of rain, particularly in summertime. Rain evaporates directly from the leaves or is siphoned back into the atmosphere via roots and the small openings on the surface of the leaves, leaving relatively little to run off the land. Root systems open a network of tiny drains that help water to percolate into the soil. There is some evidence for this change from pollen records around 5,400 years ago. This shows the advance of alder and meadowsweet – plants which thrive in wetter conditions. Later, around 5,200 years ago, the global climate changed and this saw dramatically increased rainfall in Ireland, promoting further the spread of bog, and ultimately subsuming the fields in peat.

The Céide Fields still hold many mysteries. The pace and causes of the collapse of the community in this area remain matters of debate. Did this larger, more dramatic climate event cause famine, resulting in climate refugees? Or had society begun to wind down long before that, unwitting victims of the changes they had themselves wrought on the land? Either way, it demonstrates a vulnerability to unexpected changes in the environment, which went on to undermine the viability of their whole society.

Not far to the south-west of the Céide Fields, Achill Island sits close to the Mayo mainland. Its cliff-rimmed northern shore narrows to a stony headland pointing to the vastness of the ocean. Early visitors to Achill never failed to remark on the abundance of eagles on the island; the name itself derives from the Latin word for eagle – *aquila*. In days gone by the picturesque beach at Keem was the scene of an epic battle between man and beast: the basking shark hunt. The second largest fish in the world was speared from open boats, corralled inside a hand-thrown net, and its liver cut out of the half-submerged remains. The normally docile behemoth, also known as the sun-fish for its habit of feeding close to the water surface during sunny weather, feeds by passing water through its gaping, toothless maw. The substantial liver was cut from the animal at sea and hauled ashore to be rendered down. The oil went on to light street lamps across Europe in the days before mineral oil gushed from the ground in Pennsylvania. Later the sharks would be harpooned and processed on shore. For much of the nineteenth and twentieth centuries, the population of Achill Island remained in the region of 5,000, dipping only slightly in the wake of the Great Famine in the 1840s. Farming was the main source of income but as well as the basking shark there was a thriving mixed fishery along with extensive cultivation of seaweed. In 1856 a Scot, Alexander Hector, established a site for boiling and pickling salmon at Keel, a short distance east of Keem. There were schools of mackerel and herring, depending on the season, while mussels, crabs and periwinkles were harvested for local consumption.[6] The extension of the railway to Achill at the end of the 1800s brought tourists and new routes to market for local produce. In 1902, a herring-curing station was established and new fisheries opened for lobster. In 1906, 340 fishermen were employed full-time with some sixty curraghs (traditional open rowing boats) and other vessels. The scenery, with its homesteads dwarfed by the sweep of open sky and bog, was inspiration for Ireland's best-known painter of landscapes, Paul Henry. The naturalist Robert Lloyd Praeger wrote of it in 1937: 'Achill, wind-swept and bare, heavily peat-covered, with great gaunt brown mountains rising here and there, and a wild coast hammered by the Atlantic waves on all sides but the east, has a strange charm which everyone feels, but none can fully explain.'[7]

The German writer Heinrich Böll spent time on the island between 1957 and the 1970s. He went on to win the Nobel prize for literature in 1972 and his home on the island is still used as a residency for writers. In the 1940s it was briefly home to the novelist Graham Greene, whose novels *The Heart of the Matter* and *The Fallen Idol* were partly written from the village of Dooagh.

Yet the years have not been kind to Achill. On New Year's Day 2016, the pier at Purteen Harbour near Keel was dull and cold. Broken panels detailed the now-defunct salmon fishery and basking shark bonanza that lasted until the 1970s. The rusting winches hunched over the pier are a reminder of an industry that landed an incredible 9,258 sharks between 1950 and 1956. By 1971 the catch had dwindled to only twenty-nine animals.[8] At the nearby beach at Keel, plastic debris lines the shore and bits of old fishing nets and discarded bottles are enmeshed with seaweed remains. The bogs are as desolate and as windswept as ever but these days they are also pretty lifeless. The eagles are long gone having been persecuted to extinction over 100 years ago. Once common birds like the corncrake, red grouse and curlew no longer breed on the island. Near the coast I saw land slippage and soil washing into the sea, the telltale signs of too much grazing by sheep. The bogs are susceptible to wildfires in spring, a number of which have affected large areas in recent years. The island has also suffered from built developments during the Celtic Tiger years which have not been sympathetic to the island's rich heritage.

In many ways Achill is a microcosm of the problems highlighted in this book. Extinction, overfishing, damage to habitats and people/wildlife conflicts have all left their mark. Over-exploitation of the high seas diminished the incomes from fishing while a ban on drift-net fishing for salmon came into force in 2007, depriving fishers of their livelihoods. Sea anglers once flocked here for the variety of large fish, including porbeagle, blue, thresher and mako sharks. According to Kevin McNally in his book about the island:

'the porbeagle in particular was so numerous [off Achill] that it created a serious problem for the island fishermen whose nets were repeatedly torn into useless shreds.'[9]

A British and Irish record was set here in 1932 for a 166kg porbeagle shark, caught with a rod and line by Dr O'Donnell-Browne. The record stands to this day. Alas, no one comes to fish for porbeagles any more and they are considered critically endangered in European waters.[10] Fish landings to Achill in 2015 were the lowest on record, at just 48.4 tonnes. In 2005 they were 552 tonnes, according to figures from the Sea Fisheries Protection Authority. The decline in fishing has led to local conflict between fishermen and the local seal population, as seals will take fish straight from nets or render whole catches unsellable. In 1981 the Inishkea Islands, not far to the north of Achill, was the scene of an illegal slaughter of seals, bitterly dividing the community.[11]

On land, meanwhile, much of the island lies within designated areas for the conservation of rare habitats or species. But on the ground this has little significance. In 2011 the upland habitats in one of these protected areas were surveyed by the NPWS. All of the habitats based on peat soil (bog, heath, etc.) and covering 63 per cent of the surveyed area, were found to be in 'unfavourable bad' condition. The main damaging impacts were found to be grazing by sheep, turf-cutting and erosion. During the 1980s and 1990s, like many parts of the country, too many sheep were put on the hills, resulting in overgrazing and soil erosion. While sheep densities were subsequently reduced, significantly the researchers found that erosion remains 'widespread' and that 'once exposed by removal of the vegetation, areas of bare peat may continue to erode due to climatic conditions regardless of manipulation of grazing levels'.[12] In other words, irreversible damage may have occurred which is ongoing due to the extreme levels of rainfall and the continued presence of sheep. To date no management plan has been published which hints at how these problems are to be addressed.

When the botanist Robert Lloyd Praeger visited Achill in the early 1900s he recorded bladderwort, marsh clubmoss, starry saxifrage, lesser twayblade, stiff sedge, brittle bladder fern, Foula eyebright, roseroot and the delicate maidenhair fern among his list of plants.[13] One hundred years later, a survey for the *New Atlas* of plants could find no trace of them[14] (although since then, the starry saxifrage, lesser twayblade and roseroot have happily been refound). For a number of years Mayo County Council has been working with local volunteers to eradicate swathes of the invasive giant rhubarb, a native of South America. Its enormous leaves cast a dense shade under which the native flora cannot grow. Clearing it is painstaking work that involves injecting the rough stems with herbicides but is a technique that has met with much success. Nevertheless, the threat of reinvasion is ever present. Rhododendron, another invasive plant, is also to be found on the cutaway bog. Most strikingly, the decline in the natural environment on Achill has been accompanied by a fall in the human population. Having supported a steady 5,000 people for much of the last two centuries, it fell to 2,700 in 1996 and has been dropping steadily (albeit slightly) ever since. The 2011 census found that the level of vacant homes was as high as 85 per cent in some townlands and averaged 45 per cent.[15]

The decline of nature and people is no coincidence, although as farming and fishing incomes have declined, tourism has taken up some of the slack. Achill is a popular spot in summer and it is particularly geared toward water-based activities. It lies along the Wild Atlantic Way, billed as the world's longest scenic drive. Much of the island and its offshore waters are designated as SACs or SPAs. The basking sharks have not disappeared and, although they probably have not recovered to their original abundance, they are still known to delight visitors with close encounters (search YouTube for some footage). All is not lost. Achill still has dramatic cliff scenery and great potential for outdoor activity. It retains much of its unique heritage, including deserted famine villages and prehistoric remains. Its people are as warm and welcoming as ever. To revitalise its population, perhaps it is time to start a new chapter in its remarkable history. Interestingly, to see what this new phase might look like, we don't have to travel too far.

Heading east across the short bridge that connects Achill Island with the mainland, walkers and cyclists can pick up the Great Western Greenway, a 42km off-road path that follows the route of the former railway, snaking along seashore, farmland and open bogland. It is the longest trail of its kind in Ireland and since its inception in 2010 is estimated to have earned over €1 million annually for the local economy. It has spawned a flurry of activity across the country as local authorities scramble to develop greenways of their own. In economic terms it is estimated that the project took a mere six years to pay for itself while proving that there is significant appetite among local and international visitors for contact with the countryside away from the noise and dangers of roads. The route passes close to the town of Newport but before this it intersects with a tangle of walking trails that lead north of the road into the deep valleys of the Nephin Beg mountain range. Signposted looped walks, cycle trails along little used boreens (the type of roads with grass growing up the middle) and the long-distance Bangor Trail traverse the stunning Lough Feeagh and perhaps the largest roadless tract of land in all of Ireland. Those who persevere along the dead-end road, past the Burrishoole salmon research station and beyond the lake fringes, will find themselves at a small car park and a wooden sign announcing their arrival at 'Wild Nephin'. Established in 2013 by Coillte, it is billed as 'Ireland's first wilderness area' and aims to connect the Ballycroy National Park with the wider area, stretching across 110km^2 of mountain and bog. According to the press release: 'As well as providing completely unique recreational opportunities where challenge, solitude and remoteness are the hallmarks, the project also aims to increase nature conservation biodiversity values, protect a large landscape from human artefacts while facilitating research and enhancing the status of natural ecosystems through a process of non-intervention.'[16]

The project manager for the plan added: 'This is an important day not only for Ireland but also for Europe as this agreement is a key step along the way towards the goal of setting aside 1,000,000 hectares of wilderness in Europe by 2020. Ireland is in the forefront of looking at modified landscapes, the challenges of rewilding and how these can contribute to wilderness in Europe.'

In 2015 Minister for Tourism Michael Ring (Fine Gael) was on hand to show his enthusiasm for the project and its clear potential: '[the area is] too beautiful to put into words ... There is a lot of work to be done on this but when it is finished, it will be one of the biggest areas of wilderness in the world. It will certainly be the envy of Europe.'[17]

The term 'conservation' is, by definition, associated with keeping things the same, or at least as they were at some point in the past. However, this new concept – 'rewilding' – is something totally different. Here, there would be no attempt to impose a vision of what nature should look like but rather nature would be given free rein to shape and mould the landscape as it sees fit. The concept was popularised by George Monbiot, author and journalist for *The Guardian* newspaper, in his 2013 book *Feral*. In it, he describes his interpretation of the concept: 'Rewilding recognises that nature consists not just of a collection of species but also their ever-shifting relationships with each other and with the physical environment. It understands that to keep an ecosystem in a state of arrested development, to preserve it as if it were a jar of pickles, is to protect something which bears little relationship to the natural world.'[18]

'Wilderness' is something of a subjective concept but in the popular mind it suggests a place that is, at least, subject to the rules of nature more than those of humans. It's not a thought that many would associate with Ireland, or indeed anywhere in Europe, given the thousands of years of human activity across the land and sea. It is, nevertheless, a word that appeals as much to the soul as it does to the mind. It is not so much about protecting the environment but unleashing an elemental energy that has heretofore been suppressed. It's exciting stuff. In Britain the rewilding movement (if that is the right word) has captured the public imagination.[19] It would see the return of extinct species, such as wolves and lynx, the removal of dams and weirs along rivers to allow them to flow freely, and the natural rejuvenation of woodlands long gone – particularly in upland areas.

A future for wildlife and people

A native woodland is rich in animal and plant life as well as being a pleasure to visit.

The main criticism of the approach to date has been the impression that 'rewilding' excludes humans. However, its proponents are keen to challenge this: 'Rewilding is not about excluding people ... Current land management practices [in Britain] support a tiny number of jobs per hectare. Few rural dwellers are employed directly on the land; nor do they hunt or get firewood or food from the land on which they live. This can change. But only if the environment is enriched on a scale big enough to create a host of new niches for wildlife, and new opportunities for sustainable rural employment.'[20]

They point to the example of Norway, which saw extensive deforestation through historic times so that landscapes in the south-west of that country were bald and treeless, just as they are in Scotland or the west of Ireland today. In the past 100 years the forests have returned and yet the rural population density is *greater* than it is in Scotland.[21]

In the Wild Nephin area of Mayo, today's landscape is entirely the product of human intervention. There are large tracts of conifer plantations, an extensive network of drains have been excavated to dry out the peat bog, invasive rhododendron is widespread, while the high number of sheep and deer will ensure that the regeneration of native forests will not be possible. In a conference presentation by Dr Craig Bullock it was estimated that the Mayo wilderness project could earn between €860,000 and €3.16 million per year for the local economy (not including the cost of any restoration works) but he emphasised that to attract attention the project needed to be 'unique or distinctive'. The fact that part of the area is associated with the Ballycroy National Park will help in raising its profile. The scenery is dramatic, new walkways provide access and there is no doubting the tantalising potential that this areas holds. Whether it becomes 'unique or distinctive', however, remains to be seen. Much will depend on just how far Coillte is prepared to go in putting the 'wild' into the wilderness. Today's visitors may see some ravens in the sky or encounter a herd of deer; however, other than the ubiquitous sheep, there is little wildlife. But could this change?

The mountains of Asturias and Cantabria in northern Spain at first glance offer little in comparison with Ireland. The lofty Picos de Europa mountain range make mere hills of our highest peaks. In the neighbouring province of Galicia, however, the oddly familiar sound of the *gaites* (a bagpipe-like instrument with Celtic origins) hints at deeper and more ancient connections between our two countries. Research into the origins of our wildlife reveal even more obscure connections. DNA and other analysis show the post-ice-age colonisation of a number of species of plants and animals (from badgers and spotted slugs to the Cantabrian heath shrub) can be traced, not to Britain as might be expected, but the Iberian Peninsula. So northern Spain is not as bad a place as might be imagined with which to compare the landscape of the west of Ireland.

The Picos de Europa National Park is one of a series of designated natural areas that span the north of Spain but, even on its own, it is six times larger than the Wild Nephin wilderness area. It, too,

A future for wildlife and people

has a rural population, traditionally based on low-intensity animal herding and despite the grandeur of the scenery the evidence of human influence is everywhere to be seen. During summer months the town of Cangas de Onis is chock-a-block with tourists, many of whom are Spaniards fleeing the heat south of the mountains. As in most national parks, visitors flock to a small number of scenic locations but there is ample opportunity elsewhere to feel the sense of space which the mountains inspire. Asturias is famous for its traditional, rustic cuisine and particularly its cider, which is gas free but poured at arm's length to infuse a natural effervescence. In promoting itself to the world the province places great store in the quality of its natural environment, its slogan is *paraiso natural* – natural paradise.

Shoppers for souvenirs along the streets of Cangas de Onis will not miss what makes this natural paradise so distinctive and unique, for these mountains still have the big animals that disappeared from Ireland long ago – bears, wolves, wildcats and wild boar. Their faces look out from T-shirts, mugs and key rings. Despite the fact that people exploring the mountain trails are highly unlikely to encounter any of these shy animals the very thought that they are out there, possibly around the next corner, gives a heightened sense of alertness to the would-be hiker. Even in the absence of a close encounter with any of these magnificent beasts the mountains teem with other wildlife. The protection of the top predators in the ecosystem acts as a kind of umbrella, safeguarding the health of populations of flora and fauna across the mountains. Visitors cannot miss the frequent sight of vultures and eagles circling the skies above, in summer there is a dazzling array of butterflies while it's not unusual to disturb sunbathing reptiles as they bask on the rocks.

It is true that the Cantabrian mountains, which run all along the northern coast of Spain, have a naturally more diverse range of native plants and animals than Ireland. Nor is the area without its conservation issues: the region's recovering brown bear population is sensitive to disturbance and is threatened by the development of ski resorts, etc. Wild boar is considered a nuisance, as it is throughout Europe. Even though they are widely hunted the low density of natural predators means there is insufficient check on their numbers. Wildfires are an increasing problem as the human population dwindles. But the biggest problem for wildlife is human acceptance of predators, and of wolves in particular. Centuries of cohabitation with wolves in this part of Spain has led to tolerance of a sort and its iconic value for tourism is further enhancing its public image. But in areas where wolves have not been seen in years, there is bitter conflict with farmers who see yet another threat to their already besieged way of life.

The word 'wolf' on its own is enough to make people sit up straight. Could we see the reintroduction of wolves to somewhere like Mayo and the Wild Nephin wilderness area? To date, Coillte have not commented on their plans for the reintroduction of long-lost species, other than to say that the transformation of the landscape is a long-term vision.

The history of wolves around the world is not a happy one. Persecuted, poisoned, shot and trapped, they have rarely been viewed as anything but a menace and a pest. Great energies have been expended in their outright extermination throughout their range, something that was largely successful in many countries. They are shy and adaptable animals, however, and managed to hang on in more remote mountain areas or places such as Canada and Russia where there were simply fewer people. A modern appreciation of ecology, and particularly the vital role that predators play in maintaining the health of whole landscapes, has been key in changing human perceptions and, in many places, bringing with it legal protection.

In the past fifty years the wolf has made a remarkable comeback in Europe. Its numbers are growing in most of its range and there are now breeding packs in Denmark, France and Norway, where they had been absent for decades or even centuries. In Spain the population shrank to about 400–500 individuals but is now believed to be somewhere between 2,000 and 3,000 and is finding its way to central parts of the country that haven't seen wolves in years. In 2011 a lone wolf was even caught on a camera trap in Belgium, a remarkable apparition given that to travel from the nearest breeding group in Germany meant crossing some of Europe's busiest motorways. This resurgence has come to the delight of nature lovers across the continent and is on foot of evidence, published in *Science* magazine in 2014, of a wider comeback for big, predatory animals. 'Europe's large carnivores are making a comeback ... sustainable populations of brown bear, Eurasian lynx, grey wolf, and wolverine persist in one-third of mainland Europe. Moreover, many individuals and populations are surviving and increasing outside protected areas set aside for wildlife conservation. Coexistence alongside humans has become possible ... because of improved public opinion and protective legislation.'[22]

In Europe, only in Britain and Ireland is the enthusiasm of conservationists muted as there is no chance that wolves will naturally recolonise these islands. On the upside, this geographical happenstance provides us with an opportunity: can we have a reasoned debate that would permit a controlled and publically accepted reintroduction of wolves to Ireland? Arguments in its favour include the restoration of an intrinsic part of our heritage and culture. For centuries, Irish wolves and people lived side by side, only disappearing in the late 1700s. It would make places like Wild Nephin 'unique and distinctive', bringing with it the enormous potential for tourism and the reinvigoration of the local rural economy. Is it feasible from an ecological point of view? Donor wolves would need to be found from a healthy population. The habitat would need to be available, with prey such as deer or wild boar, to support their needs. There would need to be sufficient genetic diversity to prevent future inbreeding, something that is a problem for wolves in Sweden.

Wolves are not dangerous to people but are known to kill sheep so a way of protecting flocks or moving away from sheep-rearing into other ways of land management would need to be found. Could Irish hill and sheep farmers accept wolves were they to enhance their incomes and livelihood? Evidence from Yellowstone National Park in the USA, and other studies that look at the effect of large predators on the behaviour of other animals, suggests that wolves would dramatically reduce the number of crows, foxes and deer in the landscape – nuisance species for many farmers. But it's not just farmers who need to be convinced. A lot of Irish naturalists and conservationists believe that wolf reintroduction is either not feasible or unethical. However, the evidence from the rest of Europe proves that this is not so. The main challenge is human acceptance. Can we overcome our fears of wolves to allow them return to their native land?

Across Europe, where rural communities are already living with apex predators like bears and wolves, the EU is helping to promote coexistence with people through their 'LIFE' projects. This work is focused on helping people to adapt to the presence of brown bears, wolves, lynx and wolverines. Of these 'big four', only for the wolverine is there no evidence that it was ever present in Ireland. The projects are hoping to improve the image of predators among local communities. They are promoting traditional herding or shepherding that minimise the loss of livestock and are actively seeking resolution to conflicts as they arise – and, crucially, they provide the funding to go with it. Since 1992, there have been seventy-eight projects supporting large carnivore conservation across EU members providing €54 million in aid. Then EU Environment Commissioner, Janez Potočnik, summed it up:

> Attitudes towards large carnivores vary widely from village to village, region to region, and from country to country. Some see these apex predators as powerful symbols of wild nature and natural systems, while to others they are fundamentally a threat to lives and livelihoods ...
>
> Coming from a rural community in Slovenia where people, wolves and bears have cohabited for centuries, I can testify that coexistence is not only possible but brings with it innumerable benefits.[23]

Support has not been universal and the challenge must not be underestimated. In Spain, the shooting of wolves, both legal and illegal continues. In Norway, a country traditionally associated with progressive environmental policies, authorities strictly limit the population, which stands at only about thirty-four animals according to WWF. Wolves wandering over the border with Sweden are liable to be shot. In Ireland the experience of reintroducing golden eagles, white-tailed eagles and red kites demonstrates that initial apprehension can be overcome and these projects now enjoy widespread local support. But 'rewilding' the landscape and reintroducing lost species will stretch far beyond just wolves and bears, and includes the restoration of many species lacking such contentious headwinds. This includes much of our lost birdlife, such as cranes, bitterns and osprey. Although the archaeological evidence for lynx in Ireland is slim (a single femur from a cave in County Cork) there is probably already enough woodland habitat in some parts of Ireland to support a viable population of these beautiful animals. In the rest of Europe these shy felines prey on roe deer, a species not found in Ireland. Would a reintroduced Irish lynx population prey on, and so help control numbers of, sika deer or the alien invasive munjac deer?

Naturalist Cóilín MacLochlainn thinks it might and has a vision for a rewilded landscape along the Avonmore River in County Wicklow. He sees such a project rivalling the Black Forest in Germany in terms of its extent and amenity value. The area already holds the largest amount of semi-natural woodland in Ireland. It is home to small but unique populations of the specialist birds goosander, redstart and pied flycatcher. In recent years the red kite was reintroduced and it was here that the great-spotted woodpecker chose to reintroduce itself naturally after disappearing from our landscape. MacLochlainn's vision very much includes people. Restored native woodlands would provide sustainable fuel and other resources, while a long-distance walking trail would follow the valley floor from source to sea. It would be a great forest (*Coill na hAbhann Móire*, the Irish for Avonmore, means forest of the big river) and would be inhabited by great animals – lynx, wolf, wild boar and the grouse-like capercaillie.[24] 'Rewilding' visions such as these offer hope for restoring marginal rural communities and natural landscapes.

In fact, rewilding is not a new concept and Wild Nephin is not the first project of its kind in Ireland. Although neither the words rewilding nor wilderness are to be found on its promotional material, there is a project where nature's hand has had more or less free rein for twenty years, where populations of near-vanished species are rebounding, where natural processes have been restored and where material benefits to local people are being reaped. Listed by *The Irish Times* in 2015 among the top twenty 'best days out in Ireland' the Lough Boora Parklands in County Offaly were a peat-mining wasteland in the early 1990s. Having stripped away all the turf, the state company responsible for the exploitation of Ireland's boglands – Bord na Móna – had initially attempted to develop the area for conifer plantations. When this didn't work, a group of Bord na Móna workers formed the Boora Parklands Group to develop a new approach to using the land.

That was in 1996. After turning off the water pumps which had allowed machinery to work the peat, large areas were flooded, forming new lakes and wetlands. New communities of plants and animals emerged, from birch woodlands to expanses of bog cotton and there was no attempt to recreate what had been there (probably an impossible task in any event) and no attempt was made to impose a vision of what it should look like. Flocks of ducks, geese, swans and other wetland birds settled on the lakes – over 5,000 of them during the winter of 2008–09 with a particular abundance of the threatened golden plover. In 2012 the National Biodiversity Data Centre (NBDC) organised a 'bioblitz' at the Boora Parklands (a bioblitz is an attempt to count as many different types of plants and animals as possible over a 24-hour period). During that unseasonably cold and blustery May day (I was there!), 946 species were counted – an astonishing feat which was greater than the number found at similar events in either the Wicklow Mountains and Burren National Parks the following year.

At the turn of the century the exploited bogs of Boora were the only place left in Ireland where wild grey partridges could still be seen. Once found throughout the country, by 2001 there were only twenty-one individuals left. Since then the Irish Grey Partridge Conservation Trust has been boosting the population with special conservation measures and by 2010 there were over 900 birds.[25] And it's not only the partridges that are benefiting: in 2015 BirdWatch Ireland reported that there were over 122 lapwing nests, a highly endangered bird, at Boora. On a warm spring day, this rather hidden part of Ireland is one of the best places anywhere in the country to see wildlife. Few leave for home without seeing hares or frogs, hearing the sound of the cuckoo or watching the spectacle of flocks of birds alighting on one of the many lakes.

Since its early days people have been at the heart of the plans for the Boora Discovery Park (as it is now known). Its 3,000 hectares are home to a sculpture park, bird hides, cycle trails and bike-hire facilities, walking paths and, more recently, a visitor centre and café. In 2015 it scooped the Best Environmental Tourism Innovation Award at the biennial Irish Tourism Industry Awards. According to the press release: 'Lough Boora Discovery Park is a first-class example of an amenity that emphasises harmonious integration with the environment and maximises protection of environmental resources.'

Bord na Móna controls a substantial 80,000 hectares of publically owned land throughout Ireland. Established in the 1950s to exploit peat, it is now being forced to diversify as the peat resource dwindles. According to the company itself, by 2030 there will be no more turf worth harvesting. This date may even be ambitious as concerns over climate change and the wisdom of maintaining three peat-powered electricity generating plants is called into question. For some years Bord na Móna has acknowledged the growing environmental value of peatlands and is actively involved in restoring a number of midlands raised bogs. There is greater understanding of the importance of these habitats for storing carbon, reducing the impacts of floods and providing a home for some unique wildlife.

As the peat harvesting winds down, however, there are significant questions to be asked about how this land will be used. With these questions come the tantalising opportunity that perhaps Ireland could see itself at the forefront of what could be one of the largest habitat restoration programmes ever seen. In fifteen years from now, could we see the lessons learned at Lough Boora replicated across the midland counties? Could we see the reintroduction of lost icons of our heritage, like the crane? Will we again hear the bittern boom across the vast reedy wetlands that are waiting to be created, or watch the osprey swoop on fish-filled lakes?

The amenity and tourist potential is vast – a massive fillip to communities that have suffered chronic emigration and economic decline. New self-generating woodlands could in time provide renewable sources of fuel for local people. There could be other benefits in rewilding the bogs as the peat starts to suck carbon once again from the atmosphere rather than spewing it out; creating natural reservoirs may also hold enough water during floods to prevent it entering people's homes. We need solid scientific research to answer these questions but there is some cause to be optimistic that decisions made in the coming years will maximise benefits for the environment and people. As a public asset, it must ultimately be the Irish people who decide the outcome of these decisions.

It is how we use the land which is the most important factor in determining the future of our wildlife. One of the most contentious ways in which we use land in Ireland today is forestry. Ireland has a magnificent climate for growing trees – cool and well watered. We also have a great diversity of native tree species from which to choose from, from fast-growing birch, willow and ash to slow-growing hardwoods like oak and yew. Each type of wood brings its own unique heritage and uses. Despite this, Ireland has one of the lowest levels of forest cover in Europe. According to the World

Bank, in 2013, 10.7 per cent of our land area was under trees – half that of Belgium. Only treeless Iceland has less.[26]

However, the word 'forest' is not a good description for most of these trees. Since government policy promoted the growth in tree cover in the 1950s the preference has been overwhelmingly in favour of fast-growing, non-native conifers which are alien to our landscapes and ecosystems. Their uniform stands of even-aged trees jar in our countryside and, because they are dark and tightly spaced, they leave practically no room for wildlife. Although state-owned plantations are open to public access, being dim and uniform, most are not pleasant places in which to walk. On dropping to the ground their needles change water chemistry, acidifying watercourses, while the method of felling the trees, known as 'clear-felling', washes away sediment and further damages fish habitats. Because they are monocultures, they are prone to infestations and so they need to be repeatedly sprayed with noxious fungicides and pesticides. The vast majority of the tree cover in Ireland today is of this type as only 1 per cent of the land area is under native forests.

Native forests are those where the greater proportion of the trees are composed of native tree species – such as oak, alder, ash, yew, Scots pine, rowan, hazel or willow. They are themselves diverse places and give us the ancient oak woods of Killarney, where the boughs are festooned with mosses and ferns, or the swamp forests of the Shannon and Lee rivers, with their tangle of willow and alder. These forests are full of life but are vanishingly small fragments, scattered widely. The decision to promote non-native plantations and clear-felling over permanent-cover native forests has had terrible consequences for the landscape and the environment. The social impacts have also been considerable but under-evaluated. Although providing an income to rural landowners, plantations have displaced traditional farming, breaking up communities with walls of dark, impenetrable vegetation.

As part of the EU's commitment to prevent catastrophic climate change, Ireland has signed up to legally binding targets of reducing emissions of greenhouse gases by 20 per cent by 2020 when compared to 2005. According to the EPA, in 2014 the agriculture sector was the single largest contributor to Ireland's emissions, responsible for fully one third of all greenhouse gases. This is primarily a result of the methane that is belched and farted from the 7 million-strong cattle herd. In the absence of any drive to reduce the size of the national herd the government has suggested that these emissions could be offset by restoring peatlands and planting new forests. Clearly such an approach holds a lot of potential, not only for taking carbon out of the atmosphere, but also for creating habitat, amenity and sustainable job opportunities. In its 2014 report, 'Forests, Products and People', the state envisages that tree cover will increase to 18 per cent of Ireland's land cover by the middle of the century. However, it is clear from this document that the focus is firmly on earnings from the export of timber products and there are fears that the expansion of the sector will see ever more plantations of exotic conifers.

Although the negative social impact of past forestry policies has never been fully explored, this concern was articulated succinctly by the Irish Natura and Hill Farmers Association – a group representing marginal, mostly upland, farming communities:

> statements made in relation to the role forestry can play in providing a carbon sink is of growing concern in many rural communities.
>
> It is becoming quite obvious that an agenda is being pursued that encourages large-scale afforestation in some areas which can then be used as a carbon sink (off-set) in order to facilitate an expansion in other areas ...
>
> Rural communities and especially those in disadvantaged areas have experienced a difficult eight years through business closures, emigration, unemployment and the withdrawal of services. Farming, and the money spent around it, is what now keeps many of these communities going. Farm families are often what keep the schools open, the GAA clubs going and the community alive, because for better or worse the farming activity they are involved in keeps them there.[27]

Rural communities are right to be worried. A report published early in 2016 by the state forestry research body COFORD highlighted that there are 1.8 million hectares of land 'limited for agriculture'. 'These lands have a higher proportion of difficult soils, often economically marginal for agriculture, with forestry presenting a viable alternative land use option.'[28]

In contrast, the government's commitment to planting or restoration of native woodlands is a paltry 2,700 hectares. The time is right for a change in approach. Many of these areas with 'difficult soils' coincide with our most scenic places. Policies which drive greater production of food where the land is not able for it lead to depopulation and environmental damage. In Ireland there is a remarkable absence of research being carried out into how best to farm in these marginal areas. In upland areas long deforested, nobody knows what a healthy landscape should look like. Fortunately, there are examples of new approaches that show promise for those who want to see more trees and more farming, and although we have to leave Ireland, we don't have to travel too far.

The Pontbren Project in Wales describes itself as a 'farmer-led approach to sustainable land management in the uplands', established, run and managed by a group of neighbouring farmers in the valleys of the River Severn. Since the end of the 1990s they have dramatically changed the way in which they farm: 'The key to these changes was to improve shelter by tree planting and restoring neglected woodland and hedges … [and] reinstating woodland management as an integral part of successful modern upland livestock farming systems.'

The Pontbren farmers reared sheep, as they had always done, but used traditional breeds and managed their numbers so that native woodlands could develop on the pastures. These provided shelter for the animals, making them healthier and more productive. New hedgerows separated neighbouring flocks, reducing the risk of spreading disease. New ponds replaced patches of mud and rushes, providing a supply of drinking water while reducing the chances of animals contracting foot rot and liver fluke. The emerging woodland not only benefits the sheep but provides a source of commercial timber for sawmills and woodchip for home heating and animal bedding. As the habitats improve so the wildlife has returned and farmers may have earning potential in leasing hunting rights on their land. In the space of only ten years, 120,000 new trees and shrubs were planted, 16.5km of hedges have been created or restored and woodland cover increased from 1.5 per cent to nearly 5 per cent, 'with no loss of agricultural productivity'.[29]

Seeing the potential of their work, the Pontbren farmers collaborated with academics who would add the necessary data to confirm the changes they were seeing on the ground. One of their most important findings was how the newly wooded areas helped to increase the amount of rainwater absorbed into the soil, helping to prevent flooding at the bottom of the Severn valley. In fact, the rate at which water was absorbed into land with trees was found to be a massive 60 times that of the open pasture. The restored landscape has broken up the once homogenous hillsides into a mosaic of habitats that is providing a home to rare and threatened species. Excluding sheep from watercourses has allowed tree-fringed riverbanks to emerge, altering the flow of water to a more natural state and benefiting local trout populations. The work at Pontbren continues but there is now clear evidence that landscape-scale habitat restoration can work for farmers and, by mitigating the effects of flooding, society as a whole.

In Ireland restoring large-scale native woodlands can reinvigorate rural communities, generate sustainable incomes, help alleviate flooding, create amenity and tourism opportunities, enhance water quality, absorb carbon and provide a future for our unique upland wildlife. Planting more swathes of conifer plantations to be exported and made into cheap furniture will create incomes for some, but will provide few other benefits.

Rewilding provides hope for declining communities, damaged landscapes and beleaguered wildlife populations. New models of upland farming, like that at Pontbren in Wales, maintain agricultural production but maximise the benefits nature can bring

A future for wildlife and people   309

Inis Meáin on the Aran Islands, County Galway, where High Nature Value farming is benefiting wildlife and local people.

to the community. Their success depends on engaged and open-minded farmers, supported, but never led by, local politicians and government agencies. They recognise that nature in Europe is the product of thousands of years of farming activity but, equally, that modern agriculture and centralised policy making have been bad for the environment and many rural communities. Thus the concept of High Nature Value farming has emerged. It has been around since the 1990s and so is not new, but as a concept it has been slow to take root, especially given its potential and the increasing sense of social and environmental urgency. It is associated with traditional, indeed bygone, ways of farming, requiring little or no use of artificial fertilisers, herbicides or pesticides. These systems have tended to produce less food and are frequently ill-suited to modern machinery. As a consequence they have not fared well in the global agri-industrial drive that characterises much of modern food production. Nevertheless, these traditional ways of managing land have formed the characteristic landscape of Europe that is such an essential part of the continent's heritage.[30]

In Ireland Dr James Moran from the Institute of Technology, Sligo, has been promoting High Nature Value farming and the research needed to apply it on the ground. His studies with the European Forum on Nature Conservation and Pastoralism have found that as much as 30 per cent of Ireland's agricultural land is suitable for this lower-intensity approach to farming. Unlike traditional methods of protecting nature this approach does not involve the designation of land – something that has been particularly contentious in Ireland. The Burren is perhaps the best example of High Nature Value farming in practice. It works because it utilises local knowledge and genuinely values the contribution of local farmers. Rather than imposing a 'one size fits all' policy, there is ample scope for innovation and solutions adapted to individual farms. Incomes are based on outcomes – in the Burren farmers are paid a premium based on the quality of the grassland habitat or the protection of particular features such as watercourses or archaeological remains. Although a part of the Burren is a National Park, and much of it falls under European designations for nature conservation, these are complementary. Designations have provided opportunities for financing projects, and are not seen as a threat. And the results are there for all to behold – the Burren is the best-managed landscape in Ireland. In 2014 the project was expanded to the Aran Islands and it is hoped that the Burren success can be replicated on these unique places that stand guard at the entrance to Galway Bay.

High Nature Value farming holds enormous potential for many parts of Ireland, and particularly those areas that have been bypassed by the economic boom of the 1990s and early 2000s. Yet, despite the promise it holds for the Cork and Kerry peninsulas, Connemara, the Wicklow uplands or Donegal, little has been done up to now. Much of the blame lies in the way in which farming subsidies have been

A future for wildlife and people

paid out. Since the 1990s so-called agri-environmental schemes have been used to boost farming incomes while attempting to enhance water quality or valuable habitats. The large amount of money spent on these schemes, however, has not been targeted at areas where it is most needed, but spread widely and thinly. The actions needed to qualify have been decided centrally and do not avail of farmers' knowledge and experience, or take account of local circumstances. And so, for instance, a farmer with a wet field full of rushes halfway up a hillside, in order to qualify for direct payments, must drain the field, remove any scrub and reseed with grass. The added value to the farmer may be negligible while the effects on downstream flooding, wildlife habitats, water quality and landscape are not considered. It will be a challenge for future rounds of the CAP to acknowledge the highly contrasting demands of different farming sectors and the environments in which they work. But not an insuperable one.

Greater protection for the environment and restoring landscapes in scenic areas is in many ways a no-brainer. The current economics of farming in these areas make little sense to anyone, benefit no one and erode the basis on which any future recovery relies. A tougher battle lies where, superficially at least, there is a sharper line between environmental protection on the one hand, and economic growth on the other. Getting past the marketing spin about how Irish agriculture is among the world's most sustainable, and developing profitable models of food production – particularly in the dairy sector – without the negative side effects will be difficult. But by no means is it impossible. Large farming organisations in Ireland have traditionally seen moves to improve water quality or conserve rare habitats as a threat, and have fought them vigorously. But even the most ardent conservationists do not suggest that productive farmland should be turned over to nature. According to government figures, the food and drink industry in Ireland was worth €10.8 billion to the economy in 2016, employing over 100,000 people. Nobody is advocating that we turn our back on this. Intensive agricultural grassland or tillage crops have little to offer wildlife but that doesn't mean that farms themselves cannot become more nature friendly.

In these farms, which cover approximately half of Ireland's land area and whole counties in the south and east, there are concrete measures which can be taken to boost nature, while crucially, not imposing undue restriction on farming practices or impacting the bottom line of farmers themselves. The main problems that need to be addressed are the loss of farmland birds and pollinating insects, and water pollution. The good news is that there is sufficient evidence to show what needs to done to address these issues.

Because birds are so well monitored compared to other groups of animals, the loss of familiar farmland birds has been flagged for some time. It is a phenomenon that has occurred across Europe but in Ireland has resulted in the near extinction of some species (quail and corncrake, for instance) and the outright loss of one – the corn bunting. Remedies introduced through so-called 'agri-environment' schemes, whereby farmers are paid for specified actions, have not delivered results despite the significant costs to the taxpayer. In the UK there have been extensive studies undertaken in an effort to understand why this is so.

In general, the longer an ecological experiment has been going on the more value it has and for this reason the UK is a particularly rich source of data. One recent publication has given rise to hope that a combination of measures on farms can significantly boost bird numbers, even those which are rare or declining, such as lapwing or yellowhammer.[31] The study from ornithologists at the Game & Wildlife Conservation Trust and the RSPB looked at two demonstration farms, one at Loddington Farm in Leicestershire, the other at Hope Farm in Cambridgeshire, over a twenty-year period. The farms are a mix of arable crops and grassland pasture with small woodlands and hedgerows. Over the two decades, not only were different species of birds recorded but also their abundance. This information was compared to records from the surrounding regions so that the changes to bird populations could be reliably attributed to farm management rather than broader factors.

At both farms it was found that the abundance of birds increased 'much faster' than in the surrounding countryside through measures

A future for wildlife and people

to increase the level of cover during the nesting season, and food provision during both summer and winter. At Loddington Farm the chief predators of nesting birds – crows and magpies – were controlled and once this ceased, other bird numbers fell back to what they were at the start of the experiment. At Hope Farm, however, there was no shooting of predators and here the gains were maintained. The researchers suggested this may have had something to do with the presence of mature woodland at Loddington, which in turn provided more nesting and lookout posts for the crows and magpies (although other evidence points to the benefits of well-managed hedgerows in better protecting nests from predators[32]). At Hope Farm the initial abundance of predators was less to begin with and stayed much lower than at Loddington. Most Irish farms are not adjacent to mature broadleaved woodland so for this reason the appropriately named Hope Farm may be a closer approximation to conditions here.

Different bird species were grouped so that farmland specialists, 'priority' (i.e. particularly rare or threatened) species, herbivores, grain eaters, insect-eaters and open-cup nesters were all examined separately. But across each group the results showed the same impressive increases. At Hope Farm the farmland specialists tripled in number while the 'priority' birds more than doubled. Across each group the trends in the surrounding countryside were flat or declining. So what did they do at Hope Farm to achieve these results? It mostly came down to providing good-quality habitat and, in particular, cover for wild birds, leaving a buffer around field margins, 'beetle banks' (basically a 2m-wide strip of rough grass running through an arable field which provides an invertebrate food source for young birds in particular), leaving some of the cereal crop unharvested over the winter, the creation of new ponds, ditches and hedges and sowing flower-rich seed mixes.

Many of these general measures benefit a number of bird species. One was specific to skylarks, a bird which has declined in number in Ireland by 30 per cent since 1970 and which no longer breeds in many areas of the south and east of the country.[33] This involves the creation of 'skylark plots' in the centre of arable fields, basically square patches of ground where the crop seeds are not sown. The technique leaves areas in winter-sown crops available for the birds to forage and, according to the RSPB, these threatened birds have more and healthier chicks where the plots are present.

What effect did these measures have on the basic business of the farm, i.e. the growing of crops or the rearing of animals? At Hope Farm the total area was 181 hectares while the different habitat measures combined approximately 14 hectares in any given year, under 10 per cent of the extent. Given that many Irish farmers are already participating in an agri-environment scheme and that even those who are not are required to undertake 'greening' measures in order to receive the basic subsidy payment, the implementation of Hope Farm-type measures is unlikely to affect their current levels of production. In concluding their paper, the researches say: 'Both Loddington Farm and Hope Farm demonstrate that appropriate management can generate large local increases in numbers of priority farmland birds over relatively short time periods.'

One of the principal causes of the decline in bees has been a loss of flowers in pastures and hedgerows. Many bee species emerge from hibernation early in the spring and need to feed straight away, while only re-entering hibernation late in the autumn. The range of flowering plants in a traditional hay meadow and well-maintained hedgerow provided the succession of nectar sources needed to sustain whole colonies. Modern intensive grasslands, however, may contain only one type of clover or no flowers at. Allowing common plants like thistles, dandelions and vetches to grow around field margins or planting pollen-rich trees like willow, hazel and hawthorn is great for pollinators. Leaving ivy on trees provides a particularly valuable resource because this plant flowers very late in the year. Many of these wildflowers have traditionally been seen as no more than 'weeds' so simply a change of mindset, in seeing the value of the commonplace, will go a long way.

Most Irish farms are smaller than those in England and in particular we still have a remarkably dense network of hedgerows.

But these hedges are no longer being maintained in the traditional way, resulting in lower diversity of wildlife and, without intervention, they will disappear entirely. Fewer people working on farms and the availability of cheap, electrified fencing means that maintaining hedges has become costly and time consuming. But the benefits of healthy hedgerows is far greater than just dividing fields – they allow more water to be absorbed into the soil, thereby reducing the effects of flooding; in preventing contact between neighbouring animal herds they help prevent the spread of disease such as TB; and they provide shelter for livestock, meaning more of the food they eat goes to putting on weight rather than keeping warm. These are tangible benefits to farmers so we need more promotion of the advantages and better availability of training into best-practice maintenance. Vibrant hedgerows will help farmland birds, pollinators and a host of other plants and animals, as well as retaining the character of our countryside. Giving farmers the supports they need to get the job done is public money well spent but must be paid based on the outcomes, not the actions.

A particularly intractable problem is the ongoing pollution of waterways arising from farm run-off. The EPA estimates that half our rivers and lakes are polluted and half of this is from agriculture. As the rivers run to the sea many of our estuaries are also polluted. The expansion of the dairy sector in particular presents an enormous challenge as intensively reared dairy cows produce more manure than beef cattle. The drive to reseed pastures to provide more palatable grass for the cows exposes soil to the elements, leading to the loss of silt and sediment to watercourses. As things stand, the scale of pollution in our major rivers is set to increase, even though the government agreed to achieve clean water throughout the country by 2015. Our climate is simply too wet, and the soil does not have the capacity to use the nutrients in the slurry that is being generated. But even in a scenario whereby the dairy sector continues to grow, there are things that can be done to stop nutrients and sediment escaping to rivers. Manure, after all, is a valuable resource that should not be wasted. In many European countries, farm waste contributes to the generation of natural gas. Placed in an oxygen-free container (e.g. a sealed tank) manure will generate methane, which can go on to power homes or businesses. It effectively reduces carbon emissions by displacing fossil fuels while also generating income for the farmer. Germany has built 9,000 biogas plants since 2000 while it is estimated that creating and maintaining the infrastructure across Ireland would generate 2,250 jobs and power 20 per cent of Irish homes.[34] A further by-product of biogas generation is hot water, which can be used to heat homes or farm sheds. Organic matter is reduced to an odourless residue that requires less storage space but which is just as good a fertiliser for the land. Of course, if it enters a river or lake it will cause pollution, but because it is more manageable than litres of liquid slurry, farmers can be much more targeted in applying it in the right quantities at the right time, or even exporting it if needs be. To date, however, the uptake of biogas generation has been unimpressive in Ireland; this is because fees paid for feeding the electricity into the grid are presently too low to encourage investment.

Soil is the basic foundation of any farming system and is by far a farmer's greatest asset. Despite this, tonnes of it gets washed into rivers every year where it damages fish-spawning habitats and other aquatic life. This arises when soil is exposed to rain from ploughing, reseeding or too much trampling from hooves. Yet well-researched techniques exist here too, from 'no plough' drilling of cereal crops to the creation of small wetlands at the lowest part of a field. Wetlands are excellent for trapping sediment and are also proven to remove other pollutants such as dangerous bacteria and nutrients – not to mention providing habitat for farm wildlife. Many farms already have unproductive corners of fields full of rushes which would be perfect for wetland creation.

Can we have unlimited expansion of the agricultural sector without environmental problems? Probably not. But the know-how, both modern and traditional, is readily available so that we can have sustainable farms while at the same time making, at most, marginal sacrifices in terms of productivity and profit. But we have yet to see the drive to put the right measures in place that will address some of the greatest environmental pressures we currently face.

The loss of nature at sea has more than matched that on land. In 1900 the Irish fishing fleet amounted to 6,341 boats employing over 26,000 men and boys. According to BIM, the seafood industry employed 11,000 people in 2013 and contributed €700 million to the economy. There are currently nearly 2,100 fishing vessels registered with the Department of Agriculture. Much of this economic activity is based in coastal and rural areas like Killybegs in Donegal, Kilmore Quay in Wexford or Castletownbere in Cork. However, many of the smaller ports have dwindled greatly in recent years so that activity is increasingly centred on fewer, larger ports. Ask anyone in the Irish fishing industry what the biggest problem is and they will invariably state that it is lack of control over fish in Irish waters, most of which are caught by vessels from other nations. Correcting the historic wrong of the terrible deal done by a previous Irish government is high on the agenda of fishing communities and their representatives, and has been for thirty years. But this does not mean there is nothing we can do to boost the quantity of fish and seafood, particularly in the seas close to shore. Within the so-called six-mile limit the Irish government has full authority over the management of fishing and it is out of bounds for foreign vessels.

It is within this shallow, narrow strip that centuries of fishermen and women made their living, typically in small boats using very low-impact gear. This cohort of the fishing industry has been virtually wiped out in recent decades while governance, i.e. who is allowed to do what, where and when, is practically non-existent. The system allows hydraulic dredging for razor clams (among the most destructive forms of fishing) right up to the shore, or the practice of 'pair trawling' whereby a net attached between two boats sieves the waters of an inland bay of its contents – scooping up marine life as well as the gear of smaller fishermen. It's a system that robs local fishermen of their livelihoods as well destroying a marine environment that would otherwise attract anglers, wildlife watchers and scuba divers. Many of our coastal bays and inlets are designated for nature conservation but, except in one or two areas, we have yet to see any framework that would allow conservation aims to be met. This is a travesty for all concerned, so what needs to be done?

In countries that successfully manage their fishery resource, protecting the environment is the number one priority. It is simply a fact that without taking this step the resource upon which communities depend will continue to diminish. In Norway, where this nettle has been grasped, public acceptance relied on setting rules in an open and transparent manner, in consultation with those affected, and strictly enforcing the rules once they were set. Local fishers, angling groups, and tourism-based businesses have a say in how the system is run and contribute to the science which allows for successful conservation. None of this happens in Ireland and so we are left with a system which satisfies no one. A study published by scientists from the University of York, and analysing 144 examples from around the world, found that closing off 30 per cent of the sea from any human interference would not only be good for marine life but would also be good for people.[35] So far, none of Ireland's coastline is off limits in this way. Yet such a move, to close off the sea entirely, may be drastic. More preferable would be better management, where the use of damaging fishing methods, such as dredging and trawling, is greatly restricted, certainly to beyond the six-mile limit, while the use of some gear, such as tangle nets would be banned entirely. Any method of harvesting has the potential for over-exploitation so the key is wise management.

In 2008 the small inlet of Lyme Bay, off the coast of Dorset in southern England, was closed off entirely to scallop dredging, which was destroying marine life in the area. The closure has allowed the 'dramatic recovery' of sea fans and other fauna which live attached to the sea floor.[36] However, it had the unintended consequence of driving an increase in the use of other types of fishing gear, particularly pots and fixed nets, so there is now concern that this is leading to overfishing. Fishers in the region, however, had already seen the benefits to having restrictions on the level of activity through the labelling of their catch as 'responsibly caught', something that is rewarded with higher prices, while sea anglers have reported an increase in the diversity of fish within the closed area. Before the closure, relations between the fishing community, conservationists and government authorities were characterised by mistrust and rancour. This has been overcome and there is now better cooperation between the various interested parties. The recent history at Lyme Bay shows how the issues are more complex than they first appear; however, they can be overcome to the benefit of people and the environment.

Across Ireland the solutions to a brighter future are already at hand. Whether it is rewilding our bogs, protecting our seas or restoring nature in hand with farming, we already have examples that show the potential waiting to be unleashed. Progress is painfully slow, as yet nowhere near fast enough to see a reversal in the worrying trends that face our wildlife.

The body that markets Irish food and drink, An Bord Bia, has set a strategic objective: 'To enhance the reputation, based on the principles of sustainable development, of Irish food, drink and horticulture, among consumer and trade buyers in the marketplace.'[37]

The main challenge is that while An Bord Bia is steering the ship in the right direction, the industry leaders and politicians have yet to pick up the oars and start rowing. On the one hand, we have some of the best and most dedicated scientists working in government agencies such as the Marine Institute, the EPA, Teagasc (farm research), Inland Fisheries Ireland and the NPWS. Yet this experience and expertise does not seep into the higher levels of government. Much of it is stymied by the smothering force of powerful industry lobbies so that sectoral interests win out over the public good. But while lobby groups are fully entitled to promote the interest of their constituent members, the real failing lies with the politicians, who too often abandon their duty to keep these forces in check. An overwhelming majority of Irish people believe that nature conservation is important: 97 per cent of respondents to a European Commission poll in 2015 said they agreed that 'we have a responsibility to look after nature' while fully 10 per cent of people said they 'participate as a volunteer in projects and actions dedicated to biodiversity'.[38] Yet this is not reflected in the words or actions of our politicians.

The principal agency responsible for the conservation of nature in Ireland is the National Parks and Wildlife Service (NPWS). It is a reviled organisation in many rural areas, seen as preventing traditional turf-cutting or dredging of rivers which some are convinced will solve the problem of flooding homes and farmland. In 2010 a review of the body was carried out by Grant Thornton consultants, which stated that the responsibilities of the NPWS were (among others):

- to create and update an inventory of the species of wildlife and habitats in Ireland
- to protect the most important areas for habitats and species
- control of activities throughout the country that may harm or threaten habitats or species
- the management and protection of designated sites, other sites and national assets particularly the national parks
- the development of management and conservation plans for sites, habitats and species

and

- the raising of awareness through the dissemination of information and education in relation to our environmental heritage.[39]

This is a substantial responsibility. In preparing their report Grant Thornton consulted with those outside the NPWS and found that:

- The general feeling ... is that the NPWS strategy is not clear or not clearly communicated.
- The external perception is that the NPWS takes a very reactive approach to its work and there is little or no forward planning.
- There is a perception that the NPWS is focused solely on the legislation that it works under and not on the communication of their overall wildlife conservation message.

Meanwhile, those working for the agency were also canvassed. Their views were that:

- The NPWS tends to take a very reactive approach to the various tasks that fall into its remit, rather than the proactive approach required for the successful execution of its activities.
- Communication is poor within the organisation.
- Views were expressed on the lack of staff dedicated to the management of National Parks.
- There is a perception of insufficient resources across the organisation to carry out all of the tasks and activities required to be fulfilled by the NPWS.
- The feeling among staff is that the organisation is missing out on opportunities for the generation of positive relations with the public, as it was considered that the NPWS only appears to be associated with negative stories in the media.

In conclusion, the auditors found that 'the NPWS was not performing as it should have been in some particular areas'. Yet how could it, given the low profile it has in broader government strategy? The audit report highlighted one of the biggest failings in that it doesn't even have a Chief Executive Officer, rather responsibility is diffuse and overlapping and this leads to what the report describes as 'scope to avoid accountability within management'. The NPWS itself is tucked away in the peripheral Department of Arts, Heritage, Regional, Rural and Gaeltacht Affairs, even though it would be better situated where its tasks are most relevant: as a wing of the EPA where it would enjoy a degree of independence.

The low level of awareness of the NPWS and the onerous expectations placed upon it is reflected in its annual budget. Expenditure on the agency in 2014 was €14.3 million, a paltry sum when compared to other government budgets. For instance, the state body responsible for marketing Irish food and drink as 'green and sustainable', An Bord Bia, was funded to the tune of €32.3 million in the same year.

Many will query why we should prioritise plants and animals over hospital beds, housing or schools for our children. Yet due to the sometimes-curious priorities of the state, to increase funding for nature would not necessarily require a cut in these other vital functions. In 2014 for instance, the 'horse and greyhound racing fund' (no doubt vital to the public interest) received an impressive €54.2 million – well over three times that allocated to the NPWS. In 2015 the NPWS had a staff of 283, according to figures I received from the Department of Arts, Heritage and the Gaeltacht, while the Grant Thornton report recommended that it needed at least 400 to carry out its functions. Only sixty-eight of these were actual rangers on the ground, dealing with everything from education to monitoring of rare species to policing wildlife law. The low number spreads them thinly while some counties have no full-time dedicated rangers at all. During the recession, from 2008 to 2013, the agency was gutted, with a massive 67 per cent cut to its non-staff budget.

During the Celtic Tiger years the largest habitat creation scheme in Ireland in over 200 years got under way and today provides at least 2,000 hectares of grassland, woodland and wetlands which had not previously existed. Practically fenced off from human interference,

A future for wildlife and people

these habitats are secure from spraying or ploughing, culling or poaching. Surprisingly, this new reserve falls into no specially protected area and is subject to no particular management plan to steer its development, yet its future is as secure as any national park.

The construction of over 1,100km of motorway was never intended to provide a refuge for wildlife, and indeed measures are taken to keep larger animals such as deer away lest they cause accidents. Yet, like the canals and the railways before them, motorway edges now provide a remarkably diverse range of habitats free from many human intrusions. Native woodlands have already emerged along the older sections, rocky scree with shallow soil is flush with orchids in summer, kestrels hover as they hunt for small mammals; reed-filled ponds remove pollutants from road run-off but are also home to ducks, dragonflies and swans. And these are only the features that can be glimpsed at high speed, as no study seems to have been carried out on the true diversity of nature now inhabiting these areas. Speaking to a farmer in County Meath in 2015, I was told there had never been so many birds on his land since the road was built.

Big infrastructure projects are not normally seen as being good for wildlife but today both of the canals which cut across the country from Dublin to the River Shannon – the motorways of their day – are now protected as Natural Heritage Areas. Perhaps in 100 years, when airborne shuttles make roads obsolete, the M50 and its cousins will also be protected for nature and amenity. Until recently, Ireland's quarries largely operated without regulation, planning permission or environmental controls. No doubt some of them caused pollution or loss of habitats but many today harbour remarkable populations of plants and animals, even in the ones where quarrying still goes on. In a large quarry near Dublin recently, I watched as a peregrine falcon circled its nest oblivious to the din of rock crushers beneath. This bird nearly vanished in the 1960s from pesticide poisoning but has rebounded in large part due to the availability of new cliff ledges in quarry pits. Meanwhile, our cities are becoming increasingly wildlife-friendly. City parks and gardens provide homes to badgers, safe from snares and culling programmes. The Lough in Cork city teems with bats on summer nights, Galway natives can stroll through ancient oaks in Merlin Woods while otters swim through the side channels of the River Shannon in the heart of Limerick city. Dublin Bay, meanwhile, recently became a UNESCO biosphere reserve. Increasingly local authorities are working with community groups to make parks more wildlife (and people) friendly. In my own local patch – St Catherine's Park in Lucan, County Dublin – kingfishers can be seen along the wooded banks of the Liffey while a herd of Highland cattle is helping to restore rare plant species. There are even reports that pine martens have made their way back – surely for the first time in many hundreds of years.

Most of the positive news stories for nature over the last couple of decades have come from these local community groups or small organisations – I'm thinking of the fight to save the Burren by local people and its subsequent rebirth, the emergence of the Lough Boora Parklands in County Offaly after the intervention of the Bord na Móna workers, the reintroduction of birds of prey by the Golden Eagle Trust, or the army of everyday naturalists taking small actions to save nature by taking litter from their beach, recording birds or planting native trees. These people do what they do for the love of the outdoors and the love of their local area. They know instinctively that nature is more than just nice to look at, but essential sustenance for our lives and our well-being.

But these bottom-up initiatives will only get us so far. If we want to see changes in how we fish or farm or use the landscape, it has to happen at government level. We need politicians to act with a longer-term vision, to see the opportunities that protecting nature brings and to fight the lobby groups in defence of the public interest. If I could do one thing right now to stop the haemorrhaging of nature in Ireland it would be to reboot the NPWS. This requires the will of politicians to recognise the important task it has been set and provide it with the tools to get on with the job. I would put it in a government department where its voice will be heard and respected, not marginalised and blamed.

Michael Viney is one of a small number of environmental champions in Ireland who, in his weekly column for *The Irish Times* over forty years, has quietly but consistently advocated for our wildlife. It's the first page I turn to on a Saturday over lunch – one of life's small, anchoring moments. In 2002 his musings fell on the recent acquisition of James Joyce manuscripts at a price of €12.6 million and destined for the National Library. It was, he described, 'a coup' that nourished then Heritage Minister Síle de Valera's 'particular notions of heritage'.

> I could never imagine such radiant announcement of the same millions spent on, say, acquiring the Old Head of Kinsale, or on land for a new national woodland. Nature conservation, while a big part of her department's brief, was never really her thing.
>
> What is it about Fianna Fáil, indeed, that make it so hard to visualise the words 'flower', 'bird', 'tree', 'moth' being uttered in Cabinet with anything but sheepish necessity? Even outside, looking after our natural heritage ought to merit at least public lip service; it is as worthy of commitment as health or education.[40]

It would be wrong to single out one particular political party: even the Greens, Viney pointed out 'are glad to fight elections on waste disposal issues, not wildlife'. This has not changed in the intervening fourteen years. But even now, change and restoration are possible. Do we want it? Just say the word.

Endnotes

EPIGRAPH
1 From the introduction of *Protection of the National Heritage*, Ireland's first policy document urging state action for the conservation of wildlife. An Foras Forbartha, 1969.

1. NOT AS GREEN AS WE'D LIKE TO THINK
1 'Origin Green' from An Bord Bia, 2013.
2 National Parks and Wildlife Service. 2005. *Site Synopsis: Moycullen Bogs NHA*.
3 www.connemara.ie
4 O'Dowd, P. 1993. *Down by the Claddagh*. Kenny's Bookshop, Galway.
5 James Hardiman. 1820. Quoted in *Down by the Claddagh*.
6 Earl Hodgson, W. 1906. *Salmon Fishing*. Adam & Charles Black, London.
7 Inland Fisheries Ireland. 2014. *Sampling Fish for the Water Framework Directive. Lakes 2014. Lough Corrib*.
8 www.wildatlanticway.com
9 Eurobarometer. 2015. Attitudes of Europeans Towards Biodiversity.
10 NPWS. 2013. 'The Status of EU Protected Habitats and Species in Ireland.' Habitat Assessments Vol. 2. Version 1.0. Unpublished Report, National Parks & Wildlife Services. Department of Arts, Heritage and the Gaeltacht, Dublin.
11 EPA. 2015. *Water Quality in Ireland 2010–2012*. Environmental Protection Agency, Ireland.
12 EPA. 2016. *Ireland's Environment: An Assessment*. Environmental Protection Agency, Ireland.
13 NPWS. 2013. *The Status of Protected EU Habitats and Species in Ireland*. Department of Arts, Heritage and the Gaeltacht, Dublin.
14 Krebs *et al.* 1997. *Bovine Tuberculosis in Cattle and Badgers*. UK Ministry of Agriculture, Fisheries and Food.
15 *British Wildlife*. 2014. Vol. 25, No. 3.
16 Battersby, J. (ed.). 2005. *UK Mammals: species status and population trends*. Joint Nature Conservation Committee and Tracking Mammals Partnership.
17 Roper, T. J. 2010. *Badger*. New Naturalist Series. Collins.

18 Marnell, F., Kingston, N. & Looney, D. 2009. *Ireland Red List No. 3: Terrestrial Mammals*. National Parks and Wildlife Service, Department of the Environment, Heritage and Local Government, Dublin (NPWS).
19 Yalden D. 1999. *The History of British Mammals*. Poyser Natural History.
20 Sleeman D. P. 2008. 'Quantifying the prey gap for Ireland.' *Irish Naturalists' Journal*. Special Supplement; *Mind the Gap – Postglacial Colonisation of Ireland*. Pp. 77–82.
21 Cummins, S., Bleasdale, A., Douglas, C., Newton, S., O'Halloran, J. & Wilson, H. J. 2010. *The status of Red Grouse in Ireland and the effects of land use, habitat and habitat quality on their distribution*. Irish Wildlife Manuals, No. 50. NPWS.
22 Mac Laughlin, Jim. 2010. *Troubled Waters. A Social and Cultural History of Ireland's Sea Fisheries*. Four Courts Press.
23 *The Stock Book*. 2015. Marine Institute.
24 From Nolan, P. *Following the Shoals*. The History Press, Ireland.
25 *The Stock Book*. 2015. Marine Institute.
26 http://www.msc.org/track-a-fishery/fisheries-in-the-program/certified/north-east-atlantic/Irish-pelagic-sustainability-association-western-mackerel/assessment-downloads-1/20120328_ANMT_Suspension.pdf
27 *Opinion: Ireland can become a global player in the seafood sector* by Donal Buckley. *Irish Examiner*. 8 April 2013.
28 BIM Annual Report 2013.
29 'That Fabulous Irish Fishing' by Anthony Pearson. *Fishing: Coarse and Sea Angling*. April 1968.
30 Gammon, C. 1966. *Salt Water Fishing in Ireland*. Herbert Jenkins, London.
31 See http://www.anirishanglersworld.com/index.php/bass-fishing/a-socio-economic-review-of-the-atlantic-bass/
32 Eurostat News Release. 45/2014 19 March 2014. *GDP and Beyond. Measuring quality of life in the EU*.
33 *The Economic and Social Aspects of Biodiversity – Benefits and Costs of Biodiversity in Ireland*. 2008. Department of Environment, Heritage and Local Government.
34 *Valuing Nature-based tourism in Scotland*. Scottish Natural Heritage. Unknown year.
35 *Top Free Visitor Attractions 2012*. Fáilte Ireland.
36 *Jobs lost at sea – Overfishing and the jobs that never were*. New Economics Foundation. 2012.
37 Dáil Debate, 15 December 1999.

2. THE QUEST TO CATCH THE LAST FISH

1 McKenna, C. 2009. *Fresh From the Sea*. Gill & Macmillan.

2 Mac Laughlin, J. 2010. *Troubled Waters – A Social and Cultural History of Ireland's Sea Fisheries*. Four Courts Press.
3 Ibid., p. 216.
4 Went, Arthur E. J. 1946. *The Irish Pilchard Fishery: Proceedings of the Royal Irish Academy*. Vol. 51, pp. 81–120.
5 Data from the Marine Institute's *Stock Book* (2015) give total landings from the Irish Sea in 2014 as 4,118 tonnes.
6 *Atlas of Irish Groundfish Trawl Surveys*. 2012. Marine Institute.
7 Appendix to the 1st Report to the Great Britain Commissioners on the Municipal Corporations in Ireland. 1835.
8 Muus, B. J. & Dahlstrøm, P. 1974. *Sea Fishes of Britain and North-Western Europe*. Collins.
9 Fitzgerald, Séamus. 1999. *Mackerel and the Making of Baltimore, Co. Cork 1879–1913*. Irish Academic Press.
10 Fahy, E. 1972. *A preliminary report into areas of scientific interest in County Cork*. An Foras Forbartha.
11 Sharrock, J. T. R. (ed.). 1973. *The Natural History of Cape Clear Island*. Comharchumann Chléire Teo.
12 Hiscock, K. & Hiscock, S. 1980. 'Sublittoral plant and animal communities in the area of Roaringwater Bay, south-west Ireland.' *Journal of Sherkin Island*, 1, pp. 7–48.
13 Dale, A. L., McAllen, R. & Whelan, P. 2007. *Management considerations for subtidal Zostera marina beds in Ireland*. Irish Wildlife Manuals, No. 28. NPWS.
14 MERC Consultants. 2007. *Surveys of sensitive subtidal benthic communities*. NPWS.
15 Leeney, Ruth. 2007. *Distribution and abundance of harbour porpoises and other cetaceans in Roaringwater Bay, Co. Cork*. Centre for Ecology and Conservation, University of Exeter. NPWS.
16 Went, Arthur E. J. 1948. 'The Ling in Irish Commerce.' *The Journal of the Royal Society of Antiquaries of Ireland*, Vol. 78, No. 2, pp. 119–126.
17 *Article 6 Assessment of Aquaculture and Fisheries in Roaringwater Bay*. 2013. Marine Institute.
18 Kelly et al., 2008. *The Shrimp* (Palaemon serratus *P.*) *Fishery: Analysis of the Resource in 2003–2007*. An Bord Iascaigh Mhara.
19 King, G. L. & Berrow, S. D. 2009. 'Marine turtles in Irish waters.' *Irish Naturalists' Journal*. Special Supplement 2009.
20 Read, A., Drinker, P. & Northridge, S. 2005. 'By-catch of marine mammals in US and global fisheries.' *Conservation Biology* 20: pp. 163–9.
21 Berrow et al. 1998. *Discarding Practices and Marine Mammal By-Catch in the Celtic Sea Herring Fishery. Biology and Environment: Proceedings of the Royal Irish Academy*. Vol. 98B, No.1, pp. 1–8.

22 MERC Consultants. 2007. *Surveys of sensitive subtidal benthic communities*. NPWS.
23 Stout *et al*. *SIMBIOSYS: Sectoral Impacts on Biodiversity and Ecosystem Services*. 2012. Environmental Protection Agency.
24 *EPA funded research maps feral Pacific Oyster populations in Ireland*. 2013. Press Release. Environmental Protection Agency, Dublin.
25 *Landing the Blame. Overfishing in EU Waters 2001–2015*. New Economics Foundation.
26 *Landing the Blame. Overfishing in the Northeast Atlantic 2016*. New Economics Foundation.
27 Quigley, D. 2013. 'Exceptionally large Angler Fish *Lophius piscatorius* L. captured on the Labadie Bank, Celtic Sea.' *Irish Naturalists' Journal*. Vol. 33, Part 1, p. 73.
28 Dulvy, N. K., Notarbartolo di Sciara, G., Serena, F., Tinti, F. & Ungaro, N., Mancusi, C. & Ellis, J. 2006. *Dipturus batis*. In: IUCN 2013. IUCN Red List of Threatened Species. Version 2013.2. www.iucnredlist.org. Downloaded on 8 April 2014.
29 Anon. 2012. *The Stock Book*. Marine Institute.
30 King, J. L., Marnell, F., Kingston, N., Rosell, R., Boylan, P., Caffrey, J. M., FitzPatrick, Ú., Gargan, P. G., Kelly, F. L., O'Grady, M. F., Poole, R., Roche, W. K. & Cassidy, D. 2011. *Ireland Red List No. 5: Amphibians, Reptiles & Freshwater Fish*. NPWS.
31 Valpy, R. 1848. 'The Resources of the Irish Sea Fisheries.' *Journal of the Statistical Society of London*. Vol. 11. No. 1. pp. 55–2.
32 Molloy, J. *The Herring Fishery of Ireland (1900–2005)*. Marine Institute.
33 De Courcy Ireland, J. 1981. *Ireland's Sea Fisheries: A History*. The Glendale Press.
34 From the inaugural address to the London Fisheries Exhibition of 1883 by T. H. Huxley.
35 De Courcy Ireland, J. 1981. *Ireland's Sea Fisheries: A History*. The Glendale Press.
36 *Ibid*.
37 Molloy, J. *The Herring Fishery of Ireland (1900–2005)*. Marine Institute.
38 Miller, D. D., and Zeller, D. 2013. *Reconstructing Ireland's marine fisheries catches: 1950–2010*. Fisheries Centre Working Paper #2013-10.
39 *Ibid*.
40 *The Stock Book*. 2013. Marine Institute.
41 *Ibid*.
42 Norton, T. A. & Geffen, A. J. (eds). *The Irish Sea Study Group Report. Part 3*. 1990. Liverpool University Press.
43 Roberts, C. 2007. *The Unnatural History of the Sea*. Island Press.
44 Valpy, R. 1848. 'The Resources of the Irish Sea Fisheries.' *Journal of the Statistical Society of London*. Vol. 11, No. 1, pp. 55–72.

45 *Ibid.*
46 Anon. 'British Oysters Past and Present.' 1922. *The Irish Naturalist*. Vol. 31, No. 3, p. 33.
47 Fraser, J. 1938. *Guide Through Ireland*. Fraser & Co, Edinburgh.
48 Rees, J. 2008. *The Fishery of Arklow 1800–1950*. Four Courts Press.
49 Anon. 2012. *Irish Wildlife*. Irish Wildlife Trust. Winter 2012. p. 32.
50 Roberts *et al.* 'The Irish Sea: is beauty skin deep?' Unpublished presentation given to the Pew Research Fellows, October 2013.
51 *The Stock Book*. 2013. Marine Institute.
52 Wilkins, N. P. 2004. *Alive Alive-O*. Tír Eolas.
53 *The Stock Book*. 2014. Marine Institute.
54 *Ibid.*
55 Fahy E. 2013. *Overkill! The eurphoric rush to industrialise Ireland's sea fisheries and its unravelling sequel*. Self-published.
56 *The Stock Book*. 2014. Marine Institute.
57 *The Stock Book*. 2001. Marine Institute.
58 *The Stock Book*. 2014. Marine Institute.
59 Wilkins, N. P. 2004. *Alive Alive-O*. Tír Eolas.
60 Fahy, E. & Carroll, J. 2009. 'Vulnerability of male spider crab *Maja brachydactyla* (Brachyura: Majidae) to a pot fishery in south-west Ireland.' *Journal of Marine Biological Association of the UK*.
61 *Shellfish Stocks and Fisheries Review*. 2011. Marine Institute.
62 *Ibid.*
63 Wilkins, N. P. 2004. *Alive Alive-O*. Tír Eolas.
64 ICES Advice Note. September 2016. See: www.ices.dk
65 Roberts, C. 2012. *Ocean of Life*. Allen Lane.
66 Fahy, E. 2013. *Overkill!*
67 Hardiman, J. in *Down by the Claddagh* by O'Dowd, P. 1993. Kenny's Bookshop, Galway.
68 Brabazon, W. 1848. *The Deep Sea and Coast Fisheries of Ireland*. Commissioners of Public Works in Ireland.
69 Rees, J. 2008. *The Fishery of Arklow 1800–1950*. Four Courts Press.
70 Winter *et al. 2014*. 'Razor clam to RoboClam: burrowing drag reduction mechanisms and their robotic adaptation.' *Bioinspiration & Biomimetics*. Vol 9 036009.
71 Letter from the North Irish Sea Razor Fishermen's organisation to the Department of Agriculture, Food and the Marine. 2014.
72 Gerritsen, H. D. & Lordan, C. 2014. *Atlas of Commercial Fisheries Around Ireland*. Marine Institute.
73 Pusceddu *et al.* 2014. 'Chronic and intensive bottom trawling impairs deep-sea biodiversity and ecosystem functioning.' *Proceedings of the National Academy of Sciences*. www.pnas.org/cgi/doi/10.1073/pnas.1405454111.

74 Frank *et al.* 2005. 'Trophic Cascades in a formerly Cod-dominated ecosystem.' *Science.* Vol. 308 p. 1621.
75 Pauley *et al.* 1998. 'Fishing down marine food webs.' *Science.* 279: 860–863.
76 http://www.cbc.ca/news/canada/newfoundland-labrador/newfoundland-cod-stocks-improving-but-recovery-still-years-away-researchers-1.3006663
77 Gullestad *et al.* 2013. 'Changing attitudes 1970–2012: evolution of the Norwegian management framework to prevent overfishing and to secure long-term sustainability.' *ICES Journal of Marine Science.*
78 See Section 23 of the Wildlife Act 1976, which defines 'fauna' but excludes 'fish or aquatic invertebrate animals (or their eggs or spawn or brood or young)'.
79 Fernandez-Chacon, A., Moland, E., Espeland, S. & Olsen, E. 2015. 'Demographic effects of full versus partial protection from harvesting: inference from an empirical before-and-after control-impact study on Atlantic cod.' 2015. *Journal of Applied Ecology.* Vol. 52. pp. 1206–1215.

3. THE WHITTLING-AWAY OF OUR ICONIC LANDSCAPES

1 Dudley, N. (ed.). 2008. *Guidelines for Applying Protected Area Management Categories.* IUCN.
2 Deguignet, M., Juffe-Bignoli, D., Harrison, J., MacSharry, B., Burgess, N. & Kingston, N. 2014. *2014 United Nations List of Protected Areas.* UNEP-WCMC: Cambridge, UK.
3 Chape *et al.* 2003. *United Nations List of Protected Areas.* IUCN.
4 A Natural Heritage Area (NHA) is protected under national rather than European law.
5 From *Living with Nature*, anon., unknown year, Dúchas The Heritage Service, The Department of the Environment and Local Government.
6 NPWS. 2013. *The Status of Protected Habitats and Species in Ireland.* Department of Arts, Heritage and the Gaeltacht. Dublin.
7 Anon. 1981. *Areas of Scientific Interest in Ireland.* An Foras Forbartha.
8 Quinn, A. 1971. *Provisional Survey of Areas of Scientific Interest in Co. Galway.* An Foras Forbartha.
9 Dáil Record, 18 May 1993.
10 *Tourism Facts 2015.* Fáilte Ireland.
11 *National Landscape Strategy for Ireland 2015–2025.* 2015. Department of Arts, Heritage and the Gaeltacht. Dublin.
12 Aalen, F. H. A., Whelan, K. & Stout, M. (eds). 2011. *Atlas of the Irish Rural Landscape.* Cork University Press.
13 Ripple, W. J. & Beschta, R. L. 2012. 'Trophic cascades in Yellowstone: The first 15 years after wolf reintroduction.' *Biological Conservation.* Vol. 145, Issue 1, pp. 205–213.
14 Hickey, K. 2011. *Wolves in Ireland.* Open Air.

15 Ryan, S. 1998. *The Wild Red Deer of Killarney*. Mount Eagle Publications.
16 Perrin, P. M., Mitchell, F. J. G & Kelly, D. L. 2011. 'Long-term deer exclusion in yew-wood and oakwood habitats in southwest Ireland: Changes in ground flora and species diversity.' *Forestry Ecology and Management*. Vol. 262. Issue 12, pp. 2328–2337.
17 Higgins, G. T. 2008. 'Rhododendron ponticum: A guide to management on nature conservation sites.' *Irish Wildlife Manuals*, No. 33. NPWS.
18 Barron, C. 2000. *Groundwork Rhododendron clearance in Killarney National Park 1981–2000*. Report by Groundwork.
19 Unattributed. 2013. *Observations of rhododendron in Killarney oakwood areas cleared & maintained by Groundwork Conservation Volunteers in the period 1981–2005*. Report by Groundwork Conservation Volunteers.
20 Bolwer, M. & Bradshaw, R. 1985. 'Recent accumulation and erosion of blanket peat in the Wicklow Mountains, Ireland.' *New Phytologist*. Vol. 101, pp. 543–550.
21 Evans, M., Warburton, J. & Yang, J. 2006. 'Eroding blanket peat catchments: Global and local implications of upland organic sediment budgets.' *Geomorphology*, pp. 79, 45–57.
22 NPWS. 2013. *The Status of Protected EU Habitats and Species in Ireland. Overview Vol. 1*. Department of Arts, Heritage and Local Government.
23 Perrin, P. M., Roche, J. R., Daly, O. H. & Douglas, C. 2011. *The National Survey of Upland Habitats*. Taken from the proceedings of the Conserving Farmland Biodiversity conference 2011. Teagasc.
24 Colhoun, K. & Cummins, S. 2013. 'Birds of Conservation Concern in Ireland 2014–2019.' *Irish Birds*. Vol. 9, No. 4, pp. 523–541; Holloway, S. *The Historical Atlas of Breeding Birds in Britain and Ireland 1875–1900*. T. & A. D. Poyser.
25 Balmer, D. E., Gillings, S., Caffrey, B. J., Swann, R. L., Downie, I. S., Fuller, R. J. 2013. *Bird Atlas 2007–11: the breeding and wintering birds of Britain and Ireland*. BTO Books, Thetford, UK.
26 Praeger, R. L. 1974. *The Botanist in Ireland*. Hodges, Figgis & Co.
27 Preston, C. D., Pearman, D. A. & Dines, T. D. 2002. *New Atlas of the British & Irish Flora*. Oxford University Press.
28 Wilson, F. & Curtis, T. G. F. 2008. *The Montane Flora of County Wicklow*. Report prepared for The Heritage Council and Wicklow County Council.
29 Aalen, F. H. A., Whelan, K. & Stout, M. (eds). 2011. *Atlas of the Irish Rural Landscape,* 2nd edition. Cork University Press.
30 Cross, J. 2006. 'Potential Natural Vegetation of Ireland'. *Biology and Environment: Proceedings of the Royal Irish Academy*. Vol. 106B, No. 2, pp. 65–116.
31 Tubridy, M. 2013. *A Study to Identify Best Management of Upland Habitats in County Wicklow*. A report for the Wicklow Uplands Council Ltd. www.wicklowuplands.ie

32 Robinson, Tim, 'Connemara, Co. Galway', an extract from the *Atlas of the Irish Rural Landscape,* 1st edition. 1997. Cork University Press.
33 MacGiollarnáth, Seán. 1954. *Conamara.* Leinster Leader Ltd.
34 Agricultural Statistics Bulletin, 1980–89, Central Statistics Office.
35 Colhoun, K. & Cummins, S. 2013. 'Birds of Conservation Concern in Ireland 2014–2019.' *Irish Birds.* Vol. 9, No. 4, pp. 523–541; Cummins, S., Bleasdale, A., Douglas, C., Newton, S., O'Halloran, J. & Wilson, H. J. 2010. 'The status of Red Grouse in Ireland and the effects of land use, habitat and habitat quality on their distribution.' *Irish Wildlife Manuals,* No. 50. NPWS.
36 Lockhart, N., Hodgetts, N., & Holyoak, D. 2012. *Rare & Threatened Bryophytes of Ireland.* National Museums of Northern Ireland Publication No. 028.
37 Bleasdale, A. & Sheehy Skeffington, M. 1995. *The Upland Vegetation of North-East Connemara in Relation to Sheep Grazing.* In: Jeffrey, W. D., Jones, M. B. & McAdam, J. H. (eds). 1995. *Irish grasslands – their biology and management,* pp. 110–224. Royal Irish Academy. Dublin.
38 McGurn, P. 2011. 'Developing a targeted-based programme for HNV [High Nature Value] Farmland in the North Connemara Area.' A report prepared for the Heritage Council.
39 The story of the golden eagle reintroduction is chronicled by the Golden Eagle Trust through news bulletins on their website: www.goldeneagle.ie
40 Golden Eagle Trust Facebook post, 12 November 2015.
41 www.glenveaghnationalpark.ie
42 Foot, M. J. 1871. 'On the distribution of plants in Burren, County of Clare.' *The Transactions of the Royal Irish Academy.* Vol. 24 pp. 143–160.
43 See www.burrenbeo.com
44 My summation of this saga is taken from Liam Leonard's *The Environmental Movement in Ireland.* 2008.
45 This quote, and my account of the Mullaghmore conflict is taken from Liam Leonard's *The Environmental Movement in Ireland.* 2008. Springer. p. 145.
46 Dáil record. 30 March 1995. Dáil Éireann debate. Vol. 451, No. 4.
47 Aalen, F. H. A., Whelan, K. & Stout, M. (eds). 1997. *Atlas of the Irish Rural Landscape.* Cork University Press.
48 Dunford, B. 'Beauty and the Burren'. *Irish Wildlife.* Spring 2012, p. 17.
49 Finn, J. A. & Ó hUallacháin, D. 2012. 'A review of the evidence on the environmental impact of Ireland's Rural Environmental Protection Scheme (REPS).' *Biology and Environment: Proceedings of the Royal Irish Academy.* 112B, 11–34.

4. EXTINCT: IRELAND'S LOST SPECIES
1. Went, A. 1948. 'The Status of the Sturgeon, *Acipenser sturio L.*, in Irish Waters Now and in Former Days.' *Irish Naturalists' Journal*. Vol. 9 No. 7. pp. 172–174.
2. Muus, B. J & Dahlstrøm, P. 1964. *Sea Fishes of Britain and North-Western Europe*. Collins.
3. Dickson, B. & Pinnegar, J. K. 2010. 'Skipper Newson of Grimsby – the "Sturgeon Hunter".' *British Wildlife*. Vol. 21, No. 6, pp. 416–419.
4. Praeger, R. L. 1950. *Natural History of Ireland*. EP Publishing, London.
5. Greaves, J. W. 1968. 'A Second Sturgeon *Acipenser sturio* landed at Ardglass, 1966.' *Irish Naturalists' Journal*. Vol. 16, No 2, p. 54.
6. From *The Dingle Sturgeon Story*. Broadcast November 2013 on RTÉ Radio 1.
7. Went, A. E. 1968. 'Rare fishes taken in Irish waters in 1967.' *Irish Naturalists' Journal*. Vol. 16, No 2, pp. 35–39.
8. IUCN. 2000. *IUCN Red List Categories and Criteria*. Version 3.1. Second Edition.
9. Bullock, C., Kretsch, C. & Candon, E. 2008. *The Economic and Social Aspects of Biodiversity. Benefits and Costs of Biodiversity in Ireland*. Department of Environment, Heritage and Local Government.
10. Curtis, T. G. F. & McGough, H. N. 1988. *The Irish Red Data Book. 1 Vascular Plants*. Wildlife Service Ireland.
11. Parnell, J. & Curtis, T. 2012. *Webb's An Irish Flora*. Cork University Press.
12. Fairley, J. 1981. *Irish Whales and Whaling*. Blackstaff Press.
13. http://wwf.panda.org/what_we_do/endangered_species/cetaceans/about/right_whales/north_atlantic_right_whale/
14. Note that there are records of other mammals from Ireland, such as woolly mammoth, Arctic fox, horse, reindeer, lemming, spotted hyena and Irish elk but these all date from pre-glacial times or, in the case of the Irish elk, is believed to have gone extinct before humans arrived on this island.
15. Yalden, D. 1999. *The History of British Mammals*. T. & A. D. Poyser.
16. Ibid.
17. Ibid.
18. Ibid.
19. Harris, S. & Yalden, D. W. (eds). 2008. *Mammals of the British Isles: Handbook, 4th Edition*. The Mammal Society.
20. Hickey, K. 2011. *Wolves in Ireland*. Open Air.
21. Hayden, T. & Harrington, R. 2000. *Exploring Irish Mammals*. Town House, Dublin.
22. NPWS. *The Status of Protected Habitats and Species in Ireland. Overview Vol. 1*. 2013.
23. Yalden, D. W. 2011. 'A distant history of Irish birds.' *Irish Birds*. Vol. 9. No. 2, pp. 225–228.

24 D'Arcy, G. 1999. *Ireland's Lost Birds*. Four Courts Press.
25 Ibid.
26 Ibid.
27 Ibid.
28 Ibid.
29 Balmer, D. E., Gillings, S., Caffrey, B. J., Swann, R. L., Downie, I. S., Fuller, R. J. 2013. *Bird Atlas 2007–11: the breeding and wintering birds of Britain and Ireland*. BTO Books.
30 D'Arcy, G. 1999. *Ireland's Lost Birds*. Four Courts Press.
31 Ibid.
32 Ibid.
33 Ibid.
34 Ibid.
35 Ibid.
36 Ibid.
37 Ibid.
38 Ibid.
39 Ruttledge, R. F. 1966. *Ireland's Birds*. H. F. & G. Witherby Ltd.
40 Taylor, A. J. & O'Halloran, J. 2002. 'The Decline of the Corn Bunting, *Miliaria calandra*, in the Republic of Ireland.' *Biology and Environment: Proceedings of the Royal Irish Academy*, Vol. 102B, No. 3, pp. 165–175.
41 Regan, E. C., Nelson, B., Aldwell, B., Bertrand, C., Bond, K., Harding, J., Nash, D., Nixon, D. & Wilson, C. J. 2010. *Ireland Red List No. 4 – Butterflies*. NPWS.
42 Allen, D., O'Donnell, M., Nelson, B., Tyner, A., Bond, K. G. M., Bryant, T., Crory, A., Mellon, C., O'Boyle, J., O'Donnell, E., Rolston, T., Sheppard, R., Strickland, P., FitzPatrick, Ú. & Regan, E. 2016. *Ireland Red List No. 9 – Macro-moths (Lepidoptera)*. NPWS.
43 FitzPatrick, Ú., Murray, T. E., Byrne, A., Paxton, R. J., & Brown, M. J. F. 2008. *Regional Red List of Irish Bees*. Queen's University Belfast and the Higher Education Authority.
44 Foster, G. N., Nelson, B. H. & O Connor, Á. 2009. *Ireland Red List No. 1 – Water beetles*. NPWS.
45 Alexander, K. N. A. & Anderson, R. 2012. 'The beetles of decaying wood in Ireland. A provisional annotated checklist of *saproxylic Coleoptera*.' *Irish Wildlife Manuals*, No. 65. NPWS.
46 Curtis, T. G. F. & McGough, H. N. 1988. *The Irish Red Data Book: 1 Vascular Plants*. The Stationery Office, Dublin.
47 Lockhart, N., Hodgetts, N. & Holyoak, D. 2012. *Ireland Red List No. 8: Bryophytes*. NPWS.
48 Byrne, A., Moorkens, E. A., Anderson, R., Killeen, I. J. & Regan, E. C. 2009. *Ireland Red List No. 2 – Non-Marine Molluscs*. NPWS.

49 Kelly-Quinn, M. & Regan, E. C. 2012. *Ireland Red List No. 7: Mayflies (Ephemeroptera)*. NPWS.
50 http://www.speciesrecoverytrust.org.uk/Templates/Lost per cent2olife.html
51 Edgar, G. J., Samson, C. R. & Barrett, N. S. 2004. 'Species Extinction in the Marine Environment: Tasmania as a Regional Example of Overlooked Losses in Biodiversity.' *Conservation Biology*. 1294–1300.
52 Webb, T. J. & Mindel, B. L. 2015. 'Global Patterns of Extinction Risk in Marine and Non-marine Systems.' *Current Biology*. Vol. 25, Issue 4, pp. 506–511.
53 Pimm, S. L., Jenkins, C. N., Abell, R., Brooks, T. M., Gittleman, J. L., Joppa, L. N., Raven, P. H., Roberts, C. M. & Sexton, J. O. 2014. 'The biodiversity of species and their rates of extinction, distribution, and protection.' *Science*. Vol. 344, No. 6187.
54 Quigley, D. T. G. & Moffatt, S. 2014. 'Sika-like deer *Cervus Nippon* observed swimming out to sea at Greystones, Co. Wicklow: increasing deer population pressure?' *Bulletin of the Irish Biogeographical Society* No. 38, pp. 251–262.
55 Sleeman, D. P. 2008. 'Quantifying the prey gap for Ireland.' In Davenport J. L., Sleeman, D. P. & Woodman, P. C. *Mind the Gap – Postglacial colonisation of Ireland. Irish Naturalists' Journal*.
56 Mrs Houston. 1879. *Twenty Years in the Wild West, or Life in Connaught*. John Murray, London. p. 106.
57 Lauder, C. & Donaghy, A. 2008. *Breeding waders in Ireland 2008*. BirdWatch Ireland & National Parks and Wildlife Service, Dublin.
58 Trewby, I. D., Wilson, G. J., Delahay, R. J., Walker, N., Young, R., Davison, J., Cheesman, C., Robertsone, P. A., Gorman, M. L. & McDonald, R. A. 2008. 'Experimental evidence of competitive release in sympatric carnivores.' *Biology Letters*. 4(2) 170–175.
59 Lawton, C. & Rochford, J. 2007. 'The Recovery of Grey Squirrel *(Sciurus carolinensis)* populations after intensive control programmes'. *Biology and Environment: Proceedings of the Royal Irish Academy*. Vol. 107B, No. 1, pp. 19–29.
60 Hayden, T. & Naulty, F. 2008. *Phoenix Park Grey Squirrel Project Report*. University College Dublin.
61 Sheehy, E. & Lawton, C. 2014. 'Population crash in an invasive species following the recovery of a native predator: the case of the American grey squirrel and the European pine marten in Ireland.' *Biodiversity and Conservation*.
62 Ibid.
63 O'Toole, L. 2014. *The Eurasian Crane* (Grus grus) *in Ireland – another extinct bird or a key species for an ancient belief system?* In: Sleeman, D. P.,

Carlsson, J. *Mind the Gap II: new insights into the Irish postglacial*. *Irish Naturalists' Journal*.

64 Sleeman, D. P. 2008. 'Quantifying the prey gap for Ireland'. In, Davenport, J. L., Sleeman, D. P & Woodman, P. C. *Mind the Gap. Postglacial colonisation of Ireland. Irish Naturalists' Journal* Special Supplement. 2008.

65 Carden, R. F. 2012. *Review of the Natural History of Wild Boar (Sus scrofa) on the island of Ireland*. Report prepared by Ruth Carden for the Northern Ireland Environment Agency, Northern Ireland, UK, National Parks & Wildlife Service, Department of Arts, Heritage and the Gaeltacht, Dublin, Ireland and the National Museum of Ireland – Education & Outreach Department.

66 Welander, J. 1995. 'Are Wild Boars a Future Threat to the Swedish Flora?' *The Journal of Mountain Ecology*. Vol. 3.

67 Check www.biodiversityireland.ie for the latest sightings.

68 McDevitt, A. D., Carden, R. F., Coscia, I. & Frantz, A. C. 2013. 'Are wild boars roaming Ireland once more?' *European Journal of Wildlife Research*. Published online.

69 'A total boar or a pig's pedigree?' *The Irish Times*. 4 May 2013.

70 O'Rourke, E. & Lysaght, L. 2014. *Risk Assessment of* Sus scrofa – *wild boar/feral pig/hybrid*. Published for consultation 09/05/2014 by Inland Fisheries Ireland and the National Biodiversity Data Centre.

71 'After 5,000 years Kerry red deer as Irish as can be, DNA analysis shows.' *The Irish Times*. 27 March 2012.

72 Maclean, N. 2010. *Silent Summer*. Cambridge University Press.

73 National Biodiversity Data Centre. *Irish Butterfly Monitoring Scheme 2016* newsletter.

74 WWF. 2014. *Living Planet Report 2014*.

75 Colhoun, K. & Cummins, S. 2013. 'Birds of Conservation Concern in Ireland 2014-2019.' *Irish Birds* Vol. 9, No. 4, pp. 523-544.

76 Ussher, R. J. & Warren, R. 1900. *Birds of Ireland*. Gurney & Jackson.

77 Perry, K. W. 2013. 'Rare Breeding Birds in Ireland in 2012.' *Irish Birds*. Vol. 9 No. 4. pp. 563-576.

78 Ruttledge, R. F. 1966. *Ireland's Birds*. H. F. & G. Whiterby Ltd., London.

79 Perry, K. W. & Newton, S. F. 2014. 'Rare Breeding Birds in Ireland in 2013.' *Irish Birds*. Vol. 10, No. 1, pp. 63-70.

80 Lusby, J. & O'Cleary, M. 2014. *Barn Owls in Ireland*. BirdWatch Ireland.

81 Balmer, D. E., Gillings, S., Caffrey, B. J., Swann, R. L., Downie, I. S., Fuller, R. J. 2013. *Bird Atlas 2007-11: the breeding and wintering birds of Britain and Ireland*. BTO Books, Thetford, UK.

82 Inger, R., Gregory, R., Duffy, J. P., Stott, I., Vorisek, P. & Gaston, K. J. 2015. 'Common European birds are declining rapidly while less abundant species' numbers are rising.' *Ecology Letters*. Vol. 18. pp. 28-36.

83 Crowe, O., Coombes, R. H. & O'Halloran, J. O. 2014. 'Estimates and trends of common breeding birds in the Republic of Ireland.' *Irish Birds*. Vol. 10, No. 1, pp. 23–32.
84 Went, A. 1953. 'Fisheries on the River Liffey.' *The Journal of the Royal Society of Antiquaries of Ireland*, Vol. 83, No. 2 (1953), pp. 163–173.
85 Archer, J. 1801. *Statistical Journal of County Dublin*. From Went, A. 1954. 'Fisheries of the River Liffey: II. Notes on the Corporation Fishery from the Time of the Dissolution of the Monasteries.' *The Journal of the Royal Society of Antiquaries of Ireland*, Vol. 84, No. 1 (1954), pp. 41–58.
86 Cromwell, T. 1820. *Excursions through Ireland*. From Went. A. 1954. 'Fisheries of the River Liffey: II. Notes on the Corporation Fishery from the Time of the Dissolution of the Monasteries.' *The Journal of the Royal Society of Antiquaries of Ireland*, Vol. 84, No. 1 (1954), pp. 41–58.
87 Whelan, K. 2014. 'Under Pressure: the Atlantic Salmon.' *Irish Wildlife* Winter '14. Irish Wildlife Trust.
88 *Ibid*.
89 Unknown author. 2014. *The Status of Irish Salmon Stocks in 2014*. Inland Fisheries Ireland.
90 'Hundreds protest over drift-net fishing ban' by Juno McEnroe. *Irish Examiner*. 2 November 2006.
91 'Simon Coveney: weather delayed collection of fish' by Claire O'Sullivan. *Irish Examiner*. 31 March 2014.
92 King, J. L., Marnell, F., Kingston, N., Rosell, R., Boylan, P., Caffrey, J. M., FitzPatrick, Ú., Gargan, P. G., Kelly, F. L., O'Grady, M. F., Poole, R., Roche, W. K. & Cassidy, D. 2011. *Ireland Red List No. 5: Amphibians, Reptiles & Freshwater Fish*. NPWS.
93 WWF. 2015. *Living Blue Planet Report 2015*.
94 Paleczny, M., Hammill, E., Karpouzi, V. & Pauly, D. 2015. 'Population Trend of the World's Monitored Seabirds, 1950–2010.' *PLOS One*. DOI: 10.1371/journal.pone.0129342
95 Gordon, W. J. unknown year. *Our Country's Fishes*. Simpkin, Marshall, Hamilton, Kent & Co.
96 See http://www.baskingshark.ie/
97 Taken from Fairley, J. 1981. *Irish Whales and Whaling*. Blackstaff Press.
98 *Ibid*.
99 Whelan, K. 1989. *The Angler in Ireland*. Country House.
100 'Only 12 left of Irish shark species that's 4 million years old' by Lynne Kelleher. *Irish Examiner*. 23 December 2013.
101 Brennan, D. 'Common skate and conservation.' *Sea Angler* magazine. May 1978.
102 'Angler sets new record for landing half ton shark in Ireland' by Alastair Jamieson. *The Telegraph*, 25 June 2009.

103 Hudson, A. V. & Furness, R. W. 1989. 'The behaviour of seabirds foraging at fishing boats around Shetland.' *Ibis*. 131:225–237.
104 *Actions for Biodiversity 2011–2016*. Department of Arts, Heritage and the Gaeltacht.

5. CULLING: THE URGE TO KILL ANIMALS
1 From Seigne, J. M. 1930. *A Bird Watcher's Notebook*. Philip Allan & Co. Ltd, London.
2 Seanad record. 17 July 2014.
3 'Hen harriers making farms worthless – claim' by Aidan O'Connor. *Kerry's Eye*. January 2015.
4 'Healy-Rae: Birds have brains as well' by Anne Lucey. *Irish Examiner*. 16 December 2014.
5 'Councillor in a flap over pine martens' by Eoghan O'Connell. *The Irish Times*. 18 June 2009.
6 'Introducing the pine marten' by Michael Newman. Letter section. *The Irish Times*. 20 June 2009.
7 *Irish Independent*, farming supplement, 30 August 2016.
8 Cahill, N. 'The Impact of European Environment Policy in Ireland.' Report prepared for the National Economic and Social Council.
9 *The Late Debate*. RTÉ Radio 1. May 2015.
10 Foss, P. J., O'Connell, C. A. & Crushell, P. H. 2001. *Bogs & Fens of Ireland Conservation Plan 2005*. Irish Peatland Conservation Council; NPWS. 2013. 'The Status of EU Protected Habitats and Species in Ireland.' Habitat Assessments Vol. 2, Version 1.0. Unpublished report, NPWS.
11 'Agriculture Minister Tom Hayes says experience "on the ground" stands to him' by Alison Healy. *The Irish Times*. 26 August 2013.
12 Pedreschi, D., Kelly-Quinn, M., Caffrey, J., O'Grady, M. & Mariani, S. 2013. 'Genetic structure of pike (Esox lucius) reveals a complex and previously unrecognized colonization history of Ireland.' *Journal of Biogeography*. Vol. 41, No. 3, pp. 548–560.
13 www.radiokerry.ie/kerry-farmers-losing-up-to-1.3-million-euro-annually-due-to-foxes-and-mink/
14 Cook, R. M., Holmes, S. J. & Freyer, R. J. 2015. 'Grey seal predation impairs recovery of an overexploited fish stock.' *Journal of Applied Ecology*. DOI: 10.1111/1365-2664.12439.
15 Houle, J. E., de Castro, F., Cronin, M. A., Farnsworth, K. D., Gosch, M., Reid, D. G. 2016. 'Effects of seal predation on a modelled marine fish community and consequences for a commercial fishery.' *Journal of Applied Ecology*. Vol. 53, No.1, pp. 54–63.
16 'Grey Seals – should they be managed?' by Ciaran Crummey, BIM. In *The Badger*. No. 61 Autumn 1996. p. 5.

17 Fahy, E. 2013. *Overkill!* (self-published).
18 'The Seals of Beginish' by Jacquie Cozens. In *Irish Wildlife* Winter 2004, p. 5. Irish Wildlife Trust.
19 Woodlock, J. 2015. Dead Seal Database. Irish Seal Sanctuary.
20 Dáil record. 14 May 2015.
21 The information in this section comes mainly from four sources: Conn Flynn who was Development Officer for the IWT in 2011; Bernie Barrett of Badgerwatch in Waterford; Mike Rendle of the Northern Ireland Badger Group; Kelly, F. Master's thesis. 2013. 'A Review of the policy of culling badgers (*Meles meles*) as a strategy to control the spread of bTB in cattle in the Republic of Ireland'; Roper, T. 2010. *Badger. The New Naturalist Library*, Collins.
22 'Wicklow TB rates are 10 times the national average' by Darragh McCullogh. *Irish Independent*. 28 April 2015.
23 O'Connor, R. & O'Malley, E. 1989. *Badgers and Bovine Tuberculosis in Ireland*. A report prepared for ERAD by the Economic and Social Research Institute.
24 *Ibid*. Executive Summary, p. 4,
25 Mac Coitir, N. 2010. *Ireland's Animals. Myths, Legends & Folklore*. The Collins Press.
26 Smal, C. 1995. *The Badger and Habitat Survey of Ireland*. Department of Agriculture, Food & Forestry, Dublin.
27 'Badgers tracked by GPS dragged from setts for "baiting"' by Kirsty Blake Knox. *Irish Independent*. 20 April 2015.
28 Mullen, E. M., MacWhite, T., Maher, P. K., Kelly, D. J., Marples, N. M., Good, M. 2013. 'Foraging Eurasian badgers *Meles meles* and the presence of cattle in pastures. Do badgers avoid cattle?' *Official Journal of the International Society for Applied Ecology*. Published online: 22 February 2013.
29 http://www.ecoevoblog.com/2013/05/29/do-badgers-play-friesian-tag/
30 More, S. J. Year unstated. 'Towards Eradication of Bovine Tuberculosis in Ireland: A Critical review of progress.' Report published on the website of the Department of Agriculture, Food and the Marine.
31 O'Connor, R. & O'Malley, E. 1989. *Badgers and Bovine Tuberculosis in Ireland*. Economic and Social Research Council. Reproduced from the ESRI report.
32 Krebs, J. R. 1997. *Bovine Tuberculosis in Cattle and Badgers*. Independent Scientific Review Group.
33 Kelly, F. 2013. 'A review of the policy of culling European Badgers (*Meles meles*) as a strategy to control the spread of bTB in cattle in the Republic of Ireland.' Unpublished thesis, p. 36.
34 Micheal Flynn, Co. Galway Livestock Chairman for the Irish Farmers' Association speaking to the *Connacht Tribune*, 9 April 2015.

35 Kirk, S. Opening Address. From *The Badger*. 1993. *Proceedings of a seminar held on 6–7 March 1991*, Hayden, T. (ed.). Royal Irish Academy.
36 Agriculture, Food & Marine Committee record, 8 December 2015.
37 European Commission. 2014. 'Report of an audit carried out in Ireland from 21 to 28 May 2014. In order to evaluate the effectiveness of, and progress made by the programmes co-financed by the European Union to eradicate bovine tuberculosis.'
38 'New vaccine could end badger culls' by Darragh McCullough. *Irish Independent*. 5 July 2016.
39 Byrne, A. W., Acevedo, P., Green, S. & O'Keefe, J. 2014. *Ecological Indicators* 43 (2014) 94–102, http://dx.doi.org/10.1016/j.ecolind.2014.02.024
40 Byrne, A., Sleeman, D. P., O'Keefe, J. & Davenport, J. 2012. 'The ecology of the European badger (*Meles meles*) in Ireland: a review.' *Biology and Environment*. Vol. 112B, Issue 1, p. 105.
41 Byrne, A. W., O'Keefe, J., Sleeman, D. P. 2013. 'Impact on culling on the relative abundance of the European Badger (*Meles meles*) in Ireland.' *European Journal of Wildlife Research*. Vol. 59, No. 1, pp. 25–37.
42 http://www.northernireland.gov.uk/index/media-centre/news-departments/news-dard/news-dard-020914-oneill-visits-bovine.htm
43 Interview on RTÉ Radio 1's *CountryWide* programme. 12 July 2014.
44 Brooks-Pollock, E., Roberts, G. O. & Keeling, M. J. 2014. 'A dynamic model of bovine tuberculosis spread and control in Great Britain.' *Nature*. doi:10.1038/nature13529
45 'Wild Deer are "out of control": ICMSA' by Martin Ryan. *Irish Independent*. 13 May 2015.
46 'IFA: The high levels of TB in wild deer require immediate action'. *Irish Farmers Journal*. 1 May 2015.

6. THE BATTLE TO SAVE THE BOGS

1 A third type of peat, known as fen peat, is formed from nutrient-rich groundwater, but is virtually non-existent in Ireland due to drainage and land reclamation.
2 Cabot, D. 1999. *Ireland*. New Naturalist Series. HarperCollins.
3 Turf Development Act, 1981.
4 Kohl, J. G. 1844. *Travels in Ireland*. Bruce and Wyld. p. 37.
5 Feehan, J. & O'Donovan, G. 1996. *The Bogs of Ireland*. University College Dublin.
6 *Ibid*.
7 *Ibid*., p. 8.
8 *Ibid*., p. 9.
9 *Ibid*., p. 38.
10 *Ibid*., p. 45.

11 *Ibid.*, p. 39.
12 Horner, A. 'Napoleon's Irish Legacy: the bogs commissioners, 1809–1811.' 2005. *History Ireland*. Issue 5, Vol. 13.
13 2nd Report of the Bog Commissioners. 1811.
14 Feehan, J. & O'Donovan, G. 1996. *The Bogs of Ireland*. University College Dublin, p. 13.
15 *Ibid.*, p. 122.
16 'The Bog Transformed' by Stephen Rynne. In *Ireland of the Welcomes* Vol. 18, No. 6. 1970.
17 *Ibid.*, p. 126.
18 Quinn, A. C. M. 1971. *Provisional survey of areas of scientific interest in County Galway*. An Foras Forbartha.
19 Farrell, L. 1972. *Preliminary report on areas of scientific interest in County Offaly*. An Foras Forbartha.
20 Schouten, M. G. C., Streefkerk, J. G. & Ryan, J. B. 2002. *Conservation and Restoration of Raised Bogs*. Dúchas. p. 6.
21 Crushell, P., Connolly, A., Schouten, M. & Mitchell, F. J. G. 2008. 'The changing landscape of Clara Bog: the history of an Irish raised bog.' *Irish Geography*. Vol. 41, No. 1. pp. 89–111.
22 *Ibid.*, p. 6.
23 Moore, P. D. & Bellamy, D. J. 1974. *Peatlands*. Elek Science. London. p. 198.
24 2nd Report of the Bogs Commissioners. 1811. p. 33.
25 Moore, P. D. & Bellamy, D. J. 1974. *Peatlands*. Elek Science. London. p. 199.
26 Bellamy, D. 1986. *The Wild Boglands. Bellamy's Ireland*. Country House, Dublin. p. 13.
27 Stroud, D. A., Fox, A. D., Urquhart, C. & Francis, I. S. (compilers). 2012. 'International Single Species Action Plan for the Conservation of the Greenland White-fronted Goose (*Anser albifrons flavirostris*).' *AEWA Technical Series* No. 45. Bonn, Germany.
28 Bellamy, D. 1986. *The Wild Boglands.*, p. 11.
29 Cross, J. 1990. *The Raised Bogs of Ireland – their ecology, status and conservation*. Wildlife Service, Office of Public Works. Stationery Office, Dublin.
30 *Ibid.*, p. 71.
31 O'Connell, C. 1996. 'SOS Irish Bogs.' In *The Badger*, newsletter of the Irish Wildlife Trust. No. 61. Autumn 1996, p. 6.
32 *Ibid.*
33 Síle de Valera. Dail Adjournment debate. 10 February 1998.
34 *Draft National Peatlands Strategy*. 2013. National Parks and Wildlife Service, Dublin. p. 10.
35 Síle de Valera, Minister for Arts, Heritage, Gaeltacht and the Islands in a speech to the Boston College, USA. 18 September 2000.

36 European Court of Justice Ruling C67/99.
37 'Deenihan to press turf-cutting case before the EU' by Dónal Nolan. *The Kerryman*. 14 March 2012.
38 Fernandez, F., Connolly K., Crowley W., Denyer J., Duff K. & Smith G. 2014. 'Raised Bog Monitoring and Assessment Survey 2013.' *Irish Wildlife Manuals*, No. 81. NPWS.
39 *Defending Ireland's Protected Raised Bogs from Illegal Extraction*. 2012. Friends of the Irish Environment.
40 *Draft National Peatlands Strategy*. 2013. NPWS. p. 5.
41 Viney, M. 1979. *Another Life*. An Irish Times Publication. p. 45.
42 *Ibid*., pp. 49–50.
43 NPWS. 2013. *The Status of EU Protected Habitats and Species in Ireland*. Vol. 1. NPWS.
44 *Draft National Peatlands Strategy*. 2013. NPWS. p. 30.
45 NPWS. 2013. *The Status of EU Protected Habitats and Species in Ireland*. 'Habitat Assessments Vol. 2. Version 1.0.' Unpublished report, NPWS.
46 http://www.ipcc.ie/a-to-z-peatlands/blanket-bogs/
47 *Draft National Peatlands Strategy*. 2013. NPWS. p. 98.
48 Brown, L. E., Holden, J. & Palmer, S. M., 2014. *Effects of Moorland Burning on the Ecohydrology of River basins*. Key findings from the EMBER project. University of Leeds.
49 *Draft National Peatlands Strategy*. 2013. NPWS. p. 93.
50 Ussher, R. J. & Warren, R. 1900. *Birds of Ireland*. Gurney and Jackson, London.
51 Seigne, J. M. 1930. *A Bird Watcher's Notebook*. Philip Allan & Co. London. p. 100.
52 Kennedy, Rev. P. G. 1960. *A List of the Birds of Ireland*. National Museum of Ireland.
53 Anon. 2008. *Irish Forests – A Brief History*. Forest Service. Department of Agriculture, Fisheries and Food, Dublin.
54 Conaghan, J. 2000. *Distribution, Ecology and Conservation of Blanket Bog in Ireland*. Dúchas, The Heritage Service. Dublin.
55 Ruttledge, R. F. 1966. *Ireland's Birds*. H. F. & G. Witherby Ltd. London.
56 Hutchinson, C. 1975. *The Birds of Dublin and Wicklow*. An Irish Wildbird Conservancy Publication; Whilde, T. 1978. *Birds of Galway and Mayo*. Galway branch of the Irish Wildbird Conservancy.
57 D'Arcy, G. 1981. *The Guide to the Birds of Ireland*. Irish Wildlife Publications.
58 Barton, C., Pollock, C., Norriss, D. W., Nagle, T., Oliver, G. A. & Newton, S. 2006. 'The second national survey of breeding hen harriers *Circus cyaneus* in Ireland 2005.' *Irish Birds* Vol. 8 No. 1 pp. 1–20.
59 'Dillon "does not condone" bird's killing' by Sean MacConnell. *The Irish Times*. 20 May 2003.

60 'Survey shows that forestry income can be up to 57 per cent higher than sheep farming' by Donal Magner. *Irish Farmers Journal.* 15 October 2015.
61 '500,000 acres devalued by hen harrier diktat' by Allison Bray. *Irish Independent.* 24 May 2015.
62 Press Release: Irish Farmers' Association. 28 May 2015.
63 'Hen Harrier SPA crippling farmers across East Kerry'. *The Kerryman.* 24 January 2015.
64 'Hen harriers "making farms worthless"' by Aidan O'Connor. *Kerry's Eye.* 26 January 2015.
65 'Protection of hen harriers costs farmers €22 million per year' by Wayne O'Connor. *Irish Independent.* 22 July 2015.
66 Press Release. 'An Taisce challenges controversial reallocation of >€400 million in CAP funds which has hit farmers and wildlife.' An Taisce. 12 February 2015.
67 O'Donoghue, B. G. 2012. 'Duhallow Hen Harriers *Circus cyaneus* – from stronghold to just holding on.' *Irish Birds.* Vol. 9, No. 3, pp. 349–356.
68 'Recording and Addressing Persecution and Threats to Our Raptors. 2016.' Report prepared by the NPWS.
69 Ruddock, M., Mee, A., Lusby, J., Nagle, A., O'Neill, S. & O'Toole, L. 2016. 'The 2015 National Survey of Breeding Hen Harrier in Ireland.' *Irish Wildlife Manuals*, No. 93. NPWS.
70 Hennessy, T. & Moran, B. *Teagasc National Farm Survey 2014.* Teagasc.
71 Colhoun, K. & Cummins, S. 2013. 'Birds of Conservation Concern in Ireland 2014–2019.' *Irish Birds.* Vol. 9, No. 4 pp. 523–541.

7. THE MYTH OF IRELAND'S 'GREEN' FARMING
1 Questionnaire response. In *Commonage Case Studies 2014.* Prepared by the Centre for Environmental Research Innovation and Sustainability, IT Sligo & the European Forum on Nature Conservation and Pastoralism.
2 'Origin Green' advertisement from 2012.
3 Food Harvest 2020. Executive Summary. Department of Agriculture, Fisheries & Food.
4 BirdWatch Ireland. 2011. *Action Plan for Lowland Farmland Birds in Ireland 2011–2020.* BirdWatch Ireland's Group Action Plans for Irish Birds. BirdWatch Ireland, Kilcoole, County Wicklow.
5 Campbell, S. *Ireland and the Future of Sustainability.* 2015. WWF.
6 Department of Agriculture Food and Marine. 2015. Annual Report.
7 'Official "greenspeak" masks poor show on environment' by P. Woodworth. *The Irish Times.* 29 September 2015.
8 Feehan, J. 2003. *Farming in Ireland.* University College Dublin.
9 Feehan, J. 1986. *Laois: An Environmental History.* Ballykilcoran Press.
10 *Ibid.*

11 Bruton, R. & Convery, F. J. 1982. *Land Drainage Policy in Ireland*. ESRI.
12 *Ibid*.
13 *Ibid*., p. 9.
14 *Ibid*., p. 12.
15 *The Angler's Guide*. 1948. Stationery Office, Dublin. p. 174.
16 McCarthy, D. T. 1977. *The Effects of Drainage on the Trimblestown River. No. Benthic Invertebrates and Flora*. Irish Fisheries Investigations No. 16. Department of Fisheries.
17 McCarthy, D. T. 1981. 'The impact of arterial drainage of fish stocks in the Trimblestown River.' In *Advances in Fish Biology in Ireland*. Moriarty, C. (ed.). Royal Irish Academy.
18 O'Grady, M. F. 1991. *Ecological Changes over 21 years caused by drainage of a salmonid stream, the Trimblestown River*. Irish Fisheries Investigations. No. 33. Stationery Office, Dublin.
19 O'Grady, M. F. & King, J. J. 1992. *Ecological changes over 30 years caused by drainage of a salmonid stream, the Bunree River*. Irish Fisheries Investigations No. 34. Stationery Office, Dublin.
20 Demers, A. & Reynolds, J. D. 2002. 'A Survey of the white-clawed crayfish, *Austropotamobius pallipes* (Lereboullet), and of water quality in two catchments of Eastern Ireland.' *Bulletin Français de la Peche et de la Pisciculture* (2002) 367 : 729–740.
21 Baldock, D. 1984. *Wetland Drainage in Europe*. Jointly published by the Institute for European Environment Policy and the International Institute for Environment and Development. p. 71.
22 Synnott, D. M. 1968. '*Juncus compressus* Jacq. in Ireland.' *Irish Naturalists' Journal*. Vol. 16, No. 4, pp. 92–93.
23 Martin, J. 2006. *Survey of rare/threatened and scarce vascular plants in County Meath*. BEC Consultants.
24 Faulkner, J. & Thompson, R. 2011. *The Natural History of Ulster*. National Museums Northern Ireland. p. 307
25 *Ibid*.
26 Whilde, T. 1994. *The Natural History of Connemara*. IMMEL Publishing.
27 Cabot, D. 1999. *Ireland. New Naturalist*. HarperCollins.
28 Warner, D., Linnane, K. & Brown, P. R. 1980. *Fishing in Ireland*. Appletree Press. p. 148.
29 Development of the NASCO Database of Irish Salmon Rivers – Report on Progress (Tabled by European Union – Ireland). CNL(05)45.
30 Inland Fisheries Ireland. 2015. *The Status of Irish Salmon Stocks in 2014 and Precautionary Catch Advice for 2015*.
31 International Council for the Exploration of the Sea. 2010. Report of the 2010 session of the joint EIFAC/ICES working group on eels, Hamburg, Germany, 9–14 September 2010.

32. Kelly, F. L, Matson, R., Connor, L., Feeney, R., Morrissey, E., Wogerbauer, C. & Rocks, K. 2012. *Sampling fish for the water framework directive. Easter River Basin District.* Inland Fisheries Ireland.
33. See www.floodmaps.ie
34. Bruton, R. & Convery, F. J. 1982. *Land Drainage Policy in Ireland.* Economic and Social Research Institute.
35. Environmental Protection Agency. 2011. *The EPA & Climate Change.*
36. In *A View of Ireland.* 1957. Meenan, J. & Webb, D. A. (eds). British Association for the Advancement of Science.
37. Feehan, J. 2003. *Farming in Ireland.* University College Dublin.
38. Hodgson, W. E. 1906. *Salmon Fishing.* Adam & Charles Black, London.
39. King, J. L., Marnell, F., Kingston, N., Rosell, R., Boylan, P., Caffrey, J. M., FitzPatrick, Ú., Gargan, P. G., Kelly, F. L., O'Grady, M. F., Poole, R., Roche, W. K. & Cassidy, D. 2011. *Ireland Red List No. 5: Amphibians, Reptiles & Freshwater Fish.* NPWS.
40. Feehan, J. 2003. *Farming in Ireland.* University College Dublin.
41. Tunney, H., Foy, R. H. & Carton, O. T. 'Phosphorous inputs to water from diffuse agricultural sources.' In Wilson, J. G. (ed). 1998. *Eutrophication in Irish waters.* pp. 25–39. Royal Irish Academy.
42. Champ, W. S. T. 'Phosphorous/cholorophyll relationship in selected Irish lakes.' In Wilson, J. G. (ed). 1998. *Eutrophication in Irish waters.* pp. 25–39. Royal Irish Academy.
43. McCarthy, D. T. 1988. *Fish Kills 1979–1987.* Fishery Leaflet 141. Department of the Marine.
44. Cabot, D. 1985. *The State of the Environment.* An Foras Forbartha.
45. Teagasc Dairy Enterprise Factsheet 2014.
46. EPA. 2015. *Urban Wastewater Treatment in 2014.*
47. Cabot, D. 1985. *The State of the Environment.* An Foras Forbartha.
48. EPA. 2015. *Water Quality in Ireland 2010–2012.*
49. Huxley, C. & L. 2015. *Lough Carra.* Carra Books.
50. Praeger, R. L. 1937. *The Way that I Went.* Hodges, Figgis & Co.
51. Roden, C. and Murphy, P. 2013. *A survey of the benthic macrophytes of three hard-water lakes: Lough Bunny, Lough Carra and Lough Owel.* Irish Wildlife Manuals, No. 70. NPWS.
52. NPWS. 2013. *The Status of Protected EU Habitats and Species in Ireland.* Overview Vol. 1. p. 85.
53. *Freshwater Pearl Mussel.* Project Newsletter – Summer 2014. Donegal County Council & Northern Ireland Environment Agency.
54. Foulkes *et al.*, 2013. *Hedgerow Appraisal System.* Heritage Council.
55. Wolton, R. 2015. 'Life in a Hedge.' *British Wildlife.* Vol. 26, No. 5, pp. 306–316.

56 Sullivan, C. A., Finn, J. A., Gormally, M. J., & Sheey Skeffington, M. 2013. 'Field Boundary Habitats and their contribution to the area of semi-natural habitats on lowland farms in East Galway, Western Ireland.' *Biology & Environment.* Vol. 13B, Issue 2, pp. 187–199.
57 Feehan, J., Gillmor, D. A. & Culleton, N. E. 2005. 'Effects of an agri-environment scheme on farmland biodiversity in Ireland.' *Agriculture, Ecosystems and Environment.* Vol. 107. 275–86; Copland, A., O'Halloran, J. & Murphy, J. 2005. 'Maximising the biodiversity impacts of REPS.' In *REPS3 – assisting change in farming*, pp. 33–44. Carlow. Teagasc.
58 Cabot, D. 1985. *The State of the Environment.* An Foras Forbartha.
59 *Second National Forest Inventory.* 2013. Department of Agriculture, Food and the Marine. Dublin.
60 Data from An Foras Forbartha and the Department of Agriculture, Food and the Marine.
61 EPA. 2016. *Drinking Water Report for Public Water Supplies in 2015.*
62 Redfern, M. 1995. *Insects on thistles.* Naturalists' Handbooks 4. The Richmond Publishing Co. Ltd.
63 All-Ireland Pollinator Plan. 2015. www.biodiversityireland.ie/pollinator-plan
64 Goulson, D., Nocholls, E., Botias, C. & Rotheray, E. 2015. 'Bee declines driven by combined stress from parasites, pesticides, and lack of flowers.' *Science.* Vol. 347, no. 6229.
65 See https://www.buglife.org.uk/campaigns-and-our-work/campaigns/neonicotinoid-insecticides
66 'Intense lobbying behind EU decision on banning of insecticides' by Frank McDonald. *The Irish Times.* 3 June 2013.
67 'Ireland's Farmland Birds.' BirdWatch Ireland Policy Briefing 2013.
68 McMahon, B. J. 2007. 'Irish agriculture and farmland birds, research to date and future priorities.' *Irish Birds.* Vol. 8, No. 2, pp. 195–206.
69 Estimates for 2015 from Department of Agriculture, Food and the Marine's Annual Report 2014.
70 http://europa.eu/rapid/press-release_MEMO-13-631_en.htm
71 Teagasc National Farm Survey 2014.
72 'Biodiversity schemes in need of more input from farmers' by Alan Matthews. *Irish Independent*, farming supplement, 3 November 2015.
73 *Ibid.*
74 Finn, J. A. & Ó hUallacháin, D. 2012. 'A Review of Evidence on the Environmental Impact of Ireland's Rural Environmental Protection Scheme (REPS).' *Biology and Environment.* Vol 112B. pp. 11–34.
75 Richards, K. G., Jahangir, M. M. R., Drennan, M., Lenehan, J. J., Connolly, J., Brophy, C., Carton, O. Y. 2015. 'Effect of an agri-environmental measure

on nitrate leaching from a beef farming system in Ireland.' *Agriculture, Ecosystems & Environment.* 202. pp. 17–24.
76 'Ireland's Environment 2004.' Environmental Protection Agency.

8. A FUTURE FOR WILDLIFE AND PEOPLE

1 'The first phase of wilderness project unveiled' by Adam Cullen. *Irish Independent.* 7 September 2015.
2 Feehan, J. 2003. *Farming in Ireland.* University College Dublin.
3 O'Connell, M. & Molloy, K. 2011. 'Farming and Woodland Dynamics in Ireland during the Neolithic.' *Biology and Environment.* Vol. 101B. pp. 99–128. Royal Irish Academy.
4 Most of the aforementioned details are taken from *Céide Fields Visitor Guide* published by the Office of Public Works.
5 Caulfield, Seamus. *Céide Fields & Belderrig Guide.* Unknown year. Morrigan Book Company.
6 McNally, K. *Achill.* 1975. David & Charles. Newton Abbot.
7 Praeger, R. L. 1937. *The Way that I Went.* Hodges Figgis & Co. p. 189.
8 McNally, K. 1976. *The Sun-Fish Hunt.* Blackstaff Press.
9 *Ibid.,* p. 62.
10 *European Red List of Marine Fishes.* 2015. European Commission.
11 'Fishermen claim they had to cut seal numbers' by Micheal Finlan. *The Irish Times.* 29 October 1981.
12 Roche, J. R., Perrin, P. M. Barron, S. J. & Daly, O. H. 2014. 'National Survey of Upland Habitats (Phase 1, 2010–2011),' *Site Report No. 6: Croaghaun/ Slievemore cSAC (001955), Co. Mayo (Revision).* NPWS.
13 Praeger, R. L. 1909. *A Tourists' Flora of the West of Ireland.* Hodges, Figgis & Co. Ltd.
14 Preston, C. D., Pearman, D. A. & Dines, T. D. 2002. *New Atlas of the British & Irish Flora.* Oxford University Press.
15 Figure from the Central Statistics Office.
16 Coillte Press Release. 'Ireland's First Wilderness Project Launched.' 14 March 2013.
17 'The First Phase of Wilderness Project Unveiled' by Adam Cullen. *Irish Independent.* 7 September 2015.
18 Monbiot, G. 2013. *Feral.* Allen Lane.
19 See: www.rewildingbritain.org.uk/
20 See: www.walkhighlands.co.uk/news/clearing-up-some-of-the-confusion-around-rewilding/0014181/
21 See: www.rewildingbritain.org.uk/magazine/reforestation-in-norway-showing-what per centE2 per cent80 per cent99s-possible-in-scotland-and-beyond

22 Chapron *et al.* 2015. 'Recovery of large carnivores in Europe's modern human-dominated landscapes.' *Science* Vol. 346, Issue 6216, pp. 1517–1519.
23 *LIFE and human coexistence with large carnivores.* 2013. European Commission.
24 Cóilín articulated his vision in the 2014 issue of *Irish Eagle News* (Issue 6).
25 See www.greypartridge.ie
26 See http://data.worldbank.org/indicator/AG.LND.FRST.ZS
27 INHFA Press Release 26 January 2016.
28 COFORD. 2016. *Land Availability for Afforestation.*
29 The Pontbren Project. Downloadable from: www.woodlandtrust.org.uk/mediafile/100263187/rr-wt-71014-pontbren-project-2014.pdf
30 Beaufoy, G., Baldock, D., & Dark, J. *The Nature of Farming.* 1994. Institute of European Environment and Policy.
31 Aebischer, N. J., Bailey, C. M., Gibbons, D. W., Morris, A. J., Peach, W. J., & Stoate, C. 'Twenty years of local farmland bird conservation: the effects of management on avian abundance at two UK demonstration sites.' *Bird Study* (2016) 63, 10–30.
32 Dunn J. C., Gruar D., Stoate C., Szczur J. & Peach W. J. 2016. 'Can hedgerow management mitigate the impacts of predation on songbird nest survival?' *Journal of Environmental Management.* Vol 184.
33 Balmer *et al.* 2013. *Bird Atlas 2007–11.* BTO Books.
34 Figures from the Anaerobic Digestion Association of Ireland.
35 O'Leary, B. C., Winther-Jansen, M., Bainbridge, J. M., Aitken, J., Hawkins, J. P. & Roberts, C. 'Effective coverage targets for ocean protection Running Title: Effective targets for ocean protection'. 2016. *Conservation Letters.* DOI: 10.1111/conl.12247.
36 'Lyme Bay: Six years on' by Lin Baldock. *Sherkin Comment* 2015 Issue No.59. See also http://www.bluemarinefoundation.com/project/lyme-bay/
37 *Making the World of a Difference. Statement of Strategy 2016–2018.* An Bord Bia.
38 'Eurobarometer 83.4. Attitudes of Europeans towards biodiversity'. 2015. European Commission.
39 *Organisational Review of the National Parks and Wildlife Service (NPWS).* 2010. Grant Thornton.
40 'Challenging environment for new minister' by Michael Viney. *The Irish Times.* 22 June 2002.

Further Reading

Irish Wildlife magazine is published quarterly by the Irish Wildlife Trust and contains news and views from across Ireland. The subscription fee includes membership to the IWT so you are also helping their work to protect our wildlife. See *www.iwt.ie*

Marine nature

- For a chronicle of historic changes in our sea there can be no better read than *The Unnatural History of the Sea* by Professor Callum Roberts (2007, Gaia Thinking).
- For something closer to home read *Overkill!* by fisheries scientist Ed Fahy (2012, self-published).
- Michael & Ethna Viney's *Ireland's Ocean* (2008, The Collins Press) is an excellent overview of life in our seas.
- *Cod* by Mark Kurlansky (1999, Vintage Books) recalls the importance of a fish that in Irish waters has now been abandoned to the pressures of industrial trawling.

General overview

- Although in need of updating, *Ireland* by David Cabot (1999, Harper Collins) is an excellent overview of our habitats and species by someone who was at the forefront of nature conservation for that past four decades.

Past and present

- *Wolves in Ireland* by Kieran Hickey (2011, Open Air) gives a fascinating account of the importance of wolves in Irish culture and society from the earliest times to its extinction in the late 1700s.
- For a vision of the future, *Feral* by George Monbiot (2013, Allen Lane) is as insightful as it is enjoyable.
- *Secrets of the Irish Landscape,* the TV series and book (2013, RTÉ/Cork University Press) tells the fascinating (and ever unravelling) story of the origin of our island and the plants and animals that live here.

Acknowledgements

FIRST AND FOREMOST a word of thanks to the staff of Ireland's libraries without whom I would not have been able to research this book. Their interest in my project and their friendliness and willingness to help me find some pretty obscure titles has left me in their debt. In particular I want to thank the staff of my local library in Blanchardstown, Dublin and those of the libraries of the Royal Irish Academy and the Royal Dublin Society.

I want to thank my wife, Annika, for taking the time to initially proofread my drafts, and give me her honest feedback. I am also very grateful to my sister, Eimear, a skilled graphic designer, who jumped at the opportunity to design the cover of this book.

I have to thank Mike Walker, formerly of the Pew Charitable Trusts, for leading me into the world of fisheries science about which, at our first meeting, I knew nothing. I am also indebted to my fellow 'marine group' campaigners for their guidance in navigating this complex world, including Johnny Woodlock of the Irish Seal Sanctuary, Dana Brennan in Oceana, Ed Fahy, Siobhán Egan and Sinead Cummins in BirdWatch Ireland, and Karin Dubsky in Coastwatch Ireland.

For animated discussions on present and future wildlife I thank my present and former colleagues in the Irish Wildlife Trust: Daniel Buckley, Conn Flynn, Fintan Kelly, Sean Meehan, Barbara Henderson and Kieran Flood.

I have to acknowledge that all I know about badgers and the culling programme I owe to a handful of individuals. Conn Flynn worked as Development Officer for the IWT when we relaunched our campaign in 2011. This work drew a lot from Bernie Barrett of Badgerwatch in Waterford. Badgerwatch is closely affiliated with the IWT and has been campaigning from the start to see an end to the culling programme. We were also lucky to have Fintan Kelly come on board and he dedicated his master's thesis to researching badgers and their role in spreading TB (or not as the case may be!). I also need to acknowledge the work of Mike Rendle of the Northern Ireland Badger Group. Northern Ireland also has badgers but unlike the Republic they have not had a culling programme, so it is illuminating to compare the two jurisdictions.

Finally I want to thank the staff of the National Parks and Wildlife Service. Although we are sometimes at odds, I know many individuals who remain dedicated to nature conservation despite the challenges. A number (unwittingly) helped me in writing this book by sharing their unparalleled local knowledge and expertise.

Index

Note: illustrations are indicated by page numbers in **bold**.

Achill Island 147, 168, 289–92
acidification 20, 65, 159, 235, 239, 305
Adams, Ansel 79
Agri-Environment Options Scheme (AEOS) 283
Ahern, Bertie 119
alder 287, 288, 305
algae 9, 41, 48, 110, 111, 234, 248, 270
alpine heath 97
alpine lady's-mantle 98–9
alpine saw-wort 98–9
angel shark 19, 43, 138, 167
Aran Islands 2, 38, 126, 132, 212, **309**, 310
Arctic charr 8, 36, 52, 54, 266–7
Areas of Scientific Interest (ASIs) 82–3, 224
Arklow 55, 59, 66
Arterial Drainage Acts 254, 255, 261
ash 130, 248, 304, 305
Atlantic Ocean 27, 52, 62, 69, 139, 163, 166
Australia 4, 18, 78, 127, 210, 221, 279
Avonmore River 301

badger 11, 12–15, 121, 148, 152, 184–207, 248, 273, 323
Badger Trust 12–13
Ballyconneely Peninsula 82
Ballycroy National Park 31, 85, 86, 132, 293, 296
Baltimore 37–40, 43, 45, 74, 76
Bangor Trail 293
Bantry Bay 45, 66, 165

barley 264, 265, 278
barn owl 160–61, 252, 280
Barrow, river 80, 138, 163
Barry, J. M. 195
basking shark 40–41, 166–7, 289, 290, 292
bass 29, 36, 52, 53, 58, 63
bats 16, 80, 84, 248, 283, 324
Bay of Biscay 139
bears 16, 18–19, 31, 89, 136, 139, 140, 297, 300
bees 9, 11, 125, 143, 162, 171, 179, 252, 277–8, 283, 314
beetles 143, 162, 277, 284
Belfast Lough 141
Belgium 298–9, 305
Bellamy, David 9, 220–23
Ben Bulben 126, 239
Bern Convention 203–4
Białowieża National Park 78
biogas 316
birch 100–101, 104, 211, 217, 218, 287, 302, 304
Birds Directive 80–81, 155, 176, 241
bird's-nest orchid 154
BirdWatch Ireland 11, 95, 98, 109, 147, 160–61, 250, 274, 303
bittern 19, 136, 142, 301, 304
black-necked grebe 143, 151
black scabbard 63–4
blackbird 273
Blackwater, river (Cork) 243
Blackwater, river (Northern Ireland) 259
bladderwort 99, 291
blanket bog 1, 4–5, 80, 82, 96–7, 99–100, 104,

210, 215, 217–18, 225, 231–40; see also bogs
Blasket Islands **181**, 182–3
bloody crane's-bill 125
blue shark 290
bluebell 1, 154
blunt-nosed six-gill shark 170
boarfish 28, 65
Bog Commissioners 214–15, 216, 221
bog-cotton 96, 302
Bog of Allen 212, 214–15, 222
bogs 1, 2–6, 9, 11, 21, 80–82, 96–7, 99–100, 104, 106, 110, 111, 113–17, 173, 176–7, 208–46, 251, 254–5, 263–4, 286–8, 290–91, 296, 302–4
Boora Discovery Park 302–3, 324
Bord Bia 253, 281, 319, 322
Bord Iascaigh Mhara (BIM) 27–8, 45, 181–2, 184, 317
Bord na Móna 21, 216–17, 219–20, 222–3, 238, 302, 303, 324
bovine tuberculosis (bTB) 12–15, 148, 185–8, 190–207, 275, 281, 315
Boyne, river 162, 163, 248, 256, **257**, 258–62
Brabazon, Wallop 66
bracken 4, 103, 112, 123
Brady, Vincent 127
bream 6, 54
Brennan, Des 169–70
brill 54, 60
Britain 4, 11–13, 15–17, 29, 40, 49, 58–9, 61, 78,

352

Index

135, 141, 145, 148, 152, 158, 163, 191–3, 196–8, 205–6, 235, 264, 273–4, 294–5, 299, 312–14, 318–19; *see also* Scotland; Wales
brittle bladder fern 292
brittle star 42, 67
broad-leaved cottongrass 99
brown bear 18–19, 136, 139, 140, 300
brown rat 171
Brú na Boinne 93
brucellosis 187, 281
Buckley, Daniel 153, 154
Bullock, Craig 296
Bunree River 258
Burdett-Coutts, Angela 39
burning 3–5, 98–100, 102–3, 122, 213, 233, 235–6, 245–6, 284–5, 290
Burren 1, 2, 31, 32, 34, 85, 112, 124–32, 149, 252, 302, 310, 324
Burren Action Group (BAG) 127–8
Burren Farming for Conservation Programme 129–32
Burren LIFE project 131
Burren National Park Support Association (BNPSA) 128
BurrenBeo Trust 125, 130
Bush, river 259
butterflies 1–2, 3, 125, 143, 157, 158, 162, 277, 297
buzzard 238

Cabot, David 259
Cameron, David 13
Canada 69, 70, 141, 148, 210, 298
Cape Clear island 41, 43, 45
capercaillie 113, 136, 141, 301
carbon release 97, 235, 316
carbon storage 103, 221, 275, 230, 303, 304, 306, 308

Cardigan Bay 47
carline thistle 125
Carlingford Lough 49–50, 59
Carson, Rachel 276
Castletownbere 317
cattle 5, 12, 33, 104, 108, 113, 129–31, 185–8, 190–207, 234–5, 248–9, 264–5, 271, 282, 306, 315, 324
Céide Fields 94, 286–8
Celtic Sea 27, 52, 53
Central Fisheries Board 270
China 28, 30, 126
Claddagh 6, 66
Clara Bog 219–20, 223, 228–9, **229**
Clayton, Neal 182–3
clear-felling 111, **239**, 239, 305
Clew Bay 21
Clifden 82, 269
climate change 18, 30, 65, 96, 97, 145, 159, 164, 221, 287–8, 303
Clochar na gCon 220
Clonast 216
Clonmacnoise 174
cod 6, 25, 26, 43, 54, 55, 58, 60, 62, 68, 69–70, 71–3, 76, 180
Coillte 34, 113, 239, 293, 296, 298
Cole's charr 266
Comer, John 207
Comeragh Mountains 100, 131
Common Agricultural Policy (CAP) 24, 123, 242, 281–4, 311
common dolphin 43
Common Fisheries Policy 24, 26, 47, 57, 70, 72, 74
common frog 5
common seal 41, 180
common skate 43, 53, 167, 169–70
conger eel 54, 60

conifer plantations 2, 5, **7**, 7, 10, 20, 34, 96, 99, 101, 106, 110–13, 115, 123, 146, 149, 230, 233, 238–42, **239**, 245–6, 263, 272, 296, 302, 305–7
Connemara 1–8, 23, 31, 85, 104–13, 132, 212, 218, 237, 239, 240, 310
Cooley Hills 100
Cooley Peninsula 21, 131
corals 42, 59–60, 68, 127
Cork and Kerry Mountains 100
Cork city 83–4, 134, 323–4
corn bunting 19, 136, 143, 250, 252, 280, 312
corncrake 7, 19, 156–7, 250, 252, 280, 283–4, 290, 312
cottongrass 99, 217
Coveney, Simon 24–5, 28, 185, 201–2, 207, 251
coyote 89, 147
Cozens, Jacquie 182–3
crabs 2, 6, 42, 44, 45–6, 59, 164, 289
crane 19, 136, 141, 151–2, 171, 301, 304
crayfish 6, 44, 64, 167, 256, 257, 258
Cross, John 100
crows 120, 121, 122, 161, 245, 300, 313
Crowe, Tasman 49
cuckoo 303
culling 11, 12–15, 118, 148, 173–207
Cummeragh River Bog 220
curlew 5, 19, 147, 161, 229, 246, 250, 280, 290
curly waterweed 8

daffodil 171
Daly, Clare 178
D'Arcy, Gordon 141, 142, 240
Davis, Charles 39

de Courcy Ireland, John 56
de Valera Éamon 135
de Valera, Síle 34, 128–9, 225–6, 325
Deenihan, Jimmy 116, 172, 227
deer 16, 19, 87–91, 100, 101, 103, 108, 111, 118, 146–7, 150, 152–3, 155–6, 171, 191, 206–7, 296, 300, 301, 323
deforestation 9, 18, 20, 88, 99–100, 102, 106, 111, 212, 287–8, 295
delicate maidenhair fern 292
Dempsey, Noel 82–3
Department of Agriculture 13–15, 123, 177, 185–8, 190, 194, 199–204, 242–3, 247, 265, 284
Department of Arts, Heritage and the Gaeltacht 175–6, 322
Department of Environment 22, 30
Department of Fisheries 256
Derryclare Nature Reserve 106, 110–11, 113
Diamond, Jared 17
Dillon, John 241
Dingle 135, 182, 183
Dingle Bay 236
Dingle Peninsula 21
dipper 16, 259
divided sedge 138
dogfish 2, 42, 60, 167
dolphins 43, 47–8, 76, 167, 175
Donegal Bay 139
dormouse 171
Downey, Eddie 205, 206
dragonflies 125, 157, 162, 229, 248, 323
drainage 97, 136, 209–10, 212–16, 219–22, 229, 232, 251, 254–63, 271, 276, 280, 296
Drew, David 129

Drumkelin 212
Dublin 9, 34, 94, 133, 138, 148–9, 162–3, 189, 212, 240, 324
Dublin Bay prawns 26–7, 47, 68
Dúchas 80–81
ducks 171, 302, 323
Duhallow 243–5
Dundalk 134, 142
Dundalk Bay 68
Dundrum Bay 59
Dunford, Brendan 130
dunlin 246
Durrow 142
Dutch Foundation for the Conservation of Irish Bogs 220

earthworm 137, 198
East Offaly Badger Research Project 196, 198
Economic and Social Research Institute (ESRI) 187, 192, 195, 262–3
Edenderry 213
eels 6, 54, 60, 65, 165, 261
Environmental Impact Assessments 217, 224, 227, 269
Environmental Performance Index (EPI) 10
Environmental Protection Agency (EPA) 10, 250–51, 263, 269, 277, 284, 306, 315, 319, 322
Eradication of Animal Disease Board 187
Erne, river 163, 164, 254
erosion 7, 89, 96–7, 103, 105, 110–11, 233–5, 261, 271–2, 285, 290, 291, 316
Erris Head 234
European Commission 8, 44, 97, 130, 202, 226, 230, 269, 281–2, 320

European Court of Justice 44, 82, 175, 176, 226
European eel 65
European Union (EU) 23–4, 43–4, 51–2, 57, 70, 72, 74, 80, 108, 110, 114, 123, 128–9, 131, 167, 176, 182, 200, 217, 225, 230–31, 243, 251, 256, 258–9, 272, 280–84, 300, 305–6
eutrophication 9–10, 266–71
extinctions 11, 18–20, 99, 133–72, 250, 290, 312

Fahy, Ed 25, 60, 63
Fáilte Ireland 31, 85, 87
Fairley, James 139
farming 1, 7, 10, 11, 19–21, 32–4, 100, 102–4, 107–13, 120–23, 129–32, 161–2, 185–8, 190–207, 211, 234–5, 241–3, 246–89, 299–300, 305–17
Feehan, John 213, 254–5
ferns 19, 99, 101, 103, 116, 137, 144, 292, 305
fertilisers 20, 129, 159, 162, 267, 276, 280, 309
fires see burning
fish farming 8, 27–8, 53, 54, 56, 164–5
fishing 6, 7–8, 10, 23–9, 32, 34, 36–77, **37**, 135–6, 139, 146, 162–70, 178–82, 256–8, 260–61, 266–8, 270–71, 289–91, 317–19
Fitzgerald, Séamus 40
Fitzmaurice, Michael 176
floodplains 30, 163, 251, 255, 258, 259
flounder 36, 52, 53
Flynn, Conn 14
food production 8, 248–54, 263–5, 276, 281, 309–10, 311, 319, 322

Food Harvest 2020 plan 249–51, 253, 269
Food Wise 2025 plan 269
Foot, M. J. 124
Forestry Act 238
Forestry Programme 241–2
forests see woodlands
foxes 5, 120–22, 147–8, 152, 161, 171, 179, 191, 245, 300
Foula eyebright 292
Four Areas project 196, 198
France 24, 35, 38, 59, 136, 154, 163, 168, 252, 273, 298
freshwater pearl mussel 19, 80, 259, 271–2
Friends of the Irish Environment 228
frogs 3, 5, 80, 210, 217, 248, 252, 303
fuchsia 105, 171, 273
fulmar 171
fungi 137, 145

Galtee Mountains 100
Galway Bay 2, 6, 21, 28, 64, 66, 85, 111
Galway city 1, 5, 6, 66, 84, 324
Game & Wildlife Conservation Trust 312
Garda Síochána 3, 103, 176, 227
Gearagh, the 88
geese 1, 222, 259, 302
Germany 136, 208, 212, 299, 301, 316
giant rhubarb 292
Giant's Causeway 93
Glendalough 95, 101
Glenveagh National Park 31, 85, 113–24, **115**, 132, 233–4
goats 90, 103
golden eagle 19, 22–3, 118–24, 137, 142, 147, 238, 246, 289–90, 301, 324

Golden Eagle Trust 22, 23, 119–23, 324
golden plover 246
goosander 301
Gormley, John 80, 94, 121, 175, 226–7
gorse 3, 103, 235, 273
goshawk 142
Grand Canal 212, 213
grazing 5, 7, 21, 89–91, 94, 97, 100, 102–4, 107–13, 118, 212, 219, 233–5, 252, 264, 271, 284–5, 290, 291, 296
great auk 137, 141, 171
Great Barrier Reef 78, 127
Great Famine 39, 54–5, 89, 214, 215, 254, 289
great-spotted woodpecker 137, 141, 301
great sundew 99
Great Western Greenway 31, 293
Green Low-Carbon Agri-Environment Scheme (GLAS) 242, 283
greenhouse gas emissions 234–5, 251, 305–6
Greenland white-fronted goose 222, 259
grenadier 63–4
grey heron 162
grey partridge 7, 161, 280, 303
grey seal 41, 43, 44, 180–83, **181**
grey squirrel 148–50
grey wagtail 162, 259
Griffiths, Richard 214–15, 221
Groundwork 91–3, 94
Gullestad, Peter 71
gurnet 6, 54

habitat loss 11, 18–20, 23, 97–8, 108–10, 152, 218–20, 233, 271, 278, 280
Habitats Directive 43–4, 74, 80–82, 97, 110, 155,

176–7, 225–6, 233, 243, 258
haddock 6, 54, 55, 58, 60, 63, 68, 69
Healy-Rae, Johnny 174
Heaney, Seamus 211, 231
hake 6, 43, 54, 55, 60
halibut **37**
harbour porpoise 41, 43, 44, 46
hare coursing 11, 177–8
hares 5, 11, 16, 121, 177–8, 235, 303
Haughey, Charles 48, 175
hawfinch 141
hawthorn 107, 247, 273, 314
hay 263–4, 280, 283, 284, 314
Hayden, Ashley 29
Hayes, Tom 177
hazel 129–30, 287, 305, 314
heather 2–3, 4, 11, 95–6, 99, 102–3, 108–13, 217, 235, 284
Hector, Alexander 289
hedge cutting 174, 203, 274–5
hedgerows 20, 32–3, 107, 174, 194, 198, 203, 247–8, 250, 252, 273–6, 278, 280, 283–4, 307–8, 312–15
hen harrier 98, 174, 236–46, **237**
Henry, Paul 220, 231, 289
herbicides 276–7, 279, 280, 292, 309
Heritage Council 112, 132, 275
herring 6, 25, 43, 47, 54, 55–6, 58–9, 60, 65, 66, 72, 289
herring gull 160
Higgins, Michael D. 80, 128–9, 225
High Nature Value farming 132, 309–11

holly 90, 101, 106, 107, 108, 111, 116
hooded crow 121, 122
horse mackerel 26, 27, 39
housing 5, 20, 129, 230, 248, 272
Howth 26, 46, 59, 66, 134
human effluent 268–9
human settlement 17–19, 36–7, 152, 211–12, 286–8
Humphreys, Heather 85, 230, 274
Huxley, Chris 270–71
Huxley, Linda 270–71
Huxley, Thomas 55

Iceland 26, 27, 124, 139, 141, 167, 305
Inishkea Islands 139, 291
Inishowen Peninsula 131, 134, 237
Inland Fisheries Commission 256
Inland Fisheries Ireland (IFI) 164, 179, 260–61, 319
insects 4, 20, 33, 130, 143, 157–8, 179, 273–4, 277–9; see also bees; beetles; butterflies; dragonflies; mayflies; moths
intensive farming 20, 32, 123, 129, 161, 251, 265, 271, 275–80, 311–12
intermediate bladderwort 99
International Union for Conservation of Nature (IUCN) 53, 62, 78, 127, 136, 167, 169, 170
invasive species 8, 49, 50, 91–3, 94, 112, 113, 118, 148–9, 155, 279, 292, 296
Invasive Species Ireland 148
Irish Basking Shark Study Group 166
Irish Council Against Blood Sports (ICABS) 177, 178

Irish Creamery Milk Suppliers Association (ICMSA) 186–7, 206–7
Irish elk 17
Irish Environmental Network 200, 202
Irish Farmers' Association (IFA) 128, 205, 241, 246, 274
Irish Federation of Sea Anglers 170
Irish Grey Partridge Conservation Trust 303
Irish hare 5, 16, 177
Irish Peatland Conservation Council (IPCC) 11, 220, 224–5, 230, 233
Irish Sea 25–7, 53, 55, 58–60, 62–3, 65, 66, 67–70
Irish Seal Sanctuary (ISS) 60, 183
Irish Specimen Fish Committee 169–70
Irish Wildlife Trust (IWT) 11–12, 14, 70, 91, 114–17, 154, 181, 184, 198, 200, 202, 250, 274
Iveragh Peninsula 132
ivy-leaved bell flower 99

jackdaw 248, 280
Japanese knotweed 118
jellyfish 69
juniper 211

Kane, Sir Robert 212
Kelly, Fintan 14, 198
kelp 2, 42
Kennedy, P. G. 238
Kenny, Enda 129, 178
kestrel 1, 5, 162, 323
Killarney fern 19
Killarney National Park 20, 31, 85, 86–94, 108, 132, 149, 236, 305
Killarney shad 87
killer whale 180, 181
Killybegs 166, 317
Kilmore Quay 39, 317

kingfisher 259, 324
Kinnegad 218
Kinsale 40, 169
Kirk, Seamus 199–200
Knockmealdown Mountains 238
knotweed 112, 118
Kohl, Johann Georg 86–7, 208, 211–12

Lake District 78
Land Commission 217–18
land reclamation 129, 212–15, 232, 251, 254–5
Landed Property Improvement Act 254
landslides 97, 290
lapwing 5, 147, 160, 259, 280, 303, 312
leatherback turtle 47
Lee, river 88, 305
Leonard, Liam 127
lesser argentine 63–4
lesser twayblade 292
lichens 130, 145
Liffey, river 99, 133, 138, 162–3, 164, 324
Limerick city 324
Lindenberg's featherwort 109
ling 6, 43, 54
Liscannor 168
liverworts 109, 144
Living Planet reports 158–9
lizards 217
lobster 6, 44, 45–6, 59, 164, 289
Local Nature Reserves 83
lodgepole pine 238, 241
Lost Life Project 145
Lough Barra 117
Lough Boora Parklands 302–3, 324
Lough Carra 266, 270–71
Lough Conn 179, 266
Lough Corrib 2, 8, 126, 142, 179, 266
Lough Dan 101
Lough Derg 22

Lough Derryclare 110–11
Lough Ennell 267–8
Lough Feeagh 293
Lough Funshinagh 143
Lough Gealáin 126
Lough Gur 19, 94
Lough Hyne 74
Lough Inagh 106, 110
Lough Mask 179, 266
Lough Neagh 259
Lough Sheelin 267
Lough Tay 101
Lucan 324
Lyme Bay 318–19
lynx 16, 18, 31, 89, 94, 136, 139, 140, 152–3, 294, 299, 300, 301

Mac an Ríogh, Seán 6
McCarthy, Dan 179
Mac Coitir, Niall 189
MacGiollarnáth, Seán 108, 113
McIlhenny, Henry 118
mackerel 6, 26–7, 36, 39, 40, 43, 45, 52, 54–6, 58, 60, **75**, 76, 289
Mac Laughlin, Jim 24, 38
MacLouchlainn, Cóilín 301
McNally, Kevin 290–91
McSweeney, Tom 45
maerl 42, 45, 49, 50
magpie 245, 280, 313
mako shark 290
manure 20, 215, 248, 250, 253, 266–71, 315–16
Marine Institute 25, 27, 44–5, 46, 50, 51–2, 63–4, 65, 68, 69, 319
Marine Living Resources Act 73
Marine Protected Areas (MPAs) 74–7
Marine Stewardship Council (MSC) 27, 52
Marine Strategy Framework Directive 176
marsh clubmoss 99, 291

marsh fritillary 1–2, 3
marsh harrier 142
Matthews, Alan 282–3
Maumturks 2, 105, 107, 109
Maximum Sustainable Yield (MSY) 51–2
mayflies 144, 162, 270
meadow pipit 3, 98, 117, 240, 244–5, 246
meadowsweet 288
merlin 97–8, 117, 246
Merlin Woods 84, 324
methane emissions 234–5, 251, 306
mink 179
Monbiot, George 12, 95, 100, 294
Moneer Bog 215
monkfish 36, 47, 52–3, 58, 60, 68, 69
montbretia 105
Moran, James 310
mosses 96, 99, 101, 109–11, 116, 130, 144, 211, 213, 215, 229, 256, 288, 305
mossy saxifrage 99
moths 125, 143, 157–8, 277
motorways 322–3
mountain ash 101, 107, 108, 116
mountain avens 124–5
mountain everlasting 99
mountain hare 16, 177
Mourne Mountains 100
Moycullen Bogs 4–5
Muckross House and Estate 87, 89
Muir, John 78–9
Mullaghanish Mountains 245
Mullaghmore 126–9, 131
muntjac deer 171, 301
Musheramore Mountains 245
mussels 6, 8, 19, 45, 48–9, 58, 60, 65, 67, 80, 259, 271–2, 289

narrow-leaved helleborine 99, 154

National Badger Survey 199–200
National Biodiversity Action Plan 172
National Biodiversity Data Centre 155, 278, 302
national parks 31–2, 78–80, 85–132, 159, 296–7, 310, 320–21
National Parks and Wildlife Service (NPWS) 3, 10, 14, 81–3, 92–3, 100, 103, 114, 116–18, 175–7, 190, 200, 202, 227, 229–30, 233–4, 236, 243, 245, 253, 271–2, 291, 319–22, 324
National Peatland Conservation Committee (NPCC) 219
National Peatlands Strategy 229–30, 233, 236
National Plan for the Sustainable Use of Pesticides 276–7
National Pollinator Plan 278
National Soil Survey 218
native oyster 49, 50
Natura 2000 sites 261
Natural Heritage Areas (NHAs) 4–5, 43, 80–81, 82–3, 224, 228, 230, 270, 323
Nephin Beg mountains 293
Nesbitt, Thomas 166
Netherlands 59, 136, 219, 220
nettles 259, 261, 277
New Economics Foundation (NEF) 52
New Zealand 15–16, 279
Newfoundland 70, 141
Newman, Michael 174
nightjar 19, 98, 138, 160, 246
nitrogen 159, 267
Nore, river 80, 163, 255, 272
North Atlantic hepatic mat 109

North Atlantic right whale 139, 140, 145
Northern Ireland 83, 142, 177, 194, 205, 206; see also individual locations
Norway 27, 47, 71–4, 76, 93–4, 139, 142, 295, 298, 301, 318

oak 1, 7, 20, 84, 87, 90–94, 100, 101, 106, 110–11, 113, 115, 118, 149, 211, 273, 287, 304, 305, 324
oak fern 137, 144
oats 213, 264, 265
OCEAN2012 70
O'Connell, Catherine 224
O'Donovan, Grace 213
O'Dowd, Peadar 6
Office of Public Works (OPW) 91, 128, 220, 256, 261–2, 263
Ó hEochagáin, Hector 23–5
oilseed rape 33, 278
orange roughy 63–4
orange tip butterfly 158
orchids 125, 154, 252, 323
'Origin Green' campaign 249–50, 252–3
O'Shea, Gerry 26
osprey 136, 141, 301, 304
O'Sullivan, Maureen 178
O'Sullivan, Ned 173–4
O'Toole, Lorcán 121, 151
otters 43, 44, 46, 258, 259, 324
owls 160–61, 246, 252, 280
oysters 6, 25, 45, 49–50, 58, 59, 67

Pacific oyster 45, 49–50
painted lady 277
palourde 62
parsley fern 99
Patterson, Owen 12, 13
Pauley, Daniel 54, 57–8, 69
peacock butterfly 277
peacock worm 42
Pearse, Pádraic 175

peat hags 96–7
peregrine falcon 97–8, 276, 323
periwinkles 45, 50, 289
pesticides 11, 20, 28, 98, 276–9, 280, 305, 309
Phoenix Park 148–9
phosphorus 267–8
Picos de Europa National Park 296–8
pied flycatcher 301
pied wagtail 279
pigs 153, 155, 264, 271
pike 178–9
pilchard 38–40, 43, 55
pine 2, 7, 101, 106, 108, 113, 130, 238, 241, 287–8, 305
pine marten 19, 108, 149–50, 152, 174–5, 245, 324
plaice 6, 26, 47, 51, 53, 54, 58, 63
plankton 38, 48, 166, 266–7, 272
planning regulations 22, 82, 128, 129, 217, 224, 323
poisoning 23, 108, 120, 121–2, 245, 298, 323
Poland 18, 78
pollack 6, 36, 45, 52, 54
pollen analysis 211, 287, 288
pollution 8, 9–10, 20–21, 28, 54, 65, 111, 127, 136, 145, 159, 163, 173, 210, 235, 239, 248, 250, 260, 266–72, 280, 284, 315–16
Pontbren Project 307–8
porbeagle shark 168–9, 290–91
porpoise 41, 43, 44, 46
Portal, Maurice 173
potatoes 33, 213, 214
Potočnik, Janez 300
Praeger, Robert Lloyd 98, 135, 270, 289–90, 291–2
prawns 25, 26–7, 47, 68, 69, 75

purple heather 217
purple moorgrass 4, 5, 105, 111–12
purple sea urchin 64

quail 280, 312
quaking grass 125
quarries 5, 84, 107, 323
Quirke, Bill 91

rabbit 171
ragwort 179
raised bog 9, 21, 80, 81, 176–7, 210–30, 303; see also bogs
Raidió Telefís Éireann (RTÉ) 4, 23–5, 116, 176
Randomised Badger Culling Trial (RBCT) 196–7
raven 162, 174, 296
rays 25, 53, 58, 60, 68, 76, 167–8
razor clam 65, 67–8, 317
red admiral 277
Red Data books 138, 139–40, 162, 178
red deer 16, 19, 87, 89–90, 108, 118, 152–3, 155–6
red grouse 5, 11, 21, 95, 98, 102–3, 109–10, 112, 121, 147, 160, 235, 246, 290
red kite 22, 120, 142, 301
red squirrel 19, 84, 108, 137, 140, 148–9, 172
redshank 147, 280
redstart 142, 279, 301
reintroduction 11, 19, 22–3, 31, 34, 87, 113, 118–24, 137, 142, 153–6, 172, 294, 297, 299–301, 324
rewilding 293–304, 307–9
rhododendron 91–4, 105–6, 112, 118, 123, 292, 296
Ring, Michael 286, 294
ring ouzel 97–8, 160, 246, 280
roadkill 14, 161, 190, 191

Index

Roaringwater Bay 37–51, 70, 74–6
Roberts, Callum 60
robin 162, 248, 273
Robinson, Tim 107
Roche, Dick 93
rock whitebeam 99
rodenticides 161
rook 280
Roper, Timothy 193
roseroot 99, 292
Rothamsted Insect Survey 158
round-fruited rush 259
Roundstone 82, 111
rowan 106, 305
Royal Society for the Protection of Birds (RSPB) 11, 312, 314
run-off 9–10, 20, 239, 248, 266–71, 284, 315–16
Rural Development Programme 281
Rural Environmental Protection Scheme (REPS) 275, 283, 284
rushes 144, 154, 215, 259, 263
Ruttledge, Robert F. 160, 240

Sally Gap 95, 212
salmon 7–8, 25, 27–8, 54, 56, 64–5, 162–5, 243, 257–8, 260–61, 266–7, 271, 289, 290
sand martin 259
scallops 44, 46–7, 60, 68, 75
Schouten, Matthijs 219–20
Scotland 31, 102, 109, 118–22, 130, 142, 180, 295; see also Britain
Scots pine 101, 108, 113, 3-5
Scragh Bog 220
sea bass 29, 36, 52, 53, 58, 63
sea fans 42, 68, 76, 318
Sea Fisheries Protection Authority 40, 50, 291

sea urchins 2, 64, 67
seagrass 41–2, 45, 50
seagulls 68, 160, 173–4
seals 2, 34, 41, 43, 44, 46, 47–8, 145, 167, 180–84, **181**, 291
Seas Around Us project 57–8
seaweed 2, 42, 45, 50, 213, 253, 267, 289
Seigne, J. W. 104, 237–8
Shannon, river 55, 156–7, 163–4, 208–10, 254, 256, 263, 280, 284, 305, 324
Shannon callows 147–8
sharks 19, 25, 40–41, 43, 48, 58, 61–4, 76,·138, 166–70, 289–92
sheep 7, 21, 90–91, 94, 96–7, 100, 102–5, 107–13, 179, 233–6, 249, 264, 271, 282, 290, 291, 296, 299, 307–8
short-eared owl 246
shrimp 6, 44, 45, 75
sika deer 90, 101, 146, 301
silage 162, 193, 264, 271, 278, 280
Sitka spruce 111, 238, 241
skates 36, 43, 52, 53, 54, 69, 167, 169–70
Skellig Michael 93
skylark 3, 162, 231, 313–14
Sleeman, Paddy 18, 153
Slieve Beagh Mountains 246
Slieve Bloom Mountains 100, 154–5, 237
Sligo town 122
slow worm 171
small tortoiseshell 277
snails 144
snipe 5, 246, 259
sole 6, 25, 54, 58, 60
song thrush 162
Spain 24, 71, 154, 163, 168, 273, 296–8, 300–301

Special Areas of Conservation (SACs) 43–4, 80–81, 110, 167, 225–30, 233, 270, 292
Special Protection Areas (SPAs) 68, 80–81, 241–3, 245, 270, 292
speckled wood butterfly 158
sphagnum 211, 213, 215, 229
spider crab 2, 42
spotted crake 142
spotted dogfish 167
spotted slug 87
sprat 1, 47, 54
spurdog 61–2, 63
starfish 42, 67
starling 162
starry saxifrage 291–2
stoat 1, 16, 147, 153
Stock Book 27, 51–3, 63, 69
stoneflies 256, 257
strawberry tree 87
sturgeon 133–6, **134**, 138–9, 144
subsidies 7, 20–21, 57, 72–3, 103, 110, 112, 130–32, 211, 242, 246, 248, 253–4, 256, 280–85, 310–12
Suir, river 163, 269
sustainability 1, 26–7, 31, 36, 45, 47–8, 50–54, 58, 73–7, 102, 119, 130, 159, 181, 233, 236, 249–51, 253, 311, 319
swans 259, 302, 323
Sweden 153–4, 299, 301
Synge, J. M. 98
Synnott, Donal 259

An Taisce 200, 202, 242–3, 274
Teagasc 265, 281, 282, 319
thistle 125, 277, 279, 314
thresher shark 290
Tidy Towns contest 177
tormentil 96, 217

tourism 5, 8, 23, 28–32, 76, 85, 87, 95, 105–7, 126–8, 130–31, 183, 229–30, 263, 265–6, 269, 289, 292–3, 297–9, 303–4, 308
Tralee Bay 167
Treacy, Noel 220
Trim Athboy District Angling Association 261
Trimblestown River 256–8, 261
trout 2, 7–8, 178–9, 256, 257–8, 260, 261, 266, 268, 271
tuna 36, 48, 52, 53–4
tunbridge filmy-fern 99
turbot 6, 54, 59, 64
Turf-cutters and Contractors Association (TCCA) 176
turf cutting 4, 5, 9, 11, 21, 80, 96, 113–17, 173, 176–7, 208–12, **209**, 215–20, 222–4, 246, 291, 302
turloughs 259
turtles 47, 48
Twelve Bens 2, 82, 105–7, 109–13
twite 246, 280

Ulster Wildlife 12
undulate ray 53
UNESCO sites 93–4, 174, 324
United Nations 29, 79–80
United States 4, 31, 40, 78–9, 89, 147, 148, 150, 299
Ussher, Richard J. 160

Valpy, Richard 54–5
variegated horsetail 99
Viney, Michael 231–2, 325
visitor centres 105, 106, 118, 126–9, 229, 286, 303

Wales 47, 95, 307–8; *see also* Britain

Walker, Mike 70
Warren, Robert 160
water beetles 143, 162
Water Framework Directive 176
Waterford city 134, 141, 184
Went, Arthur 135
Western Way 107
wetlands 1, 7, 21, 33, 151, 162, 251, 255, 259, 284, 302, 304, 316
Wexford Harbour 269
whales 47–8, 139, 140, 145, 175
wheat 263–4, 265
Whelan, Ken 167
whelk 64, 67
white-clawed crayfish 256, 257, 258
white ray 167
white skate 53
white-tailed eagle 22, 23, 87, 108, 120, 137, 142, 173, 301
whiting 6, 25, 39, 54, 58, 60, 62, 69
whooper swan 259
Wicklow Mountains 11, 31, 85, 95–104, 131, 132, 212, 218, 236, 302
Wicklow Uplands Council 103, 104
Wild Atlantic Way 8, 131, 292
wild boar 89, 94, 136, 139–40, 152–5, 297–8, 301
Wild Deer Association of Ireland 207
Wild Nephin 293–6, 298, 299, 302
wildcat 89, 94, 136, 139, 140, 152–3, 297
wilderness 31, 79, 286, 293–304
Wildfowl Refuges 270
Wildlife Act 43, 73, 82, 103, 167, 180, 202, 207, 224–5, 235, 274

willow 3, 217, 218, 273, 287, 304, 305, 314
wind energy 5, 97, 123, 230
wolf 16, 19, 22, 31, 88–9, 91, 94, 136, 140, 146, 150, 152–4, 171, 179, 294, 297–301
wolverine 300
wood mouse 16
wood rush 154
wood white butterfly 158
Woodchester Park 192, 193
woodcock 160
woodlands 1, 7, 10, 18, 20, 33–4, 84, 87–94, 99–108, 110–11, 113, 115, 118, 130, 148–50, 152–5, 198, 203, 211–12, 238, 273, 287–8, 294–6, **295**, 301–2, 304–8, 312–13, 323–4
woodlark 143
Woodlock, Johnny 60
woodpigeon 171, 248, 280
Woodworth, Paddy 253
World Wide Fund for Nature (WWF) 11, 128, 139, 158–9, 165, 220, 251, 258, 301
worms 42, 137–8, 198

yellow wagtail 143
yellowhammer 250, 280, 312
Yellowstone National Park 31, 89, 150, 299
yew 304, 305
Yosemite National Park 78–9
Youghal 269
Yucatan Peninsula 126

zebra mussel 8
Zoological Society of London 158–9